Interdisciplinary Collaboration

An Emerging Cognitive Science

Interdisciplinary Collaboration

An Emerging Cognitive Science

Edited by
Sharon J. Derry
University of Wisconsin–Madison

Christian D. Schunn
University of Pittsburgh

Morton Ann Gernsbacher
University of Wisconsin–Madison

LEA
LAWRENCE ERLBAUM ASSOCIATES, PUBLISHERS
2005 Mahwah, New Jersey London

Lawrence Erlbaum Associates, Inc., Publishers
10 Industrial Avenue
Mahwah, New Jersey 07430
www.erlbaum.com

Cover design by Kathryn Houghtaling Lacey

Library of Congress Cataloging-in-Publication Data

Toward a cognitive science of interdisciplinary collaboration / edited by Sharon J. Derry, Christian D. Schunn, Morton Ann Gernsbacher.
p. cm.
Includes bibliographical references and index.
ISBN 0-8058-3633-0 (cloth : alk. paper)
1. Cognitive science. 2. Interdisciplinary approach to knowledge.
I. Derry, Sharon J. II. Schunn, Christian D. III. Gernsbacher, Morton Ann.
BF311.T656 2005
153—dc22 2004050626
CIP

Books published by Lawrence Erlbaum Associates are printed on acid-free paper, and their bindings are chosen for strength and durability.

Printed in the United States of America
10 9 8 7 6 5 4 3 2 1

Contents

Preface

In this volume, we call attention to a serious need to study the problems and processes of interdisciplinary inquiry, to reflect on the current state of scientific knowledge regarding interdisciplinary collaboration, and to encourage research that studies interdisciplinary cognition in relation to the ecological contexts in which it occurs. Relatively little has been written about interdisciplinary work, education, and research as objects of empirical and theoretical research. This volume helps fill this gap. It contains reflections and research on interdisciplinarity found in a number of different contexts by practitioners and scientists from a number of disciplines. Several contributors to this volume, including its editors, have been allied with a 25-year-old interdisciplinary enterprise—cognitive science—a field that attempts to promote cross-disciplinary integration of concepts, methods, epistemologies, language, data, and infrastructures for research and education on cognition. Several chapters represent attempts by cognitive scientists to look critically at the cognitive science enterprise itself. All of the seven disciplines listed in the official logo of the Cognitive Science Society and its journal—anthropology, artificial intelligence, education, linguistics, neuroscience, philosophy, and psychology—are represented. We believe that the experience of those involved in cognitive science, as well as the theoretical and empirical base of cognitive science broadly defined, can shed significant light on the nature and complexity of interdisciplinary work.

VOLUME OVERVIEW

Our volume is presented in three parts: Theories and Frameworks, Studies of Interdisciplinarity in the Wild, and Studies on Cognitive Science.

Part I: Theories and Frameworks

Chapters in Part I set the stage by providing three broad overviews of literature and theory on interdisciplinary research and education. In "Ethnocentrism of Disciplines and the Fish-Scale Model of Omniscience" (Campbell, chap. 1, this volume), the late Donald Campbell argues that successful interdisciplinarity begins with academic hiring and that universities should form a continuous texture of narrow specialties that overlap. With tongue in cheek, Campbell named his theory the fish-scale model of omniscience. This theory points to important connections between intergroup organization, group processes, and the institutionalization of interdisciplinarity. If the goal of interdisciplinary scholarship is a comprehensive, integrated multiscience, then the obstacle to overcome is the ethnocentrism of disciplines. In chapter 2 (Klein, this volume), "Interdisciplinary Teamwork: The Dynamics of Collaboration and Integration," Julie Thompson Klein discusses the larger history of interdisciplinary science and technology. The emphasis is on natural groups in interdisciplinary problem-focused research, with examples from health care and education. In closing, chapter 2 offers a model for interdisciplinary collaboration and suggests an agenda for research. The final chapter (O'Donnell & Derry, chap. 3, this volume) in this section reviews research on characteristics of effective work groups and origins of difficulties within such groups. However, much research on group work has focused on groups in the laboratory, which differ from interdisciplinary groups in important ways. The study of interdisciplinarity must focus on natural groups that experience special difficulties because of the combinations of disciplines and institutional contexts represented and the implications of disciplinary allegiance to problem representations and solution strategies. In chapter 3, O'Donnell and Derry examine several examples of interdisciplinary teamwork. Methodologies for studying interdisciplinary interaction are suggested.

Part II: Studies of Interdisciplinarity in the Wild

The contributions in Part II are from cognitive science researchers who have examined varied forms of interdisciplinarity in situ. In the now-classic "Seeing in Depth" (Goodwin, chap. 4, this volume), Charles Goodwin unpacks the ways in which scientists from different disciplines must work together on an oceanographic research vessel. Juxtaposition of theory and laboratory practice creates unique possibilities for synergy as members of one discipline make use of the tools of another for different scientific goals. Particular attention is paid to how various scientists coordinate their work and use alternative tools to organize perception in ways appropriate to complementary tasks. In "Disrupting Representational Infrastructure in Conversations Across Disciplines" (Hall, Stevens, & Torralba, chap. 5, this volume), Rogers Hall, Reed Stevens, and Tony Torralba analyze the cognitive processes of making and using generalizations in interdisciplinary work—entomologists working with a statistician and architects working with structural engineers and historical preservationists. Through careful analysis of video records of gesture and speech, Hall et al. show how people try to bridge interdisciplinary differences in representation, categorization, and tools. Hall et al. build on work by Susan Leigh Star (chap. 6, this volume) reported in her chapter titled "Categories and Cognition: Material and Conceptual Aspects of Large-Scale Category Systems." Chapter 6 unravels the organized mysteries found in formal and bureaucratic classifications systems that span disciplinary and organizational boundaries as in the World Health Organization's disease classification system and race classification in South African apartheid. As infrastructural tools, classification systems often create a means of communicating across traditional divisions, which may take place across a wide geographical scale or long periods of time, even decades. It relies not on consensus but on interpretation of the formal system and local modifications suited to the practicalities of the users. In the final chapter (DuRussel & Derry, chap. 7, this volume) in Part II, "Schema Alignment in Interdisciplinary Teamwork," DuRussel and Derry examine a case study of one interdisciplinary research team whose interactions were unsuccessful. DuRussel and Derry's observations are consistent with the hypothesis that particular dimensions of cognitive misalignment, as well as the team's lack of metacognitive awareness of misalignments and how to overcome them through communication, were responsible for the team's inability to perform.

Part III: Studies on Cognitive Science

In this part of our volume, noted cognitive scientists reflect on their experiences and turn the analytical lenses of their own discipline to the critical examination of cognitive science itself as a case study in interdisciplinary collaboration. In "Cognitive Science: Intradisciplinary and Interdisciplinary Collaboration," John Bruer (chap. 8, this volume) discusses the opportunities for cognitive scientists to engage in both interdisciplinary and intradisciplinary collaborations that could position cognitive science as the basic science for the "decade of behavior," reaching both up into applications like educational practice and down into the neuroscience of cognitive processes. In "Making Interdisciplinary Collaboration Work," Susan Epstein (chap. 9, this volume) provides interview data on many successful collaborations in cognitive science and attempts to identify the attributes that made them, in the judgment of their investigators and external arbiters, particularly successful. Consideration is given both to relatively obvious features, such as discipline and topic, and to less obvious ones, including institutional environments and funding agencies. In the third chapter in Part III, "Interdisciplinarity: An Emergent or Engineered Process?," Rogers, Scaife, and Rizzo (chap. 10, this volume) argue that interdisciplinarity is very difficult to achieve and examine alternative ways of advancing understanding, comparing interdisciplinarity with multidisciplinarity as different ends of a continuum. Rogers et al. point out how many successful breakthroughs have come about through approaches that modify and reappropriate existing frameworks and concepts rather than those that mix and match concepts to create new ones.

In "Cognitive Science: Interdisciplinarity Now and Then," Schunn, Crowley, and Okada (chap. 11, this volume) report a historical analysis of the field of cognitive science through analyses of the Cognitive Science Society and its journal. Analyses of departmental affiliations, training backgrounds, research methodology, and paper citations suggest that the journal *Cognitive Science* and the Annual Meeting of the Cognitive Science Society are dominated by cognitive psychology and computer science rather than being an equal division among the constituent disciplines, although 30% to 50% of the work is interdisciplinary. In the chapter, Schunn et al. discuss factors that may have led to the current state of affairs. The final chapter is "Being Interdisciplinary: Trading Zones in Cognitive Science," by Paul Thagard (chap. 12, this volume). Thagard hypothesizes that interdisciplinary research requires "trading zones" in which people from different fields are able to achieve at least partial communication. In this chapter, Thagard discusses some specific trading zones that make interdisciplinary work in cognitive science possible, focusing on analyses of the trading zones between artificial intelligence and psychology and between philosophy and psychology.

This theme of trading zones, a conceptual space and language in which researchers communicate even before they fully understand what meanings their communications carry for themselves and others, occurs repeatedly throughout this volume, represented in some form in nearly every chapter. For example, it is the basis for the "pidgin" that according to Thompson Klein (chap. 2, this volume) is an emergent form of communication in all interdisciplinary work. Star's (chap. 6, this volume) cognitive categories as boundary concepts, Hall et al.'s (chap. 5, this volume) generalizations, and and Derry's (chap. 7, this volume) cognitive misalignments represent different variations and operationalizations of the trading zones observed in interdisciplinary cognition revealed as communicative acts. In this sense, our volume represents alternative views and analyses of both successes and failures in establishing interdisciplinary trading zones. The volume advances theory and begins mapping a research agenda for studying interdisciplinary communicative behavior and, as Bruer (chap. 8, this volume) suggests, mapping behaviors to models of interdisciplinary cognitive process. Although the volume raises as many questions as it answers, we hope and believe it will help foster interest and promote deeper and more informed research and discourse on interdisciplinary science and education.

ACKNOWLEDGMENT

Preparation of this book was funded in part through the National Institute for Science Education, a partnership of the University of Wisconsin–Madison and the National Center for Improving Science Education, Washington, DC, with funding from the National Science Foundation (Cooperative Agreement No. RED-9452971). However, the ideas expressed herein are not endorsed by and may not be representative of positions endorsed by the sponsoring agencies.

Interdisciplinarity: A Beautiful but Dangerous Beast

Sharon J. Derry
University of Wisconsin–Madison

Christian D. Schunn
University of Pittsburgh

Interdisciplinarity—the integration of concepts, philosophies, and methodologies from different fields of knowledge—is pervasive. It permeates all educational levels and is critical to problem solving within government, industry, and academe. It crosses the humanities, social sciences, science, and technology. At its best, it engages participants in collaborative dialog, including debate and conflict, which both transforms the understandings of individual participants and produces new knowledge, new solutions, and even new disciplines that would not be possible without such dialogue. Such collaboration is especially needed when complex, real-world problems cannot be understood or solved with the tools and perspectives of only one discipline. The Web site of the GoodWork® interdisciplinary study project based at Harvard University describes the importance of interdisciplinary collaboration in the following terms:

> Decisive shifts in knowledge production characterize the turn of the 21st century. The alliance of medical doctors, engineers, computer scientists, and molecular biologists is revolutionizing medical care through new, minimally invasive surgery technologies and artificial human tissue development. Pressing social issues like globalization, poverty, terrorism, and environmental survival challenge scientists, historians, psychologists, and artists alike to converge on solutions that defy disciplinary boundaries. Interdisciplinary understanding (i.e., the ability to integrate knowledge from two or more disciplines to create products, solve problems, or produce explanations) has become a hallmark of contemporary knowledge production and a primary challenge for contemporary educators. (http://www.p2.harvard.edu/Research/GoodworksIS.htm)

Although there is much promise in interdisciplinarity, much need for interdisciplinarity, there are also many dangers. In this chapter, we briefly describe two very important facets of interdisciplinary work—research funding and university program development—that help illustrate the beauty and barbs of interdisciplinary collaboration.

FUNDING TRENDS

Since the mid-1990s, there has been a noticeable trend in U.S. (and elsewhere) research funding policy in cognitive science-related fields to emphasize interdisciplinary research. Larger than normal pots of money are placed on the table along with the restriction that many researchers from very different disciplines must come together to work on the grant. Learning and Intelligent Systems, Knowledge and Distributed Intelligence, and the Interagency Educational Research Initiative are just a few examples from the National Science Foundation (NSF). The U.S. military has the Multidisciplinary University Research Initiative program and the more recent Augmented Cognition program. Even larger are various center grants: Math and Science Partnerships, Teaching and Learning Centers, and Science of Learning Centers at NSF, and the Interdisciplinary Behavioral Science Centers for Mental Health at the National Institute of Mental Health.

Given that these programs are typically quite large (often around $1 million per year per grant and as high as $7 million per year per grant), one could (and should!) ask several questions. Is this money well spent? Are such interdisciplinary projects more likely to result in new knowledge than within-discipline projects, especially on a per dollar basis? Interdisciplinary research has many advantages and many disadvantages. Interdisciplinary research has different enabling factors than intradisciplinary research. In this introductory chapter, we just mention a few relevant issues. They are just a tip of the iceberg. In the chapters of this book, readers will find much more detailed information of relevance.

In favor of these large multidisciplinary grants, one might argue that bringing together researchers who do not normally talk to one another on issues that they have in common is a powerful technique for bringing very new insights or revealing important theoretical gaps between adjacent disciplines (see Bruer, chap. 8, this volume, and Thagard, chap. 12, this volume). To bring those researchers together requires some additional incentives such as larger than normal grants. Fewer large grants may be easier than many small ones for funding agencies to monitor because some management work associated with larger grants is offloaded to principal investigators (PI) and their institutions. Moreover, one might argue that complex research problems require interdisciplinary cooperation to make significant progress.

On the negative side, one can point to many problems. From the researcher's perspective, considerably more effort is required to put together an interdisciplinary proposal. New perspectives need to be learned and new relationships forged. As the money goes up for a proposal, so does the number of people applying, sometimes more than proportionally. Thus, the odds for winning the proposal can be lower. With many people on the proposal, the amount of money per investigator can be lower than what could be obtained with a traditional single PI grant. Once the grant is won, then there is the challenge of managing an interdisciplinary team and budget and maintaining good working relationships across disciplines and institutional partners. How is status across disciplines and institutions negotiated (as in the battles of qualitative and quantitative disciplines; Thompson Klein, chap. 2, this volume)? How are disciplinary differences in publications standards and research evaluation criteria and authorship and good versus bad venues negotiated? Finally, on top of intellectual distance between disciplines, there is typically physical distance. Researchers from different disciplines are usually in different buildings or even different campuses, and it is well known that physical distance per se is a strong barrier to collaboration (see Hinds & Kiesler, 2002; Olson & Olson, 2001). Although it can be argued that new Web-based technologies can help alleviate communication difficulties, such technologies must be standardized across multiple institutional structures, are acquired, learned, and maintained at substantial expense, and are not always successful (Cummings & Kiesler, 2003).

From the funding agency perspective, there is a question about whether the researchers will be able to successfully work together on the project (see DuRussel & Derry, chap. 7, this volume). Relative to traditional intradisciplinary collaborations, interdisciplinary collaboration requires a different management style for the leader of the group and different work styles between members of the group (Thompson Klein, chap. 2, this volume). The forces against interdisciplinary work can be strong, on top of the sheer inertia of researchers simply continuing to work within their discipline (see Campbell, chap. 1, this volume). Is it wise for the funding agencies to give money to researchers who have no experience and training in interdisciplinary collaboration and limited knowledge of how to bridge disciplines or how to coordinate and manage large-scale collaborations and grant budgets? Few graduate programs prepare future academics for such major responsibility.

One could argue that much of the money invested in interdisciplinary programs might have been better spent and that future interdisciplinary programs in the cognitive sciences should pay greater attention to what interdisciplinary research actually requires. In past competitions, PIs on the grants have not had to demonstrate prior experience or skill in leading interdisciplinary collaborations. The work plan did not have to specify how the gaps (conceptual, terminological, work practice) between the disciplines would be spanned (see Hall, Stevens, & Torralba, chap. 5, this volume). The timelines did not have to acknowledge or account for the typically long period of familiarization required between members across disciplines (see Epstein, chap. 9, this volume). As a result, many projects did not get off the ground or simply separated into component discipline subprojects without ever coming back to the issues of integration that were presumably at the heart of the originally proposed work. In this book, readers will find several concrete examples and analyses of these kinds of problems in interdisciplinary collaboration.

UNIVERSITY PROGRAMS

Complementing trends in interdisciplinary funding are the programmatic changes taking place within university and college settings. Recognizing their responsibility to both participate in and prepare students to participate in a world in which problems defy disciplinary boundaries, many institutions of higher education encourage interdisciplinary scholarship through their policies, hiring practices, and curricular offerings.

Current examples of interdisciplinary hiring initiatives include the University of Wisconsin–Madison's cluster hiring program, which includes, for instance, a plan to support cross-department hires in cognitive science. Other universities, such as Northwestern and Indiana University, have hired to build and maintain interdisciplinary graduate programs in the learning sciences, a branch of cognitive science that focuses on problems of teaching and learning in classrooms and other educational settings. Researchers in this field draw on multiple theoretical perspectives and research paradigms to understand complex learning environments at different levels of analysis, and graduate students trained in such programs experience interdisciplinary collaboration first hand. Another type of graduate training initiative is exemplified by the Graduate Research Training Program in the School of Education at the University of Wisconsin–Madison funded by the Spencer Foundation. Students in this program apprentice with faculty working from different research traditions and attend multidisciplinary seminars. The program is designed to insure that future educational researchers are broadly trained to employ multiple interpretive traditions and methods in their work. Finally, many undergraduate interdisciplinary programs designed to promote dialogue between science and the humanities, such as San Francisco State's NEXA Program funded by the National Endowment for the Humanities, have also emerged. NEXA's Web site (http://www.sfsu.edu/~nexa/) describes its mission as offering "a curriculum that demonstrates the historical, philosophical, and ethical interactions among humanities, arts, business, and the physical and social sciences."

Clearly, many interdisciplinary initiatives are well underway. Yet, there is much in university culture that mitigates against them, making interdisciplinary hiring and teaching very difficult. On many campuses, interdisciplinary hiring programs strain to overcome the limits of historical, social, administrative, and physical institutional boundaries that are highly entrenched and make meaningful interdisciplinary collaborations difficult if not impossible. The outcomes can be comical if not disastrous and may include "bridging" faculty members who find themselves appointed simultaneously in two departments and shuttling between two offices and two sets of faculty meetings located several miles apart. Similarly, faculty who participate in both interdisciplinary scholarly programs and communities (such as cognitive science) and the activities of their home department (such as computer science or psychology) may find the demands on their time nearly doubled. Faculty from one discipline (such as educational research) who are "dropped in" to another (such as a college of engineering) can be cut off from mentorship and meaningful daily interactions with their own disciplinary cultures. The demands of "tribalism" (Campbell, chap. 1, this volume) in the internal and external relations of university departments and national organizations take their toll on interdisciplinary good intentions. Faculty with interdisciplinary interests typically find they must still "fit" conventional standards and norms of scholarship for their home departments, which may reject or redirect candidates who are doing innovative work at the boundaries of a discipline as locally conceived. Those hired specifically for less stable interdisciplinary programs may find themselves adrift. (Even as we go to press, it has been announced that San Francisco State's NEXA program may be eradicated for lack of funding.) Thus, interdisciplinary scholarship is a risky business, especially for junior faculty.

CONCLUDING COMMENTS

In sum, although large amounts of human and monetary capital are now invested in interdisciplinary research and teaching enterprises within university settings, it is also clear that old systems and ways of doing things are holding fast, sometimes preventing new ones from struggling into being (Thompson Klein, chap. 2, this volume). Thus, yet another motivation for our volume was to raise awareness among funding agencies, faculty, and administrators of what is required for successful interdisciplinary collaboration. A body of literature based on "wisdom of practice" has emerged in recent years, but this is not enough. Importantly, researchers are beginning to conduct scientific investigations that examine how social and cognitive processes interact to drive intellectual growth and construction of intellectual products within these natural interdisciplinary groups, shedding light on what is required to support and evaluate such processes and outcomes. Programs such as the interdisciplinary study component of the Goodwork® Project examine exemplary collaborations. However, many studies in this book examined instances of more "typical" efforts and in greater detail. Although a few of these efforts succeeded, others only partly succeeded or almost entirely failed. Studies of both successes and failure are required to help shape the picture of what good interdisciplinary collaboration requires.

The emerging picture indicates that successful integration of disparate disciplines is a complex and difficult process. Interdisciplinary work requires skillful management as well as openness to and ample time for learning new fields through collaboration. It is fraught with and builds on conflict and misunderstanding. It can be hindered and constrained not only by time and resources and entrenched disciplinary-based systems of evaluation, but also by physical space and proximity as well as available methods of communication and the skills involved in utilizing them. Due to the historical, institutional, and physical contexts that constitute universities and other organizations that host disciplines, due to the way interdisciplinary teams are assembled, and due to the lack of time and knowledge and managerial skills required for interdisciplinary work, many initiatives are likely to fail either partly or completely in terms of their interdisciplinary goals. In fact, we may wish to question whether interdisciplinarity is an impossible holy grail in most current university contexts. In chapter 10 in our volume, Rogers, Scaife, and Rizzo argue that major breakthroughs result not from interdisciplinary collaborations but from multidisciplinary ones in which limited tools and concepts are shared and borrowed across disciplines while disciplinary alliances are maintained. We hope this and other chapters in our volume will help readers assess whether their program and project designs might more reasonably aim for multidisciplinary rather than interdisciplinary outcomes.

Yet, interdisciplinary collaborations exist. Interdisciplinary work must be done to solve certain problems with sufficient funding and adequate management, some projects are successful (Cummings & Kiesler, 2003). What we attempt to make salient in this book is what goals are realistic and some problems that must be tackled if interdisciplinary work is to succeed. The research reported here confirms that the conceptual systems drawn from cognitive science, broadly defined, can supply useful analytical lenses for studying and helping improve interdisciplinary work. This is an important goal not only because complex world issues require integration of multiple knowledge bases and approaches but also because organizations are today investing in very large-scale interdisciplinary efforts.

As editors of this book, we hope that three key questions will arise in readers' minds as they read the following chapters:

1. What are the problems associated with interdisciplinary work both in general and in cognitive science in particular?
2. How does successful interdisciplinary work proceed (through work structure, graduate preparation, leadership, team makeup, etc.)?
3. What concepts and ideas drawn from the cognitive sciences, broadly conceived, can help both managers and researchers with-

in interdisciplinary teams and researchers of interdisciplinary collaboration address these issues?

With this information in hand, hopefully we will be in a better position to develop more appropriate interdisciplinary grant competitions, write more plausible interdisciplinary grant proposals, and conduct and evaluate successful interdisciplinary work.

REFERENCES

Cummings, J., & Kiesler, S. (2003). *KDI initiative: Multidisciplinary scientific collaborations*. Retrieved Jan 4, 2004 from http://netvis.mit.edu/papers.NSF_KDI_report

Gardner, H., & Mansilla, V. (PIs). *The GoodWork Project®: Interdisciplinary study*. Retrieved January 4, 2004, from http://www.pz.harvard.edu/Research/GoodWorkIS.htm

Hinds, P., & Kiesler, S. (2002). *Distributed work*. Cambridge, MA: MIT Press.

Olson, G. M., & Olson, J. S. (2001). Distance matters. In J. M. Carroll (Ed.), *Human–computer interaction in the new millennium* (pp. 397–417). New York: ACM Press.

I

Theories and Frameworks

1

Ethnocentrism of Disciplines and the Fish-Scale Model of Omniscience[1]

Donald T. Campbell

This chapter is a preliminary exercise in the sociology of science—an exploratory application of principles of groups and intergroup organization to group processes in the institutionalization of science. The goal in this book is a comprehensive, integrated multiscience. The obstacle described in this chapter is the "ethnocentrism of disciplines," that is, the symptoms of tribalism or nationalism or ingroup partisanship in the internal and external relations of university departments, national scientific organizations, and academic disciplines. The "fish-scale model of omniscience" represents the solution advocated, a solution kept from spontaneous emergence by the ethnocentrism of disciplines. The slogan is collective comprehensiveness through overlapping patterns of unique narrownesses. Each narrow specialty is in

[1]From *Interdisciplinary Relationships in the Social Sciences* (pp. 328–348), by M. Sherif and C. W. Sherif (Eds.), 1969, Chicago: Aldine. Reprinted by permission of the estate of the late Donald T. Campbell. The author is deceased. At the time of his death he was Professor Emeritus at Lehigh University

this analogy a "fish scale." Figures 1.1 and 1.2 illustrate the title. Our only hope of a comprehensive social science or other multiscience lies in a continuous texture of narrow specialties that overlap with other narrow specialties. Due to the ethnocentrism of disciplines, what one gets instead is a redundant piling up of highly similar specialties leaving interdisciplinary gaps. Rather than trying to fill these gaps by training scholars who have mastered two or more disciplines, we should be

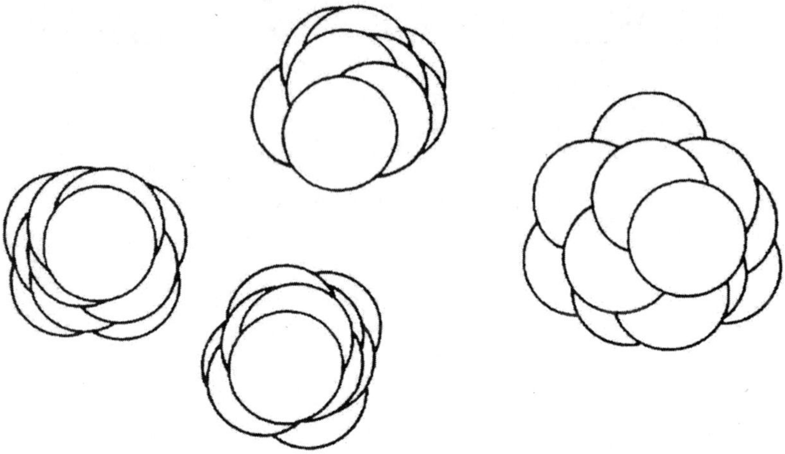

FIG. 1.1a. Present situation: Disciplines as clusters of specialties, leaving interdisciplinary gaps.

FIG. 1.1b. Ideal situation: Fish-scale model of omniscience.

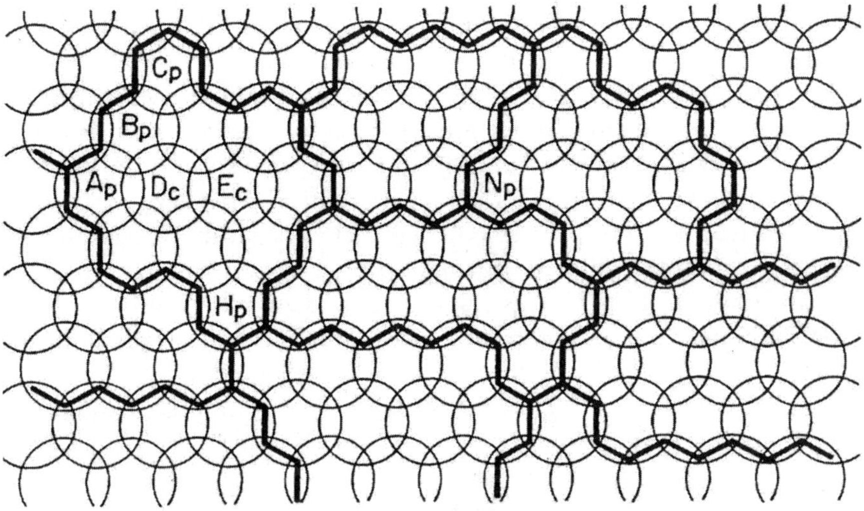

FIG. 1.2.a. Hypothetical pattern of specialty overlap at the time of superimposition of arbitrary "departmental" boundaries (heavier lines).

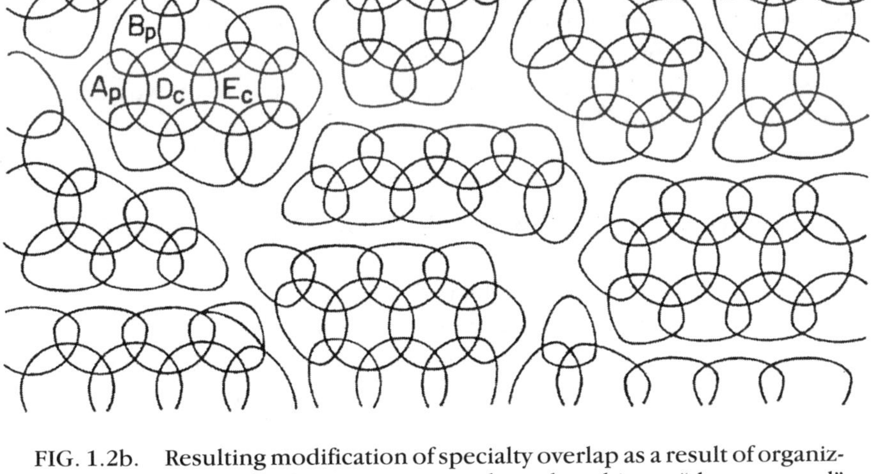

FIG. 1.2b. Resulting modification of specialty overlap as a result of organizing decision making and communication along the arbitrary "departmental" lines. ("Departmental" boundaries omitted to facilitate inspection of specialty overlap pattern.)

making those social-organizational inventions that will encourage narrow specialization in these interdisciplinary areas.

The diagrams of Figure 1.1 and 1.2 are of course an oversimplification, an analogy in two dimensions of what should be *n* dimensional. Particularly likely to be dyscommunicative are the exaggerated gaps between the disciplinary clusters. The real situation is perhaps more often one of unrecognized overlap. The disciplinary clusters may at their edges overlap other clusters, but as ships that pass in the night, they fail to make contact. The clusters, as it were, may overlap but lie on independent planes. Such an alternate diagramming might be possible, and the reader is invited to substitute it in his or her mind's eye in what follows. The issue, too, is not one of total absence but of relative density of interdisciplinary specialties.

COMPREHENSIVE TRAINING AND THE LEONARDESQUE ASPIRATION

Too often in discussions of interdisciplinary training one hears calls for breadth, for comprehensiveness. Too often we attempt the production of multidisciplinary scholars, professionals who have mastered two or more disciplines, rather than interdisciplinary specialists. This orientation I parody as the "Leonardesque aspiration": the goal of creating current-day Leonardos who are competent in all of science. As a training program, it is bound to fail in one of two directions. At its worst, it produces a shallowness, a lowest common denominator breadth, an absence of that profound specialization that is essential for scientific productivity. At best, it is evaded in the direction of the interdisciplinary narrowness here advocated.

BURGEONING LITERATURE AND THE OBLIGATIONS OF INTERDISCIPLINARIANS

One of the several background facts that lies behind my emphasis is the enormous past and burgeoning present literature. Speaking for myself, any volume such as this raises guilt feelings in that it acquaints me with the existence of scientific literatures of obvious relevance to my work that I have neglected. William McGuire's paper (1969), for example, lies in an area of high relevance to my work, and yet I have not read or read at even half of his citations and was not at all aware of the existence of another sizable proportion. However, it is not only multidisciplinary conferences that have this effect. Unidisciplinary conferences remind me that I am failing to keep up in areas once central—and so does the arrival of one of the few journals I take (and read only 10% of), or chance inspection of one of the many journals equally relevant to which I do not subscribe.

What seems to me essential is that moving into an interdisciplinary problem area not increase this obligation and guilt—that for every new literature we pick up, we are excused from or drop some other literature so that the interdisciplinarian is free to remain as narrow, as specialized, as any other scholar.

MYTH OF UNIDISCIPLINARY COMPETENCE

Lying behind many models of interdisciplinary competence is an unrealistic notion of unidisciplinary competence—the image of scholars competent in one discipline. It will clarify the discussion of interdisciplinary competence to recognize at the outset that there are no such persons. What we have instead is a congeries of narrow specialties, each one of which covers no more than one tenth of the discipline with even a shallow competence. Yet individual disciplines do have some integrity, some comprehensiveness—at least in comparison with social science as a whole or with specific interdisciplinary areas. What must be recognized is that this integration and comprehensiveness is a collective product not embodied within any one scholar. It is achieved through the fact that the multiple narrow specialties overlap and that through this overlap, a collective communication, a collective competence and breadth, is achieved. This approach is our only hope for a unified and complete behavioral science. The present social organization of science impedes it.

LOCUS OF SCIENTIFIC KNOWLEDGE IS SOCIAL

Philosophy of science and epistemology have not yet assimilated the fact that the problem of knowledge must, in the end, be stated at the social level—although Charles Sanders Peirce and James Mark Baldwin, for example, were making this point at the turn of the century. Moving the problem of knowledge from a solitary viewer's vision to language is a step, but the implicit model is still usually a single native speaker with perfect knowledge of a stable language. Sufficient attention is not yet given to the social and incomplete conditions of language learning, to the fundamental idiosyncrasy and errorfulness of functional individual lexicons, to the very partial distribution of words that are still somehow "in" the language, to the effective redundancy that makes imperfect language as competent as it is. When these have been assimilated, the locus of "truth" and "knowledge" will have clearly shifted from individual "minds" to a collective social product only imperfectly represented in any one mind. Similarly, in the philosophy of science, the competence, the discipline, the verification, and the integration are all in the end social products, imperfectly and incompletely represented in the work of any one scientist. Michael Polanyi (1966) wrote to this point in *The Tacit Dimension* identifying the locus of scientific authority as the "Society of Explorers" itself:

> ... the *principle of mutual control.* It consists, in the present case, of the simple fact that scientists keep watch over each other. Each scientist is both subject to criticism by all others and encouraged by their appreciation of him. This is how *scientific opinion* is formed, which enforces scientific standards and regulates the distribution of professional opportunities. It is clear that only fellow scientists working in closely related fields are competent to exercise direct authority over each other; but their personal fields will form *chains of over-lapping neighborhoods* extending over the entire range of science. It is enough that the standards of plausibility and worthwhileness be equal around every single point to keep them equal over all the sciences. Even those in *the* most widely separated branches of science will then rely on each other's results and support each other against any laymen seriously challenging their authority. (p. 72)

PRESENT DISCIPLINES AS ARBITRARY COMPOSITES

Although it is probably not essential to the perspective, it is certainly relevant that the present organization of content into departments is highly arbitrary, a product in large part of historical accident.

Thus, *anthropology* is a hodgepodge of all novelties that struck the scholarly tourist's eye when venturing into exotic lands—a hodgepodge of skin color, physical stature, agricultural practices, weapons, religious beliefs, kinship systems, language, history, archeology, and paleontology.

Thus, *sociology* is a study of social man in European industrialized settings, a hodgepodge of studies of institutional data in which persons are anonymous—of individual persons in social settings, of aggregates of person data losing both personal and institutional identity, and of interactions that are neither persons nor groups.

Thus, *psychology* is a hodgepodge of sensitive subjective biography, of brain operations, of school achievement testing, of factor analysis, of Markov process mathematics, of schizophrenic families, of laboratory experiments on group structure in which persons are anonymous, and so forth.

Thus, *geography* is a hodgepodge of land-surface geology, of industrial development, of innovation diffusion, of social ecology, of political territoriality, and of visual perception of areal photographs, and of subjective phenomenology of mental maps.

Thus, *political science* is a hodgepodge of political entities as actors and persons as actors, of humanistic description and scientific generalization, of history, and of social psychology.

Thus, *economics* is a hodgepodge of mathematics without data, of history of economic institutions without mathematics or theory, and of an ideal model of psychological man.

There are no doubt many natural divisions within the domain of the social or behavioral sciences—but they are not employed in the allocation of content to disciplines. A hierarchy of levels of analysis exists in which the focus of differential description at one level becomes the assumed undifferentiated atoms of the next: This is the atom-molecule-cell-organ-organism-social group and so forth model. On this hierarchy, sociology, political science, geography, and anthropology are all mixed across the individual and group levels and so is experimental social psychology. The experimental laboratory work of Sherif, Lewin, Lippitt, and Bavelas in many instances has represented psychologists doing experimental sociology, experimenting with social structure, developing laws about social norms in which persons are treated as undifferentiated atoms and in which the resulting laws relate social structural and group-product variables.

Another natural division is between the descriptive humanistic on one hand and the scientific on the other. On this dimension too our departments and disciplines are mixed, with all save history having both strong scientific and strong descriptive-humanistic factions.

There are ways of describing the internal logic and coherence of disciplines, but this cannot be done with singular principles and still capture both the reason for the content within one discipline and the reason for the same content's appearance or exclusion in others. Certainly the dimensions so used would be only a partial sample of potential classificatory criteria.

The specialties within disciplines are more coherent, and eventually such specialization takes over, each scientist allowing the congeries of irrelevancies within his or her own disciplinary knowledge to atrophy, journals to go unread, subscriptions to lapse, and so forth. The temporary disciplinary breadth transiently achieved in graduate school is of course not undesirable—the objection here is rather to the repetitious duplication of the same pattern of breadth to the exclusion of other breadths equally relevant but organizationally unsupported.

Effects of Organizing Specialties into Decision-Making Units

Consider what would happen if we took a large domain of specialties and aggregated adjacent dozens into "departments" as collective decision-making units. In this hypothetical example, suppose that the aggregation has been arbitrary except for the requirement of adjacency, that all specialties are equally well staffed to begin with, and that any specialty can belong to only one department. We are now interested in the effect of this second-level organizational structure, this superimposition of departmental boundaries on the specialty boundaries. We are particularly interested in differential effects on the future growth of the "central" specialties versus the peripheral or marginal specialties, this centrality or marginality being in this hypothetical case a purely arbitrary by-product of where the administrative boundaries happened to fall. Figure 1.2 is an effort to portray this starting point. (In Figs. 1.2a and 1.2b, A_p, B_p, and C_p, are examples of peripheral specialties, and D_c and E_c of central ones.)

Consider first purely internal decision making. For most issues, there are differences in priority and preferences that are associated with specialty points of view, and overlapping specialties are apt to have overlapping preferences. The situation will also be that collective decisions will be achieved only by a consensus of a plurality of specialties within the department. The accidentally central specialties have more natural allies within the department and find it easier to achieve support for their concerns. The natural allies of the peripheral specialties lie in other of the arbitrary departments and are organizationally prevented from effectively presenting their consensuses. The incidentally central specialties are also more frequently the compromise candidates, and when an ideology is needed to rationalize the historically arbitrary departmentalization, it is the concerns of the central specialties that are chosen as epitomizing the true common denominator as the essence of the initially arbitrary aggregate. Centrality becomes reinterpreted as common root, trunk, and fountain head when initially, it meant only remoteness from the boundaries with other departments.

Here are some considerations that would illustrate these purely internal dynamics. The selection of a chairperson in a peaceful department will follow these lines—only if a peripheral is a compromise between two strong factions, each with some central specialties, will he or she be a chairman. Deciding on a core curriculum to be required of all students, the minimum essential to their being sound, well-grounded *X*-ologists, will go in this direction. Pressure there will be to require one course from each specialty. Demands for time to meet the needs of specializing will preclude this in any but the smallest departments. The process of choosing a smaller set will involve the elimination of peripheral specialties because inherently, in the arbitrary organizational structure, there are fewer other specialists in the department who deem them important. This tautology may of course be expressed ideologically as "that's not really *X*-ology—it even comes close to being *Y*-ology." The setting of qualifying exams and dissertation committees will exercise a constant centralizing bias on the training of graduate students in peripheral specialties. Going across departmental lines in study programs is wasted effort as far as these important hurdles are concerned. The peripheral specialist himself or herself will be anxious that his or her students show up well on the central core content and will be willing to see equally or more relevant cross-department content be neglected because no punishment is involved in its neglect, no institutional reward in its achievement. This is a minor problem if these departmental hurdles are low. However, there is great pressure on departments to achieve excellence, and there is a perverse tendency to see this as implemented by high standards in the achievement of passive regurgitative mastery of past achievements in the literature. Under the institutional decision-making arrangements here described, the inevitable effect of higher standards in training is greater neglect of peripheral and cross-departmental content, and this is without the by-product of increasing the profundity of specialized training for any but the central specialties.

Deciding who in the department merits a raise or is ready for promotion, whose competing offer it is essential to meet and whose the department has not the funds to match, whose specialty needs additional staff and space—all show the effect of the arbitrary location of departmental boundaries on what becomes defined as central or peripheral. In my professional career at four universities, I have repeatedly seen men who were of exceptional creativity and competence and who were absolutely central as far as social science or behavioral science was concerned (occupying positions like those designated as H or N in Fig. 1.2a), be budgetarily neglected and eventually squeezed out because the departmental organization of specialties made them departmentally peripheral. That in some cases these were arrogant men does not explain the result because their detractors were also arrogant men but more centrally located. However, a selective feedback may well be at work; it may be that of all those who are initially attracted to peripheral, boundary-crossing specialties, only the arrogant persist in bucking the institutional pressures that would otherwise move them to more centrally defined specialties.

The dynamics just described are internal to our hypothetical departments. These arbitrary budgetary units compete with each other as budgetary entities for budget increases, space increases, and personnel increases. The specialties within the arbitrary departments thus come to share common fate and become joint actors in competition with the other arbitrary aggregates of specialties. This common fate, although arbitrary in its initiation, is real enough in practice to provide the basis of an ingroup identification against competitive outgroups. There also develops implicit or explicit competition for the most talented students and indoctrination procedures designed to maintain the loyalty of those who have tentatively joined. Characteristic of ingroup–outgroup relations in other settings, these indoctrination procedures not only emphasize ingroup virtues and ideology but also contrasting outgroup faults. Philosophy and sociology departments have frequently maintained internal solidarity by teaching about the wrongness of behavioristic psychology. Sidney Aronson (1969) has documented the manner in which historians build ingroup morale by deprecating sociologists. (It is even symptomatic of ethnocentrism that the first illustrations that come to his and my mind are ones in which our own departments are being wrongly disparaged by outgroups rather than vice versa.)

If I add to my hypothetical example the feature that on most university campuses, the arbitrary aggregations into departments be parallel as to which specialties are combined, further institutional pressures emerge. Departments' students must be prepared to appear adequate to centrally dominated hiring committees at other universities. The new faculty appointments to the department must be ones that inspire admiration on the part of the parallel departments of other universities.

Effect of Departmental Organization on Scientific Communication and Specialist Competence

In this topic, I come to the most direct effect of departmental organization on scientific knowledge. Each scientist's competence, his or her participation in the collective activity of science, is based on communication. The hypothetical departmental organization under consideration affects communication patterns in many ways. In my hypothetical example, suppose that those specialties aggregated into an arbitrary department be housed adjacently but that departments be scattered at random. Incidental oral and paper-passing communication links thus become predominantly intradepartmental, and the extradepartmental aspects of peripheral specialties suffer great relative neglect in comparison with their intradepartmental overlaps. Shop talk, reading of dissertations, reading of each other's preprints and reprints, and looking at laboratory setups and research instruments all illustrate primary modes of communication seriously warped. The bias extends also to books and journals read. One is rewarded socially for shared detailed reading of exciting new developments. No such reward occurs for unshared reading, and thus, the literature in the cross-departmental aspects of a specialty loses ground to the reinforced intradepartmental reading.

With a parallel arbitrary organization occurring at other universities and in national and international disciplinary organizations, still other boundary effects occur. Professional organizational membership and their journal discounts to members lead to stereotyped patterns of journal subscription, with most members limiting themselves to journals within one field—these are invariably so voluminous that one excuses himself or herself for not reading in other fields by noting that this would be foolish when he or she is not even able to keep up in "his or her own field." "Own field," needless to say, tends to become defined in terms of these arbitrary departments whenever it goes beyond the narrow specialty. Abstracting sources, annual reviews, handbooks, and so forth, further the redundant repetition of intradepartmental patterns of breadth at the expense of equally relevant cross-departmental ones.

The annual reviews and handbooks do another disservice from the fish-scale model. Even within departments, each scholar's coverage must be partial, collective competence being assured by different partialities for each. Annual reviews and handbooks become crutches leading large numbers of scholars to the same partiality and the inevitable neglect of other partialities. The degree of this partialness is generally underestimated, impressed as one must be at the long bibliographies appended.

The departmental grouping of communicators allows unstable language to drift into unintelligibility across departments. A basic law is that speakers of the same language, once isolated into separate communities, drift into local idiosyncrasies and eventually unintelligibility once the discipline of common conversation is removed. This tendency produces departmental linguistic idiosyncrasy even for shared contents and referents. Furthermore, as Edmund Leach and others have noted, such idiosyncrasy may be exaggerated as an ingroup solidarity device. What is despised as jargon by the outgroup may be the shibboleth of adequate professional training by the ingroup.

RESULTING ETHNOCENTRISM OF DISCIPLINES

Figure 1.2b shows the results of these dynamics. A few isolated specialties have been lost, and the peripheral specialties have allowed their cross-departmental edges to atrophy, producing departmental discreteness and interdepartmental gaps.

What started out as a hypothetical case has occasionally in the previous paragraphs drifted into the description of current actualities, perhaps blurring the point being made: Even if the true nature and the historical starting point had been one of a homogeneous texture of overlapping specialties, the organizing of specialties into departments and disciplines for decision-making and communication purposes would have produced disciplinary discreteness, cohesiveness, and interdisciplinary gaps such as exist at present. These features therefore should not be judged to justify as natural the existing arrangements of specialties.

The historical origins of departments and disciplines are of course quite other than those of my hypothetical example. The actual histories add historical depth to the tribal myths of origin and no doubt provide a greater true commonality to departments than in this hypothetical instance. Yet no matter what the degree of valid core and discreteness, the dynamics described here would inevitably have the effect of artificially enhancing it.

A MISCELLANY OF REFORMS

A New Ego Ideal for the Scholar as Student

At the present time the ego ideal of scholars calls for competence, for complete knowledge of the field they claim as theirs. In their everyday interaction with fellow specialists, scholars tend to feel guilty when they find that they have not read what others have read. Although scholars inevitably learn to live with such guilt feelings, this ego ideal spells the direction of their guilt. If scholars take to heart the notion that scientific competence can never be embodied in single minds—that their guilty neglect is not their unique shame but the inevitable predicament of all and that science is somehow achieved in spite of this—they may come to substitute a quite different ego ideal, a quite different focus of guilt. Rather than praying, "May I be a competent and well-read *X*-ologist, may I keep up with the literature in my field," a scholar will pray, "Make me a novel fish-scale. Let my pattern of inevitably incomplete competence cover areas neglected by others." Each scholar would then try to have a pattern of journal subscriptions unique to his or her department, university, or profession. Noting that the scholar and a colleague were reading the same set of journals, the scholar would feel guilty and vow to drop one of these in favor of some other. Recognizing that the interdisciplinary links in the collaborative web of knowledge are the weakest, the scholar would give up some ingroup journal in favor of an outgroup one. The scholar would feel guilty if he or she did not cut attendance at ingroup conventions to attend relevant outgroup ones, and so forth.

There is a secondary payoff in cross-disciplinary reading and conventioning. Scholarly reading for the true scholar is ideally recreational, something the scholar enjoys doing and would choose to do reactionally even if he or she had some other profession. The social system of science, particularly in graduate school and in the first stages of a scientific career, associate such activities so strongly to the reward and punishment system of competitive evaluation that they cease to be relaxing or effectively recreational. In the current disciplinary organization, the scholar will often find that the journals and conventions of a neighboring discipline can still serve this recreational function—particularly if the scholar accepts his or her role as smatterer and does not assume the obligation of "mastering" that literature. If my fish-scale model were to become the norm, such reading would of course tend to become more obligatory and run the risk of ceasing to be recreational.

While on the themes of recreational reading and the duplication of fish scales, it seems appropriate to deplore the tendency of social scientists to feel that they all should read current newspapers, particularly the *New York Times.* Certainly the collective perspective would be better if most spent the equivalent time with newspapers of other epochs or with historical, anthropological, archeological, or literary descriptions of quite other samples of social milieus. Rather than the ego ideal of keeping up with the current worldwide social developments, the young scholar should hold the ideal of foregoing current informedness for some infrequently sampled descriptive recreational literature. Too often our ego-ideals call for uniform omniscience, knowledge of both past and present, of both here and there, and too often we settle for the same pattern of compromise all our colleagues are settling for. Compromise from the Leonardesque aspiration there must be, but even in leisure reading, one can hold as ideal the achieving of unique compromises.

An Ego-Ideal for the Scholar as Teacher

Under the ideology of disciplinary competence, a department feels that its staff should be able to provide competent guidance in the PhD programs it offers, that students should bend their interests to the locally available specialties or go elsewhere, that to train a student properly, the faculty should be more competent than the student in the student's area of study. Combining this with the organizational advantages of one-headed decision-making units in dissertation direction, and so forth, a strong tendency toward duplicating of identical fish scales within departments results.[2]

Under the fish-scale ideology, professors would feel guilty when they turned out "chips off the old block," PhDs who showed the same pattern of overlap as they did. The goal instead would be to encourage each new PhD to select such a novel specialty that each one could indeed within his or her graduate training become one of science's leading experts, a fully contributing specialist. Note the difference between this approach to instant expertness and that of exhaustive mastery of an exceedingly narrow

[2]For total careers, there is, of course, considerable freedom to redefine personal specialties within disciplines. This, combined with the needs for autonomous personal identity within face-to-face departments, reestablished the needed texture of overlap within disciplines.

realm within one specialty. The latter asks graduate students to be narrower than their mentor, to subspecialize within their mentor's range. The former asks the graduate student to achieve a novel range, not necessarily broader or narrower than the mentor's. One greatly needed implementation of this goal is to encourage the PhD to give up some traditional intradisciplinary subfield in favor of mastery of a cross-disciplinary one of relevance. A common reason given for rejecting this is that for such a content, we *X*-ologists would have no way of checking the PhD's competence—a typically ethnocentric reaction, illustrating again the way in which, given our present organization into departments, concern over evaluation of competence decreases the crossing of departmental bounds in specialization.

Using the Advantages of Smallness and Bigness

Generally speaking, the larger the university, the wider the variety of specialists available and the more likely that the full range of possible specialties be represented by actual persons. This being so, one might expect the interdisciplinary gaps to be less and interdisciplinary collaboration to be more frequent. In actual practice, the contrary is more apt to be the case (although we need studies to verify this), and the laws of group organization and size applied to departmental organization explain this. The larger the department, the more obligatory relationships there are intradepartmentally and the more the totality of obligatory and informal relationships is predominately intradepartmental. The larger the department, the more required courses there are for graduate students within the department (Aronson, 1969, cites a study by Sibley to this effect) and the less opportunity and the greater jeopardy for cross-departmental study. In the smaller department, the loyalty demands are less, informal communication and friendship links are more frequently cross-departmental, collaboration across departmental lines is less apt to involve loss of intradepartmental esteem, and graduate students are more apt to feel sufficient mastery in their home departments to have time to explore outside. Northwestern University has had an exceptionally productive period of interdisciplinary collaboration in recent decades, at a time when most universities have been finding such relationships increasingly impracticable. The explanation lies in part in the chance accumulation of a few key leaders with this orientation, such as Richard Snyder in political science, but more significant, I believe, has been the smallness of its departments combined with low teaching loads and a full commitment to research and graduate training. Most places this small are focused on undergraduate training. Most places focused on research and graduate training combine this with largeness. In contrast stands the University of California at Berkeley where interdisciplinary contacts have steadily decreased as departments grew and where once interdepartmental institutes have become annexes of single departments.

The reprise is Bigness increases the isolation of departments and decreases the interdepartmental fish scales unless organizational reforms are devised to prevent this.

Ad Hoc Interdisciplinary Training Programs

Even at Northwestern, what crossing of disciplines we achieve is mostly at the faculty level. My own graduate students are as unidisciplinary as any and rarely do more than a required course or two outside of psychology. Were I to push them to a real mastery of some relevant-to-them, cross-disciplinary specialty—be it time-series analysis in economics, analysis of ideology in sociology, or child rearing customs in anthropology—I would be adding to an already inhibiting burden of requirements. However, if I could at the same time relieve them of the need to master some physiological psychology, or the sensory discrimination literature, or the like, such programs would be possible. They are also much needed, not because they would be better than the present mix but because they individually might be as good and because the combination of some of these plus some of the standard would be collectively stronger than the present all-of-one-type.

At a place with relatively good (although minimal) interdepartmental contact like Northwestern, one should be able to train hybrid specialists such as these with an ad hoc assembly of core courses, fields for prelims, and dissertation committees tailor-made for each student. The adviser and student would assemble an ad hoc training advisory faculty and schedule of courses. Perhaps a divisional review committee would have to check the program to assure that it was as exacting and as coherently specialized as the standard programs required in the overlapping departments. Such a committee would also conduct written and oral qualifying exams and guide the dissertation. If the possibility of use to evade difficult requirements emerged, the program might be restricted to only the top half of entrants. If perceived as a vehicle for inadequate professors, the prerogative of advising on such programs could be limited to an elite of proven unidisciplinary capability, and so forth. To coordinate with the labels of the disciplines nationally, the PhD would be designated according to the department of his or her senior advisor. If departments objected to thus anointing an inadequately trained *X*-ologist, a dual labeling could be adopted—PhD in *X*-ology (divisional) for the new type, PhD in *X*-ology (departmental) for the old. In the present academic market, we should have no difficulty placing such PhDs. Thus, with no such drastic reorganizations as creating new rigid interdepartments, with no new budgetary units, and with no new staffs, a congeries of new specialties each as coherent and narrow as our present PhD programs could be achieved.

Paralleling such ad hoc training programs for graduate students, ad hoc decision groups substituting for departments in deciding on raises, promotions, and tenure, might well be established for interdepartmental faculty appointments with some divisional funds allocated for such purposes separate from departmental budgets.

Organizational Alternatives for Journals and Conventions

Our academic professional organizations publish journals and offer them to their members at reduced rates. There results a repetitious patterning in the several journals each scholar takes. If our associations would make these same rates available to the members of other disciplinary associations (as indeed some do), novel journal sets would become more frequent. In the creation of new journals, broad interdisciplinary scope should be eschewed for novel narrowness. The *Journal of Verbal Learning and Verbal Behavior* sets an excellent example. It is much narrower than the *Journal of Experimental Psychology*, which it overlaps heavily, but it juxtaposes work by experimental psychologists and by linguists so as to eventually nurture some novel fish scales.

Our professional conventions are too large, have too many simultaneous meetings, cover too wide a range of specialties, and last too long. If our conventions instead were held at different times by each specialty (e.g., by divisions within the American Psychological Association), and if the scholar did his annual conventioning by attending several of these shorter ones, he would be much more likely to cross disciplinary lines.

SUMMARY

Interdisciplinary programs have been misled by goals of breadth and multidisciplinary training. Even within disciplines, disciplinary competence is not achieved in individual minds but as a collective achievement made possible by the overlap of narrow specialties. This fish-scale model of collective omniscience is impeded in interdisciplinary specialty areas by the ethnocentrism of disciplines, by the organization of specialties into departments for decision making and communication. For an integrated and competent social science, we need to invent alternative social organizations that will permit the flourishing of narrow interdisciplinary specialties.

Donald T. Campbell: A Tribute

Donald T. Campbell was "a nimble-minded social scientist who left his mark on half a dozen disciplines and helped revolutionize the fundamental principles of scientific inquiry common to them all" (Thomas, 1996). Campbell, a social psychologist, did his major work at Northwestern University, but his last university post was at Lehigh University where he was designated a "university professor" with faculty appointments in the departments of psychology, sociology, anthropology and education.

They could have easily thrown in biology, the philosophy of science and market research. For a generation, virtually no respectable researcher this side of the chemistry lab has designed or carried out a reputable scientific study without a thorough grounding in what Dr. Campbell called quasi-experimentation, the highly sophisticated statistics-based approach he invented to replicate the effects of the truly randomized scientific studies that are all but impossible in the slippery and unruly world of human interactions (Thomas, 1996).

Dr. Campbell argued that valid inquiry in the social sciences required the sophisticated interaction among many methods. This was the theme reflected in his famous paper with Donald W. Fiske, "Convergent and Discriminant Validation by the Multitrait-Multimethods Matrix." More than 300,000 copies of "Experimentation and Quasi Experimental Designs for Research," a book written with Julian Stanley, have been sold (Thomas, 1996), and this book is yet to be replaced as the research bible for field-based social intervention studies.

Campbell, who had studied how academic departments often stifle the advance of knowledge, coined the tongue-in-cheek phrase, "fish scale model of omniscience" in reference to a serious theory about the distributed nature of knowledge. Permission to reprint this piece is given by his son, Martin Campbell, who recalls his father as a personal genius as well as an intellectual. "My memory is that Dad affectionately referred to this one as perhaps chief among his crack-pot articles.... Yes, please include my father's piece with my and the family's blessings."

ACKNOWLEDGMENT

The preparation of this chapter was facilitated by the Council for Intersocial Studies, which operates under a grant from the Ford Foundation to Northwestern University.

REFERENCES

Aronson, S. H. (1969). Obstacles to a rapprochement between history and sociology: A sociologist's view. In M. Sherif & C. W. Sherif (Eds.), *Interdisciplinary relationships in the social sciences* (pp. 292–304). Chicago: Aldine.

Campbell, D. T., & Fiske, D. W. (1959). Convergent and discriminant validation by the multi-trait-multimethod matrix. *Psychological Bulletin*, 56, 81–105.

Campbell, D., & Stanley, J. C. (1963). *Experimental and quasi-experimental designs for research*. Boston: Houghton Mifflin.

McGuire, W. J. (1969). Theory-oriented research in natural settings: The best of both worlds for social psychology. In M. Sherif & C. W. Sherif (Eds.), *Interdisciplinary relationships in the social sciences* (pp. 21–51). Chicago: Adline.

Polanyi, M. (1966). *The tacit decision*. New York: Doubleday.

Thomas, R. McG., Jr. (1996, May 12). Donald T. Campbell, master of many disciplines, dies at 79. *The New York Times* (Sunday late Edition).

2

Interdisciplinary Teamwork: The Dynamics of Collaboration and Integration

Julie Thompson Klein
Wayne State University

Interdisciplinarity has became a major topic in discussions of research, education, and problem solving. For many, the word *interdisciplinary* is synonymous with teamwork. It is not. Heightened interest in collaborative and integrative skills, though, has reinforced the connection. In this chapter, I explore the nature of interdisciplinary teamwork. I survey practices in science, industry, and government and then look more closely at the examples of team teaching and research in developing countries. The major foci are organizational structure, task and activities, leadership, team dynamics, problems, enabling factors, and tools, skills, and models. In the conclusion, I offer suggestions for future research.

HISTORICAL AND ORGANIZATIONAL CONTEXTS

A *team* is a specialized group with a performance objective or goal that requires coordinating activities. Teams exhibit the general characteristics of groups. They have a definable membership, consciousness of membership, a sense of shared purpose, and the ability to act in a unitary manner. In teams, though, the sense of shared purpose is stronger, and interdisciplinary teams in particular are distinguished by the presence of members of different knowledge fields (Knowles & Knowles, 1972, pp. 39–40; Davis, 1995, pp. 77–78).

Historical Background

World War II was a watershed in the history of interdisciplinary teamwork highlighted by the Manhattan Project and the beginnings of operations research. The formation of laboratories, centers, and institutes to solve military problems accustomed academic administrators to having large-scale, collaborative projects in short time cycles on campuses. By the 1960s, interdisciplinarity had become a recognized force in space research and by the 1970s and 1980s in areas of keen international economic competition, especially manufacturing, computer sciences, biomedicine, and high technology (Klein, 1990, pp. 121–139, 1996, pp. 173–208).

The current heightened visibility of interdisciplinarity is linked with the changing scale of collaboration in science. During the 1960s, the most prevalent unit of research production in the United States was a professor with two to six students working in a single discipline. By the 1970s, scientists were more amenable to accepting externally set objectives, working in groups, and using highly specialized, complicated equipment. By the 1990s, a greater variety of research units existed including large groups with many graduate students, nontenure-track researchers, postdoctoral fellows, and technicians working under one principal investigator. Today, large projects often involve multidisciplinary or interdisciplinary teams, heightening the need for collaborative skills and experience (*Federally Funded Research,* 1991, pp. 33–35, 219–226; *Scientific Interfaces,* 1986, pp. 188, 195).

The most dramatic changes have occurred in the fields of physics, astronomy, space research, and oceanography. Many projects require expensive facilities and sophisticated instruments such as particle accelerators, observatories, space telescopes, and ocean research vessels. In particle physics, for example, physicists must collaborate to gain access to particle accelerators. Experiments typically involve from 10 to 200 collaborators or more. Astronomers conducting interplanetary research must also collaborate to use satellites. Biology and chemistry have been slower to change. In chemistry, research tools still tend to be in the mid range, and in botany, collaborations tend to consist of 2 to 10 people. Some projects though, notably the Human Genome Initiative, exhibit features of "big science." Moreover, some fields, such as evolutionary biology and biochemistry, have traditionally been collaborative (Stine, 1992, as cited in Smithsonian, 1991–1992, pp. 399–406).

Interdisciplinary task force management has also been a feature of military operations, civilian affairs, engineering projects, feasibility studies, and industrial research and development. During the 1970s, "interdisciplinary attack" was a tag phrase for combining talents to solve socio-techno-economic problems such as urban decay and environmental pollution (Bass, 1975, pp. 2–7). International competition in key industries heightened the demand for more participatory and collaborative research models, reinforcing the rhetoric of "networked," "clustered," "nonhierarchical," and "horizontal" work organizations (Davis, 1995, p. 77). In addition, intraorganizational and cross-functional projects have been appearing more frequently under the names "project" and "venture," "team" and "working group," and "ad hoc committee" and "task force" (Dermer, 1988). The increasing reliance on temporary groups documents a growing gap between the current scale and complexity of industrial processes and traditional capabilities of industrial, governmental, and academic structures (Dahlberg, 1986, p. 14).

Organizational Structures

The organizational structures that facilitate interdisciplinary collaboration range from informal clusters within existing organizations to autonomous institutions. Matrix structures are a familiar form within organizations. A *matrix* is a program structure superimposed on an existing hierarchy. Matrices have long facilitated cross-disciplinary and cross-functional research and development in the pharmaceutical industry and in engineering projects (Pearson, Payne, & Gunz, 1979, p. 114). Centers and institutes are familiar structures on campuses. The Engineering Research Centers of the National Science Foundation (NSF), for instance, brought together engineering and scientific disciplines to address fundamental research issues crucial to the next generation of technological systems. NSF's Science and Technology Centers institutionalized hybrid communities of students, scientists, and engineers from academe, the nonprofit sector, and industrial and federal laboratories. The goal was to assure a requisite pool of scientists capable of meeting changing needs, dealing with complex research problems, and improving the economic posture of the country.

Other familiar and newer structures facilitate collaboration as well. Science and research parks, government laboratories, and experiment stations are recognized sites of interdisciplinary work in the nation's research system. In recent decades, a variety of new alliances have also emerged. They include offices of technology transfer, industrial liaison programs, joint mergers and ventures, high technology partnerships, small entrepreneurial firms, research networks, and consortia that bridge academe and industry. New alliances produce a variety of end products from social science knowledge to genetic materials. Their emergence marks an important trend. A significant number of research groups in academic science are now operating as "quasi-firms" exhibiting characteristics of research groups in private business. The industrial model enhances productivity while allowing more complex problems to be addressed by large teams with specialized responsibility, a shared infrastructure, and a principal investigator (*Federally Funded Research*, 1991, p. 35). Conflicts between the commercial values of industrial culture and the academic values of academic culture persist, but they have eased in a number of important sectors (Etzkowitz, 1983, pp. 213–215, 219–220).

These examples are not only interdisciplinary, they are interinstitutional. Interdisciplinary collaboration, by definition, crosses the boundaries of disciplines, professions, and interdisciplinary fields. In science and technology, it also blurs traditional divisions of basic and applied research, research modes (experimental, computational, and theoretical), and social sectors (industrial, government, and academic). The term *hybrid community* designates a group of researchers, politicians, bureaucrats, and representatives of different groups who come together to formulate a research program. Organizations such as the Work Research Institute, a state institute established in Norway in 1965, are homes for hybrid, problem-solving communities (Mathiesen, 1990, pp. 411–413). Newly allied scholars and policy makers constitute a transformative alliance that encompasses the organizational frameworks of policy making and social research as well as the new hybrid discourse in which problems are defined, investigated, and handled (Hagendijk, 1990, pp. 58–59).

The question of integration remains, though. What exactly makes a collaboration interdisciplinary?

COMPLEXITY OF INTERDISCIPLINARY TEAMWORK

Interdisciplinary teams have similarities to other teams, but they operate in a more complex environment. McCorcle's (1982) distinction between an interdisciplinary team and a more conventional, homogeneous group highlighted two major reasons. First, an interdisciplinary team is an open, not a closed, system. It often owes its existence to an external agent who may make demands in an unpredictable sequence. Second, it has a more heterogeneous, although interconnected, membership driven by the presence of individuals from different fields.

Like all collaborations, interdisciplinary teamwork emerges for different reasons. Many teams are driven by an agenda of practical problem solving, but they also form in the interest of creating knowledge. The search for grand unifying theory has brought astronomers and physicists together, and research in artificial intelligence has required establishing multidisciplinary teams (Stine, as cited in Smithsonian, 1991–1992, p. 404). The norms of a particular culture, discipline, or organization may be a strong factor. In universities, interdisciplinary collaboration often results from self-generated meetings of faculty members. In national laboratories and in industry, it usually emanates from a directorate and management (*Interdisciplinary Research*, 1990, p. 12).

Task and Activities

Tasks vary. Projects with directed goals target specific and well-defined ends such as mapping the human genome or executing a human mission to Mars. Dynamic goals aim at achieving broader states or conditions such as a coordinated federal environmental research and development effort (Carnegie Commission, 1992, p. 22). Van Dusseldorp and Wigboldus (1994) further distinguished "narrow" and "broad interdisciplinarity." Narrow interdisciplinarity is exemplified by a team composed of agronomists, soil scientists, and climatologists, or a combination of biologists, chemists, and physicists. They use more or less the same paradigms and methods and share a similar knowledge culture. Involvement of few disciplines and location in the same organization simplify communication. Broad interdisciplinarity is exemplified by a team composed of agronomists, soil scientists, economists, and social scientists. Involvement of many disciplines, dispersal across different organizations, and the presence of different paradigms, methods, and knowledge cultures complicate communication (Van Dusseldorp & Wigboldus, 1994, p. 96). In education, Kelly (1996) made a comparable distinction between narrow and wide interdisciplinarity based on epistemological similarities and differences among participating disciplines.

Variation occurs in activities as well. In his manual on the operation of interdisciplinary teams, Bass (1975) identified five categories. Each has advantages and disadvantages. Type I, unoriented and unstructured activities, may be a starting point for developing specific concepts. They may foster better understanding and liaisons across disciplines. Rarely though do they lead directly to specific benefits. Time and energy may be too diffuse. Type II, oriented unstructured collaborations, encourage interaction, initiative, and creativity. Yet, the focus of activity may not be well clarified. Contributions and timing may also be uncertain. Type III, oriented structured programs without constraints, provide channels for diverse inputs and a common focus that encourages communication. Yet, leadership decisions may not be enforceable, and members may diverge from the schedule and planned objectives. Type IV, oriented structured projects with constraints, foster consensus by stimulating direct communication and interaction. Yet, a program may be restricted by the selected specialized inputs and the leader's managerial style. Adherence to an original concept may also deter innovation. Type V, projects under centralized executive control, lead to direct accomplishments under centralized administrative and operational control. Yet, results depend on the ability of the leader, and initiative and creativity may be limited.

Variation extends to the level of a single meeting. Bass (1975) cited a joint discussion of surgeons and engineers to consider an integrated approach to producing a new prosthetic device (Type II). The task may require general review of the diversity of skills needed (Type I). From this discussion, a consolidated program outline with a specific goal (Type III) might emerge. This effort, in turn, may be converted into a project proposal for outside support (Type IV). The activity will also include definition of objectives, justification of utility, designation of a project leader and other team members, a structured program, target date, and budget. In the end, an entrepreneurial manager will be needed with the authority to coordinate implementation (Type V).

Team Types

All teams need a results-driven structure, clear roles, an effective communication system, methods of monitoring performance and giving feedback, and a means of recording and making fact-based judgments (Davis, 1995, p. 92). All types of teams are not the same, however. Drawing on experience in futures research, Sandi (1990, p. 153) identified two major types, stable and ad hoc teams. Researchers on stable teams tend to acquire necessary research skills rather than recruiting collaborators who possess them. They have the advantage of common views, language, attitudes, and familiarity with the same bibliography. Yet, stable teams may be superficial in dealing with problems that require specialized knowledge. They may also fall into a stultifying routine and experience conflicts in interpersonal relations. Ad hoc teams have flexible organizations that correspond to the research skills and needs of ongoing projects. They have the advantage of a fresher approach and chances for insight and creativity. Yet, they may take a longer time to establish communication and to overcome barriers.

Drawing on the track record of interdisciplinary teamwork at Honeywell, a high-technology company, Sackett (1990) identified three dominant styles of carrying out projects. An "ace with consultants" worked alone on small interdisciplinary projects, soliciting expertise from within the same functional section. A "super ace (ace teams)" was the traditional approach to large interdisciplinary research projects. Work and assignments were partitioned, and role negotiations took place mostly on a one-to-one basis. Reports were submitted to the super ace who put them together into a team output. Lower than desired win percentages on proposals at Honeywell led to the third model. "Consensus teams," dubbed "Boiler Room" teams because of the noise coming out of work rooms, embody "truer" teamwork. A proposed leader is in charge, but the entire team works through every key issue in a proposal. Before individual assignments are made, everyone agrees on key elements and the selling message.

Team Size and Leadership

The composition of teams differs by size and physical proximity of members; age, gender, racial, and cultural composition; and participating disciplines, professions, and functions. Even in the same realm of knowledge or problem setting, teams differ because collaboration is an adaptable research strategy. Every combination creates a unique dynamic and a different organizational structure. The ideal size is identified as 4 to 12 members. Taylor (1975) advocated no fewer than 5 and no more than 12, with an optimum range of 7 to 9. Stankiewicz (1979) found optimum size to be 5 to 6. Overall measures suggest that small groups with stable membership are the most integrative. Large numbers complicate information processing. Large groups also tend to inhibit creativity and provoke a tendency to work at the level of the smallest commonly agreed on denominator. Large groups with stable membership and clear divisions of labor have had positive results. They can be divided into small working groups and an open-channel of communication used. However, they will encounter more difficulty achieving integration. Responsibility is also more difficult to allocate.

A leader is responsible for three major areas—managerial, relational, and target scope. Managerial functions include forecasting and planning, implementing interdependent schedules, focusing on core tasks, handling budgets, training and development, sales and public relations, and technical/professional plus administrative coordination. Relational factors are embodied in the leader's work as evaluator, motivator, and director. Target scope extends across external parties and multiple organizations plus superiors, subordinates, peers, and self (Barth & Manners, 1979, p. 52). In addition to skills of management, communication, and problem solving, leading interdisciplinary teams requires some familiarity with pertinent specialties and disciplinary paradigms.

An interdisciplinary leader has been called a ringmaster, gatekeeper, boundary agent, ombudsman, polymath, dynamo, metascientist, specialist/generalist, strong entrepreneur, and a bridge scientist. Anbar's (1973) concept of a bridge scientist highlights the need to move beyond multidisciplinary translation of a problem to interdisciplinary integration. Studies have suggested that a democratic leader tends to generate more fully integrated research than an authoritarian or a laissez-faire leader. In his studies of a wide range of projects, though, Birnbaum (1979) found that different situations call for different behaviors. Initiating behavior emphasizes tasks, authority, control, and structure. Consideration behavior emphasizes the social needs of a team, relationships among members, and a democratic decision process. In projects less than 4 years old, leaders who exhibited both initiating and consideration behaviors encouraged mutual trust and support. In older projects, task-oriented behavior is more important to ensure coordination. Members become increasingly independent as they develop expertise and a sense of security. Pearson et al. (1979) also found that a leader who delegates authority is called for in an urgent project (see also MacDonald, 1982, pp. 16–18; Klein, 1990, pp. 132–133).

PROBLEMS IN INTERDISCIPLINARY TEAMWORK

The problems of interdisciplinary teams are legion. Some derive from disciplinary territoriality and turf battles. Social and psychological impediments include resistance to innovation, mistrust, insecurity, and marginality. Participants may also lack integrative skills, systems thinking, and familiarity with interdisciplinarity. Shortfalls of integration occur when forums for exchange are inadequate, and participants settle for reductive solutions. Even strong groups may be undermined by unstable membership and unwillingness to take risks. In addition, projects encounter constraints of time and access to equipment, rigid budgetary and administrative categories, and restrictive legal mandates and policies. From beginning to end, progress may be deterred by lack of incentives and an inadequate reward system as well.

Disciplinary defaulting is widely reported. In the absence of shared understanding, members may maintain orthodox expert roles. Disciplinary value judgments privilege one's own goals while minimizing others. Collaboration, in contrast, is a consultative mode that requires mutual granting of power and surrender of some individual degree of control. If potential collaborators hold back during the early phase of a project, the prospect of arriving at a shared or interfacing cognitive framework is hindered. If one individual assumes the primary task of articulation, care must be taken to insure that his or her cognitive framework does not dominate. The cognitive framework of interdisciplinary work must not belong exclusively to any discipline or attempt to imperialize the others.

Conflict is endemic. Members of interdisciplinary teams exhibit many of the same fighting and thwarting behaviors as other groups. Conflict is associated with both technical issues (definition of problem, research methodologies, and scheduling) and interpersonal issues (leadership style and disciplinary ethnocentrism). Interdisciplinary teams, especially, must overcome the "boundaries of reticence" that disciplinary socialization creates. People also bring "excess organizational baggage" including perceptions of others, status in the organization, preconceived ideas of their roles, and differing understandings of the problem. Furthermore, conflict does not go away. It occurs throughout the lifetime of a project, task, or course as participants work through their differences and attempt to resolve them in the interest of achieving a common goal (MacDonald, 1982, pp. 34, 41; Fischer, 1992, as cited in Smithsonian, 1991–1992, p. 125).

Status

Status is a particularly tenacious problem. Interdisciplinary teams are status systems that reflect external hierarchies and disciplinary chauvinism. In the absence of a strong alternative, a team will tend to follow the status system of the world outside a project. Even a strong alternative will not eliminate status ambiguity and clashes in career goals, professional styles, and epistemologies. Status conflicts arise for many reasons including variables of gender, race, and cultural background. Disciplinary pecking order is a major obstacle. A prestigious person or discipline may dominate, inhibiting others from speaking, impeding role negotiation, delaying communal work, and creating social and cognitive dependence. The theory of "status concordance" holds that organizational success is related to matched and equal ranks among members' age, sex, academic rank, highest degree obtained, and discipline (Klein, 1990, pp. 127–128). Rarely, though, are perfect matches possible.

In her pioneering study of working relationships among psychologists, psychiatrists, and sociologists in mental health projects, Luszki (1958) found that disciplines imported to help with a project tended to be in subordinate power positions. Initiating disciplines tended to be in primary position. In studying members of an alcoholism rehabilitation organization, Fry and Miller (1974) discovered that rehabilitation counselors, as state employees, had the right to authorize payment for services to patients; because patients were clients of the state department of rehabilitation, they were technically patients of the rehabilitation counselor. In medical settings, the greater prestige of doctors over nurses and other professionals, such as social workers and therapists, is a constant source of status conflict. A large body of case studies affirms the importance of establishing a democratic system of communication that empowers team members with lesser status (Klein, 1990, pp. 143–145).

Quantitative Versus Qualitative

The quantitative–qualitative split is an added obstacle. Studies of technology assessment projects have revealed dissension within teams over quantification including near contempt of some physical scientists for social scientists and humanists. Some participants were unwilling to venture beyond the data in hand to predict the future. Others, especially economists, tended toward overuse of jargon, modeling, and unrealistic data requirements (Klein & Porter, 1990, p. 16). The list goes on. Scientists exhibit disdain for engineers, mathematicians for physicists, pure scientists for applied scientists, and physical scientists for social scientists and humanists.

Simon and Goode's (1989) chronicle of limits encountered as anthropologists working on a policy research project is a striking case in point. The project focused on the efforts of laid-off employees and union leaders to save jobs in the supermarket industry. The dominance of an economic perspective and quantitative model of research restricted the anthropologists' role to supplying "background" context from interviews. They were reduced to a contracting mode, not a full partnership. Archaeologists and geographers echo Simon and Goode's lament, complaining of being brought in as subordinate data suppliers rather than being consulted on overall design, implementation, integration, and evaluation.

The four models of collaboration Simon and Goode (1989) identified are potential levels of interaction in all interdisciplinary projects:

1. *Background or context information*, an additive step that can be supplied separately from contributions of other researchers and may only appear as an appendix or separate case study.
2. *Elaboration or explanation of findings* from quantitative components, still limited to an additive role that typically produces a concluding chapter valued as descriptive detail, not findings.
3. *Definition of important variables or categories* for quantitative study, a step that sometimes occurs at the outset or prior to finalization of research design, structured instruments, or analytic approaches.
4. *Creative combination of ethnographic and multivariate approaches in research, analysis, and interpretation*, a rare occurrence of integration in which fundamental questions are refined using the participants' approaches on a mutually illuminating basis. (pp. 220–221)

ENABLING INTEGRATION

Lest the litany of disincentives and obstacles suggest that interdisciplinary teams never succeed, the list of enabling factors is equally long. Mutuality and interdependence are the keys to developing "teamness." Both require mutual learning and power sharing. Stone's (1969) distinction between secondary and primary group relations applies. Young teams exhibit secondary-group relations. Members are self-protective of themselves, thinking in terms of "I." Primary-group relations are characterized by dedication to a common task, thinking in terms of "we." Even in the latter case, though, teams must still work to find a middle ground between, at one end of a spectrum, inefficiency and debating each decision endlessly and at the other end, groupthink and setting assignments prematurely before alternatives are considered (Davis, 1995, p. 94).

Koepp-Baker (1979) likened an interdisciplinary health care team to a polygamous marriage. The team is launched by an announcement of intentions, an engagement, considerable publicity, a honeymoon, and finally, the long haul, which is inevitably threatened by ennui (Koepp-Baker, 1979). Making it through the long haul requires identifying where difficulties lie, where and by whom goals are clarified and roles defined, what the levels of communication are inside and outside the group, how the group builds and maintains its identity and sense of purpose, what its capacity for change is, and how and by whom points are assessed and achievements measured (Logan & McKendry, 1982, p. 884).

A team trying to eradicate pollution of a lake experiences different conditions of practice than a team designing a new aircraft, a team investigating artificial intelligence, or a team teaching an interdisciplinary course on the environment. They encounter generic variables of interdisciplinary collaboration. Yet, the task at hand creates a local dynamic of collaboration and integration. Two extended examples illuminate both common and unique conditions of practice.

Team Teaching

In the kindergarten through Grade 12 system, the history of interdisciplinary teams dates to the early decades of this century to the advent of core curriculum in junior high schools. At the postsecondary level, team teaching is a major priority in interdisciplinary curriculum reform. In the first comprehensive study of the topic in higher education, Davis (1995) provided a detailed picture on macroscopic and microscopic scales based on results of a national survey and a close analysis of five courses from the University of Denver. Davis's metaphor of "inventing the subject" is an apt description of what all interdisciplinary teams go through. Participants create the boundaries of an interdisciplinary course, a program of care, a research project, or a product innovation.

Optimal integration requires high levels of collaboration. "The greater the level of integration desired," Davis (1995) admonished, "the higher the level of collaboration required" (p. 44). The four major areas of collaboration in team teaching are planning, content integration, teaching, and evaluation. They are not necessarily equal. A course may have a high degree of collaboration in planning and in content integration but a low degree in teaching and in evaluation. Planning is the area of greatest collaboration. The major considerations are the amount of contributions, responsibilities, and decision-making patterns. Some teams settle into a stable collaboration quickly. Others always seem to be in the planning phase, whereas others never truly gel.

In content integration, all participants participate in elaborating goals. The critical question from the standpoint of integration is whether disciplines provide different lenses for viewing the same phenomena or they examine different phenomena separately. Cognitive framework and organizational structure are interrelated. Intellectual integration requires a collaborative effort to produce a new way of thinking about the substance of a course, not separate serial inputs. A tension often exists between teaching "my interest and expertise" and the demands of inventing a subject collaboratively. Dealing with this tension may require team members to teach each other their subjects. The process of mutual learning can be time consuming, but it is crucial to forging a truly collaborative relationship. Generally speaking, it is also difficult to produce a highly integrated subject until a course has been taught at least once, and all faculty get a more concrete sense of what they are trying to achieve.

Teaching is typically the area of lowest collaboration. Established traditions, lack of time, and lack of imagination are formidable deterrents. Teaching in the classroom at the same time instead of relying on separate, serial presentations raises the level of integration. Collaborative decision making about teaching strategies, readings, and other materials is also needed, and testing and evaluation must be collaborative too. Decisions about how to measure learning outcomes and student achievement and how to weight varied components of testing and evaluation should be made together. And, collaboration should extend to responsibility for writing and grading exams and papers. (See "Guide to

Interdisciplinary Syllabus Preparation" for the difference between solitary and collaborative activities at different levels of integration.)

One of the governing realities of interdisciplinary collaboration is that members of the same research, teaching, or problem-solving team tend to lack formal consensus on a definition of interdisciplinarity. Moreover, they rarely engage in philosophical discussion. Different operational and implicit definitions usually emerge from pragmatic discussions. In the curriculum, they center on what topics to cover, books to read, issues to raise, and sensibilities to develop in a particular course or degree program. Themes and problems, principles and theories, and questions are the most common frameworks at the level of course delivery (Klein & Newell, 1997, pp. 404–408).

In contrast to team teaching, interdisciplinary collaboration in research and problem solving in developing countries is an understudied topic.

Research and Problem Solving in Developing Countries

The complications of interdisciplinary research in developing countries are daunting. Members of teams come from not only different academic disciplines and professions but different institutes and governmental bodies. The latter have their own organizational cultures, styles of communication, and traditions of collaboration. A team leader must also be able to negotiate with stakeholders in the community, and infrastructure is an added complication. Working transport, electronic communication, and telephones are not always available. Shortfalls of integration occur because not all disciplines may be available at the same time, and stages of planning and implementation may occur in isolation. Irrigation schemes, for example, may not be used or are underutilized because school and health facilities may be unavailable. Introduction of new production technologies may bog down because neither the market system nor the physical infrastructure of roads, storage, and facilities is adequate.

The International Development Research Center (IDRC) has actively promoted research in developing countries. In 1993, the Center channeled half of its program resources into six themes including problems of food, health, the environment, and development. In one project in Egypt, a local community, governmental agencies, and a university came together in an effort to promote community participation in health care. Project teams included 10 physicians, two social scientists, an anthropologist, a teacher, an agronomist, and an engineer. From the beginning, all parties agreed to work as equal partners. Community leaders held informal and formal meetings with the local population, the ultimate beneficiaries. Problem definition was outlined by the local population and reviewed by researchers at the university. Priority areas for research were also agreed on with the local community.

Reflecting on this and other projects under management, Kapila and Moher (1995) devised guiding principles for interdisciplinary research. Kapila and Moher affirmed the importance of the three "Cs" of collaboration, cooperation, and communication that have been widely noted in the literature. Continuous recognition of a common goal, regular communication, consultation, exchange of data, a common effort to achieve provisional conclusions, and a strong commitment to teamwork are basic requirements for success. The quality of interdisciplinary research also depends on attention to preparation, disciplinary inputs, the process of interaction, and the final synthesis. In developing countries, multidisciplinary cooperation is often more feasible where monodisciplinary research is the norm. Preparation for interdisciplinary research can be promoted by encouraging regular net- working among academics and researchers from different disciplines, policymakers, and the broader community.

Drawing on their experience in policy/action-oriented research and regional planning, Van Dusseldorp and Wigboldus (1994) addressed the practical realities of achieving integrated analysis when there is a separate period of monodisciplinary research. As a project moves from general research processes to specific research problems, subprojects need to be integrated. In the preparation stage, policy problems must be translated into research problems to be operationalized. A work program capable of integrating discipline-based research programs must also be prepared. In the fieldwork stage, research problems and the work program may be adjusted as regular consultations occur and tentative findings are reached. In the synthesis stage, disciplinary findings are presented and integrated through regular meetings.

In formulating a work program, time relations of disciplinary inputs must be designated. If, for example, agronomists do not receive timely information from climatologists and soil scientists, they cannot indicate the agricultural production potential of a region. Exchange of tentative data and information is equally important. In regional planning, some disciplines have to contribute information at an early stage. The input of climatology, hydrology, geology, and soil science, for instance, enables other disciplines to focus their own research and planning. Agronomy and animal husbandry provide information for sociology and economy. Comparable results are crucial. The outcome of monodisciplinary research and analysis must be accessible to other disciplines. If not, one discipline has significantly more information than another, and a bias is created.

Departure of team members is an added complication. Some participants may leave the project or the field early. Hypothetically, they could be stationed nearby for team meetings, but that is an expensive arrangement. Ideally, all team members should at least have opportunities to make suggestions about how integration will occur. In the case of a large team, a nucleus team may take initial responsibility for combining information provided by participating disciplines into an integrated, dynamic analysis. During the final integration stage, though, all team members need to be able to follow the way their inputs are being used.

In both areas—team teaching and research and problem solving in developing countries—the importance of continuous learning and synthesis cannot be overstated. Repeated interactions build up common language, sensitivity to disciplinary assumptions, and openness and respect for each other's disciplines. The word *trust* appears often in descriptions of collaboration. Building trust requires an atmosphere of honesty, openness, consistency, and respect (Davis, 1995, p. 94). When stakeholders are involved, different perspectives need to be integrated and synthesized on an ongoing basis as well through regular communication. Finally, Van Dusseldorp and Wigboldus (1994) echoed Kapila and Moher's (1995) concern for quality of disciplinary inputs. "An excellent integration of disciplinary contributions of poor or heterogeneous quality," they warned, "delivers a poor product" (Van Dusseldorp & Wigboldus, 1994, p. 106). The quality of outcomes depends in no small part on the quality of the building stones.

TOOLS AND SKILLS

All interdisciplinary teams generate a host of raw documents that reflect the process of knowledge negotiation. Both technological and nontechnological tools aid in managing information flow.

Tools

The taxonomy of tangible knowledge includes target products (conceptual papers, team reports), instrumental products (graphs, data matrices), and ephemeral products (transitory representations, whiteboard diagrams or lists on a flip charts). Technological tools for managing information include shared text editors and drawing media, recording devices such as audiotapes or videotapes, and computer-based tools that facilitate group decision making (chap. 7, this volume). Relational input–output system diagrams are useful for articulating conceptual models. Specialists in modeling techniques may also act as third-party facilitators.

Computers are valuable for managing aggregate data sources, databases, and archives. Teams must be careful, though, not to force all disciplines to translate their information into a mathematical language that suits the computer (Van Dusseldorp & Wigboldus, 1994, p. 130). In describing technology assessments of oil and gas operations on the U.S. outer continental shelf, Sharp (1983) depicted how technological and nontechnological approaches were combined. Information was recorded in a computer model of inputs available to a team comprised of specialists from engineering and the natural and social sciences. To ensure integration, assignments were rotated among team members. They conducted extensive and intensive internal reviews and rewrites. Internal and external reviews were particularly helpful in resolving communication problems. After several internal iterations, outside consultants, an oversight committee, and representatives of varied parties of interest were involved. By relying on a system of review procedures, the core team was able to produce papers that became the basis for an integrated rather than serial report.

Iteration entails reading and critiquing one another's work to achieve a coherent, common assessment. Assumptions can be checked on a repeated basis. Returning to earlier stages encourages critical appraisal of individual contributions and collective resolution of differences. Peer editing of drafts increases the likelihood of moving beyond multidisciplinary juxtaposition in a contracting mode to interdisciplinary integration in consulting and partnership modes. Iteration fosters a higher level of interrelation and open discussion of disagreements. "Disregarding differences," Kapila and Moher (1995) cautioned, "may accelerate consensus in the short run but create more fundamental conflict in the long run" (p. 11).

Even powerful software, electronic communication, and proven techniques such as Delphi method do not guarantee synthesis. Integration is a human action. It is negotiated, situationally dependent, and contingent on participants. Role clarification and negotiation help team members to assess what they need and expect from each other, from consultants, and from clients, patients, or students. Team members need to clarify differences in disciplinary language, methods, tools, concepts, role conceptions, and professional identities and ideologies. Allowing individuals to articulate their separate understandings in verbal or written form is a particularly useful technique (Ackerman et al., 1983).

Skills

No matter what context, participants utilize familiar skills in doing interdisciplinary work. The basic skills are differentiating, comparing, contrasting, relating, clarifying, reconciling, and synthesizing. Integration is neither formulaic nor automatic. Whether working on a teaching team or on a technology project, team members must be able to structure a workable framework that is flexible enough to allow for shifting groupings. Mutual learning also requires knowing how to recognize one's ignorance of a particular area and then soliciting and gathering appropriate information and knowledge. The task at hand requires analyzing the adequacy, relevancy, and adaptability of discrete pieces or elements. In the process, depth of disciplinary/professional contributions is balanced with breadth of perspective. Iteration enables clarification and presentation of results for mutual revision (Klein, 1990, p. 183, 1996, p. 224).

Certain ideal characteristics for interdisciplinary collaboration have been identified. Team members should be flexible, patient, willing to learn, sensitive toward and tolerant of others, and willing to venture into uncharted waters. These are ideals, however. Not everyone arrives at teamwork with these qualities or a willingness to subordinate thwarting behaviors. White (1975) suggested that people who tend toward Taylor's (1975) category of divergent thinkers are probably more likely to find interdisciplinary research compatible and enjoyable. Certain steps necessary to solving ill-defined problems seem to require divergent thinking, although later steps requiring rigor call for convergent thinking, which psychologists correlate with well-defined problems (White, 1975, p. 384). Mead (1977) suggested that "digital" thinkers may be too narrow in focus to deal with broad, cross-cutting issues and that "analogic" thinkers can better perform integrative tasks. All interdisciplinary teams, no matter what context, depend on the willingness of individuals to subordinate their individual interests to a common objective. For that reason, veterans of IDR recommend early training in group interaction skills. Even scientists, though, who are better socialized for collaborative work than humanists and many social scientists, may be unprepared for the unique demands of interdisciplinary collaboration.

MODELS OF INTEGRATION

Models of interdisciplinary integration are based in operational theory, studies of human behavior, sociocultural and sociotechnical theories of group interaction, communication theory, decision theory, and the hybrid psychologies of social, cognitive, educational, organizational, and industrial concerns. Two particular perspectives lend insight into the dynamics of collaboration and integration. The first is stage and process models; the second is linguistic and communication models.

Stage and Process Models

Klein (1990), Newell (1999), and Sjolander (1998) have designated particular steps or stages in integration. Klein's (1990, pp. 188–189) observations are based on a general description of interdisciplinary process, Newell's on the link between interdisciplinarity and complexity, and Sjolander's on the experience of teams at the University of Bielefeld's Center for Interdisciplinary Research. Systems engineering utilizes a stage model of integration. The key to implementation in systems engineering has been powerful, low-cost personal computers and accessible word processing and software for data management. Computers aid in data management and provide microscale and macroscale testing of models through simulation. Specially designed computerized systems facilitate interdisciplinary work at different levels of complexity by constructing network representations of relations among particular ideas and elements. The context may be a single discussion, an entire project, or a field of knowledge.

Within the aerospace industry, to cite a major example, systems engineering entails management of research and development and production of complex high technology, defense, and aerospace systems. It is a client-centered activity that relies on a structured, top-down, iterative approach to problem solving. A hypothesis is formed and tested or a problem is solved in six steps:

1. Identification of goals and objectives.

2. Identification of alternative approaches that are ranked comparatively.
3. Synthesis of alternatives to discover new or better solutions.
4. Selection of the "best" solution.
5. Definition of steps to implement the chosen solution.
6. Definition and documentation of the process that led to the decision. (Mar, 1988)

The final step is not always taken. However, good interdisciplinary work requires a strong degree of epistemological reflexivity.

Gold and Gold (1985) recommended proceeding backward: starting from articulation of a group goal then working back to determine what is needed to satisfy that goal. In the process, common understandings are identified and developed. The tangible outcome is twofold: a shared conceptual model of the system being studied and the process by which the group goal is to be achieved. DeWachter's (1982) model of an interdisciplinary approach to bioethics formalizes the gap between ideal models and the realities of practice. A bioethical problem forms the basis of a global question for all team members. The ideal model of integration starts with the assumption that individual team members will suspend their disciplinary/professional worldviews from the beginning in favor of a global question based on the problem to be solved. Realistically, though, participants are usually unwilling to abstain from approaching a topic in terms of their own worldviews. The best chance of succeeding, DeWachter advised, lies in starting by translating a global question into the specific language of each participating discipline and then working back and forth in iterative fashion, constantly checking the relevance of each answer to the task at hand. That way, no single answer is privileged.

Comparably, my own global model of an interdisciplinary approach to problem solving (Fig. 2.1) moves beyond earlier models of a linear sequence of steps to acknowledging the messier realities of integration. The model was devised initially in the context of design, planning, and policymaking and then later generalized as a generic model of integration. Both the "Core Steps and Types of Knowledge" indicated on the left-hand side and the "Conceptual Framework and Skills" specified on the right hand are generic to interdisciplinary process.

Synthesis does not derive from simply mastering a body of knowledge, applying a formula, or moving from Point A to Point B. It requires ongoing

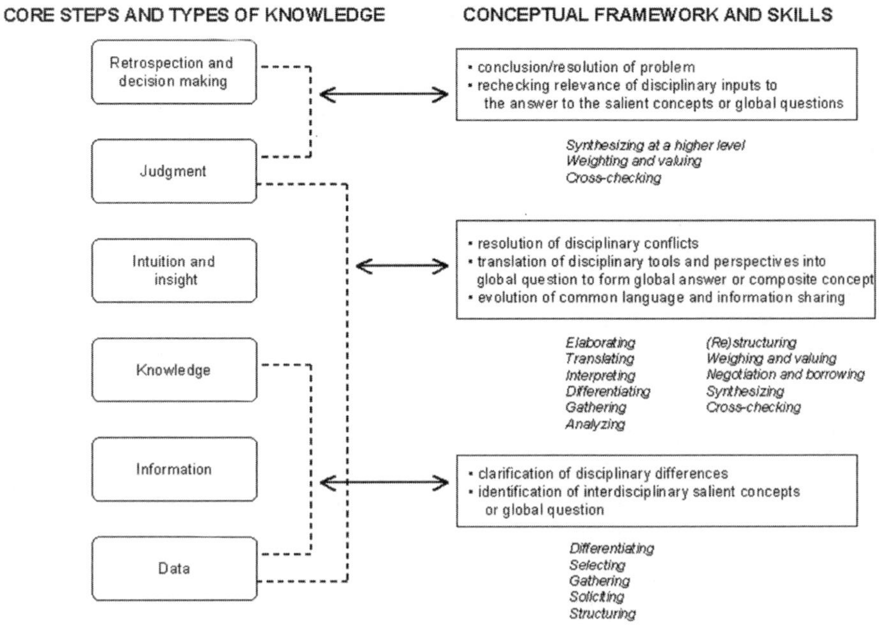

FIG. 2.1. Framework for interdisciplinary process. The core steps on the left occur iteratively rather than linearly. Arrows connecting boxes on the right align core steps and types of knowledge with clustered skills and actions in the conceptual framework.

triangulation of depth, breadth, and synthesis. Breadth connotes a comprehensive approach that draws on multiple variables and perspectives. Depth connotes competence in pertinent disciplinary, professional, and interdisciplinary approaches. Synthesis connotes creation of an interdisciplinary outcome through integrative actions. Insight develops through exploring and applying familiar techniques to new situations. Objectives and approaches are modified in light of ongoing accomplishments, not a fixed definition or formula. Because achieving a working relation between differentiation and unification is an ongoing task, boundaries between stages also blur. Synthesis is not reserved for a final step. The possibilities are tested throughout, moving in zigzags and fits and starts as new knowledge becomes available and new possibilities and limits arise (Klein, 1990–1991, p. 52, 1996, pp. 222–223).

This global model is premised on another set of assumptions. Interdisciplinarity entails communicative action.

Linguistic and Communication Models

Linguistic and communication models highlight the centrality of language in collaboration. Disagreements, in fact, often boil down to disputes over language (Davis, 1995, p. 50). Desertification, to illustrate, is a complex problem in the field of natural resources. It is ripe for interdisciplinary collaboration among climatology, soil science, meteorology, and hydrology as well as geography, political science, economics, and anthropology. Yet, the concept of desertification is defined differently in disciplinary, national, and bureaucratic settings. Individual aspects are emphasized in each setting deriving from conflicting special interests of climate, human factors, animals, soils, natural vegetation, and range management. The literature on desertification has contained no less than 100 definitions leading to miscommunication among researchers and policymakers. Any interdisciplinary effort, then, requires analyzing terminology to improve understanding of phenomena and to construct an integrated framework with a common vocabulary (Glantz & Orlovsky, 1986, p. 215; Bennett, 1986, p. 347).

The crux of the matter is difference. Misunderstandings, animosities, and competitions must be taken seriously, not mitigated or glossed over. Vosskamp (1994) and Klein (1996) have treated interdisciplinarity as communicative action. Vosskamp proposed that the agreement–disagreement structure necessary for all communication shapes the possibility of interdisciplinary dialogue. Consent–dissent (*Alteritaet*) requires accepting the unforeseeable and productive role of misunderstanding from the outset. A sense of the new and the surprising is decisive in mutual exchange and dialogue. The result is not necessarily consensus or unity. Dissent will remain a thorny issue because even if negotiated and mediated, differences do not go away.

Interdisciplinarity conceived as communicative action rejects the naive faith that everything will work out if everyone just sits down and talks to each other. Decades of scuttled projects and programs belie the hope that status hierarchies and hidden agendas will not interfere, or the individual with the greatest clout or the loudest voice will not dominate. The ideal speech situation assumes lack of coercion and equal access to dialogue at all points. The ideal, Habermas (1987) urged, is a valid critical standard against which apparent consensus can be called into question and tested. However, the shamming of communicative relations occurs in disciplinary relations as much as the interpersonal and social relations on which Habermas based his theory of communicative action.

Other related problems may haunt the effort. The fallacy of eclecticism is the naive faith that partial methods will add up to a complete picture of the phenomena, not a microcosm of disciplinary struggles to colonize phenomena. Creating an integrated product, solution, or perspective, Fuller suggested, requires moving from lower level translation of disciplinary perspectives by bootstrapping up to higher levels of conceptual synthesis. Linguistic models are not imported intact from metamathematics, set theory, symbolic logic, or any paradigm. They evolve in the creation of a trade language that may develop into a pidgin, an interim tongue, or a creole, a new first language among a hybrid community of knowers (Fuller, 1993, p. 42; Klein, 1996, pp. 216–222). Bilingualism is a popular metaphor of interdisciplinary work. However, mastery of two complete languages rarely occurs. Pidgin and creole are the typical forms of interdisciplinary communication. Studies of interdisciplinary communication reveal that everyday language is usually combined with specialist terms. Interdisciplinary discussions, Frey (1973) found, typically take place on a level similar to that of the popular scientific presentation. They become more precise as individuals acquire knowledge of other disciplines. At a higher level of conceptual synthesis, new and redeployed terminology form the basis of a working metalanguage.

Communicative competence is a necessary, although not sufficient, condition for interdisciplinary teamwork. The quality of outcomes cannot be separated from development of a language culture and its richness. Most misunderstandings, Frey (1973) found, are the result of using the same words with different meanings. In this respect, conflict is a defining condition as well. Luszki (1958) found that members of mental health teams paid a price for congeniality. By not dealing with conflicts in disciplinary definitions of such core terms as *aggression*, for instance, they reduced the number of creative problem-solving conflicts that would have promoted high-level, shared concepts. Difference, tension, and conflict are not barriers that must be eliminated. They are part of the character of interdisciplinary knowledge negotiation. For that reason, to reiterate, interdisciplinary process is grounded in social learning. Indeed, Frank and Schulert (1992) suggested, "Changing one's perspective is like entering another culture" (p. 235).

FUTURE RESEARCH NEEDS

A lot is known, but many questions remain unanswered. A number of important research tasks have been identified including finer detailed studies of local contexts of practice and collecting the raw documents of integration. Three additional areas merit further research.

First, evaluation and criteria are the most understudied aspects of interdisciplinarity. Scattered discussions of evaluation in the literature on problem-focused research and an emerging discussion of assessment in education have emphasized the inappropriateness of a single standard or a conventional measure. Excellence is not to be measured in terms of fidelity to an a priori disciplinary or professional standard but interdisciplinary originality. Disciplinary accuracy and clarity are important, but so is the clarity and strength of interdisciplinary communication and the solution to a problem or the creation of new meaning.

Second, the effect of collaboration on knowledge production merits further study. Collaboration has ill-understood effects on the research enterprise. There is no reliable method of measuring impact on productivity in various fields. Relatedly, the relation between specialization and collaboration needs attention. Collaboration may actually promote overspecialization or impose a hierarchical decision-making structure. It may also undercut gains in dismantling barriers of race, religion, gender, and age by reerecting barriers within the invisible domain of informal and personal networks. The full range of social, political, and ethical choices that confront sponsors and participants needs to be considered (Bordon, as cited in Smithsonian, 1991–1992).

Third, the relation between interdisciplinarity and creativity holds particular promise for building a cognitive model of collaboration. Most definitions of creativity, Sill (1996) found in the first extended study of the link between interdisciplinarity and creativity, have emphasized the appearance of something new. Creativity is heuristic, not algorithmic. It relies on rules of thumb or incomplete guidelines to drive learning and discovery, not mechanical rules. It also draws from the richness of the subconscious in relying on nonlogical and nonlinear thought processes. Preinventive structures in the subconscious provide raw material for creative combinations. These structures can be ideas, images, or concepts. Their ripeness encourages creative insights that tend to be ambiguous, novel, meaningful, incongruent, and divergent. They also contain emergent features.

Koestler's (1964) model of bisociation, Sill (1996) suggested, provides a model for understanding creativity as a form of synthetic thought. Disciplines represent independent matrices. Bisociative thought works at the intersection of separate matrices, some cognitive and some social. When they contradict or conflict, tensions must be resolved through creation of a new order. Barthes's (1977) notion of a "mutation," Paulson's (1991) notion of "noise," Davis's (1995) notion of "inventing the subject," and Hursh, Haas, and Moore's (1983) emphasis on "disequilibrium" in interdisciplinary learning all underscore the role of novelty in interdisciplinary work. By its very nature, creativity violates the present order. Novelty brings about instability. Interdisciplinarity unsettles existing assumptions. Something that makes sense in one matrix may not make sense in another. Making the subject or idea problematic makes it "absurd." It no longer makes sense.

Ultimately, interdisciplinary teamwork is not simply a practical arrangement. It raises profound epistemological questions. Is the production of new knowledge only a temporary, isolated phenomenon? Or does it represent a new paradigm? In the current system of knowledge production and in education, more than one system is often operating. In a business setting, interdisciplinary teams are typically involved in two separate but complementary managerial systems. One is focused on organization and the other on activities. For individuals, the organizational system provides a framework of policies and practices. The activities system is the route through which they apply skills to problem solving (Davis, 1995, pp. 112, 132). In education, an older system often holds fast while a new one struggles into being. The new one has implications for all components of the current system from philosophy, organizational structure, management style, institutional culture, curriculum, and instruction to scheduling, tracking and sequencing, budgets, certification and licensure, teacher education, inservice training, and professional development (Clarke & Agne, 1997). The present condition underscores the importance of developing both disciplinary expertise and interdisciplinary capacity. Nothing less is capable of equipping members of interdisciplinary teams for the tasks and the problems they address.

REFERENCES[1]

Ackerman, L. et al. (1983). Role clarification: A procedure for enhancing interdisciplinary collaboration on the rehabilitation team. *Archives of Physical Medicine and Rehabilitation, 64,* 514.

Anbar, M. (1973, July). The bridge scientist and his role. *Research/Development, 24,* 30–34.

[1]In addition to works cited here, the growing literature on transdisciplinary problem solving in the realm of sustainability is an excellent source of insights and case studies on collaboration and integration. For an entry into this literature, see Klein et al. (2001). For a continuing updated bibliography, see http://www.transdisciplinarity.ch

Barth, R. T., & Manners, G. E., Jr. (1979). Some behavioral, managerial, and research perspectives on the organization and management of interdisciplinary research: An overview. In R. T. Barth & R. Steck (Eds.), *Interdisciplinary research groups: Their management and organization* (pp. 50–59). Vancouver: Interdisciplinary Research Group on Interdisciplinary Programs.

Barthes, R. (1977). From work to text. *Image, music, text* (Stephen Heath, Trans., pp. 155–64). New York: Hill and Wang.

Bass, L. W. (1975). *Management by task forces: A manual on the operation of interdisciplinary teams*. Mt. Airy, MD: Lomond.

Bennett, J. (1986). Summary and critique: Interdisciplinary research on people–resources relations. In K. Dahlberg & J. Bennett (Eds.), *Natural resources and people: Conceptual issues in interdisciplinary research* (pp. 343–372). Boulder, CO: Westview.

Birnbaum, P. (1979). Academic interdisciplinary research: Problems and practice. *R&D Management Journal, 10,* 17–22.

Carnegie Commission. (1992). *Enabling the future: Linking science and technology to societal goals: A Report of the Carnegie Commission on Science, Technology, and Government*. New York: Author.

Clarke, J., & Russell, A. (1997). *Interdisciplinary high school teaching: Strategies for integrated learning*. Boston, MA: Allyn & Bacon.

Dahlberg, K. (1986). The changing nature of natural resources. In K. Dahlberg & J. Bennett (Eds.), *Natural resources and people: Conceptual issues in interdisciplinary research* (pp. 11–35). Boulder, CO: Westview.

Davis, J. R. (1995). *Interdisciplinary courses and team teaching: New arrangements for learning*. Phoenix, AZ: American Council on Education and Oryx Press.

Dermer, J. (1988). Intraorganizational interdisciplinary projects: Design issues. *INTERSTUDY Bulletin, 9,* 19.

DeWachter, M. (1982). Interdisciplinary bioethics: But where do we start? A reflection on epoche as method. *Journal of Medicine and Philosophy, 7,* 275–287.

Etzkowitz, H. (1993). Entrepreneurial scientists and entrepreneurial universities in American academic science. *Minerva, 21,* 198–233.

Federally funded research: Decisions for a decade. (1991). Washington, DC: Office of Technology Assessment, United States Congress.

Frank, A., & Schulert, J. (1992). Interdisciplinary learning as social learning and general education. *European Journal of Education, 27,* 223–237.

Frey, G. (1973). Methodological problems of interdisciplinary discussions. *RATIO, 1,* 161–182.

Fry, L., & Miller, J. P. (1974). The impact of interdisciplinary teams as organizational relationships. *Sociological Quarterly, 51*(3), 417–431.

Fuller, S. (1993). *Philosophy, rhetoric, and the end of knowledge*. Madison: University of Wisconsin Press.

Glantz, M., & Orlovsky, N. (1986). Desertification: Anatomy of a Complex environmental process. In K. Dahlberg & J. Bennett (Eds.), *Natural resources and people: Conceptual issues in interdisciplinary research* (pp. 213–229). Boulder, CO: Westview.

Gold, H., & Gold, S. (1985). Implementation of a model to improve productivity of interdisciplinary groups. In B. Mar, W. T. Newell, & B. O. Saxberg (Eds.), *Managing high technology: An interdisciplinary perspective* (pp. 255–267). Amsterdam: North-Holland.

Guide to Syllabus Preparation. (1996). *Journal of General Education, 45,* 170–173.

Habermas, J. (1987). *The theory of communicative action: Vol. 2. Lifeworld and system: The critique of functionalist reason* (T. McCarthy, Trans.). Boston: Beacon. (Original work published 1981)

Hagendijk, R. (1990). Structuration theory, constructivism and scientific change. In S. Cozzens & T. Gieryn (Eds.), *Theories of science in society* (pp. 43–66). Bloomington: Indiana University Press.

Hursh, B., Haas, P., & Moore, M. (1983). An interdisciplinary model to implement general education. *Journal of Higher Education, 54,* 42–59.

Interdisciplinary research: Promoting collaboration between the life sciences and medicine and the physical sciences and engineering. (1990). Washington, DC: National Academy Press.

Kapila, S., & Moher, R. (1995). *Across disciplines: Principles for interdisciplinary research.* Ottawa, Ontario, Canada: International Development Research Centre.

Kelly, J. (1996). Wide and narrow interdisciplinarity. *Journal of General Education, 45,* 95–113.

Klein, J. T. (1990). *Interdisciplinarity: History, theory, and practice.* Detroit, MI: Wayne State University Press.

Klein, J. T. (1990–1991). Applying interdisciplinary models to design, planning, and policy making. *Knowledge and Policy, 3*(4), 29–55.

Klein, J. T. (1996). *Crossing boundaries: Knowledge, disciplinarities, and interdisciplinarities.* Charlottesville: University Press of Virginia.

Klein, J. T., & Newell, W. (1997). Advancing interdisciplinary studies. In J. Gaff & J. Ratcliff (Eds.), *Handbook on the undergraduate curriculum* (pp. 395–415). San Francisco: Jossey-Bass.

Klein, J. T., & Porter, A. (1990). Preconditions for interdisciplinary research. In P. H. Birnbaum-More, F. Rossini, & D. Baldwin (Eds.), *International research management: Studies in interdisciplinary methods from business, government, and academia* (pp. 11–19). New York: Oxford University Press.

Koepp-Baker, H. (1979). The craniofacial team. In K. R. Bzoch (Ed.), *Communicative disorders related to cleft palate.* Boston: Little Brown.

Koestler, A. (1964). *The act of creation.* New York: Macmillan.

Knowles, M., & Knowles, H. (1972). *Introduction to group dynamics.* New York: Cambridge University Press.

Logan, R. L., & McKendry, M. (1982). The Multidisciplinary team: A different approach to patient management. *New Zealand Medical Journal, 95,* 883–884.

Luszki, M. B. (1958). *Interdisciplinary team research methods and problems.* Washington, DC: National Training Laboratories.

MacDonald, W. (1982). *The management of interdisciplinary research teams: A literature review* (A report for the Department of the Environment and the Department of Agriculture). Government of Alberta (January).

Mar, B. (1988). *System engineering.* Unpublished manuscript.

Mathiesen, W. C. (1990). The problem-solving community: A valuable alternative to disciplinary communities? *Knowledge: Creation, Diffusion, Utilization, 2,* 410–427.

McCorcle, M. (1982). Critical issues in the functioning of interdisciplinary groups. *Small Group Behavior, 13,* 291–310.

Mead, M. (1977). *Can research institutions accommodate interdisciplinary researchers?* Paper presented at 143rd annual meeting of the Association for the Advancement of Science, Denver, CO.

Newell, W. (1999, October). *A theory of interdisciplinary knowledge.* Paper presented at annual meeting of the Association for Integrative Studies, Napierville, IL.

Paulson, W. (1991). Literature, complexity, interdisciplinarity. In N. K. Hayles (Ed.), *Chaos and order* (pp. 37–53). Chicago: University of Chicago Press.

Pearson, A. W., Payne, R. L., & Gunz, H. P. (1979). Communication, coordination, and leadership in interdisciplinary research. In R. T. Barth & R. Steck (Eds.), *Interdisciplinary research groups* (pp. 112–127). Vancouver, British Columbia, Canada: International Group on Interdisciplinary Programs.

Sackett, W. T. (1990). Interdisciplinary research in a high-technology company. In P. Birnbaum-More, R. Rossini, & D. Baldwin (Eds.), *International research management: Studies in interdisciplinary methods from business, government, and academia* (pp. 60–72). New York: Oxford University Press.

Sandi, A. M. (1990). Interdisciplinarity and future research. In P. Birnbaum-More, R. Rossini, & D. Baldwin (Eds.), *International research management: Studies in interdisciplinary methods from business, government, and academia* (pp. 146–54). New York: Oxford University Press.

Scientific interfaces and technological applications. (1986). Washington, DC: National Academy Press.

Sharp, J. M. (1983). A method for peer group appraisal and interpretation of data developed in interdisciplinary research programs. In S. R. Epton, R. L. Payne, & A. W. Pearson (Eds.), *Managing interdisciplinary research* (pp. 211–219). Chichester, England: Wiley.

Sill, D. J. (1996). Integrative thinking, synthesis, and creativity in interdisciplinary studies. *Journal of General Education, 45,* 129–151.

Simon, E., & Goode, J. G. (1989). Constraints on the contribution of anthropology to interdisciplinary policy studies: Lessons from a study of saving jobs in the supermarket industry. *Urban Anthropology, 18,* 219–239.

Sjölander, S. (1985). Long-term and short-term interdisciplinary work: Difficulties, pitfalls, and built-in failures. In L. Levin & I. Lind (Eds.), *Inter-disciplinarity revisited: Re-assessing the concept in the light of institutional experience* (pp. 85–92). Stockholm: OECD, SNBUC, Linköping University.

Stankiewicz, R. (1979). The effects of leadership on the relationship between the size of research groups and their performance. *R&D Management, 9,* 207–212.

Smithsonian. (1991–1992). Series on collaboration. *Knowledge:* Part I on "The Arts." (1991). *Knowledge, 13*(2). Part II on "The Sciences." (1992). *Knowledge, 13*(4). And Part III on "The Humanities and Social Sciences." (1992). *Knowledge, 14*(1).

Taylor, J. B. (1975). Building an interdisciplinary team. In S. Arnstein & A. Christakis (Eds.), *Perspectives on technology assessment* (pp. 45–60). Jerusalem: Science and Technology Publishers.

Van Dusseldorp, D., & Wigboldus, S. (1994). Interdisciplinary research for integrated rural development in developing countries: The role of social sciences. *Issues in Integrative Studies, 12,* 93–138.

Vosskamp, W. (1994). Crossing of boundaries: Interdisciplinarity as an opportunity for universities in the 1980s? *Issues in Integrative Studies, 12,* 43–54.

White, I. L. (1975). Interdisciplinarity. In S. Arnstein & A. Christakis (Eds.), *Perspectives on Technology Assessment* (pp. 380–387). Jerusalem: Science and Technology Publishers.

3

Cognitive Processes in Interdisciplinary Groups: Problems and Possibilities

Angela M. O'Donnell
Rutgers, The State University of New Jersey

Sharon J. Derry
University of Wisconsin at Madison

Groups are ubiquitous phenomena in our lives and the relationship between individuals and various types of groups has been of interest in a wide variety of fields such as social psychology, sociology, medicine, computer science, and education. Many facets of groups have been the subjects of research including the nature of group interaction, the discourse among group members, leadership style, and group effectiveness. In this chapter, we are particularly concerned with interdisciplinary task-oriented groups. Many work teams, policy-making groups, and research teams are made up of individuals who represent a variety of disciplines. These teams are usually constituted for the purposes of framing problems or policies and solving problems that transcend disciplinary boundaries. Because these groups have members with different skills and perspectives, they have the potential to tackle broad issues, select problems that go beyond the confines of any one discipline, answer complex questions, frame problems with greater accuracy and breadth of understanding, combine resources and capitalize on differing skills in pursuing solutions to problems, and develop innovative solutions to problems.

Groups, however, do not necessarily operate in productive ways. Conflict will inevitably arise between group members when working on complex problems. For example, there may be differences in opinions about strategies for achieving group goals. Unresolved conflict between group members can erode effectiveness, and the time taken to negotiate solutions can also be deleterious to group functioning. Interdisciplinary teams can experience problems with developing shared understanding of the problem at hand (Journet, 1993). Conflicts may arise as a result of assumed or actual status differences among participants (Meeker, 1981) and differences related to decisions about appropriate courses of action and methodologies for implementing such plans.

Cognitive processing by individuals or by the members of a group as a whole is very important, as group members must come to understand the goals of the group, represent the problem under discussion by group members, identify strategies for achieving goals, and decide on strategies for achieving goals. The group's ability to do these activities effectively depends on the necessary information being available to the group. The analysis of cognitive processes in interdisciplinary groups is expected to identify mechanisms and strategies for facilitating the effective use of interdisciplinary teams.

Studies of the interaction of groups provide keys to the cognitive and social processes that represent effective group functioning or are predictive of effective group outcomes. With the increased specialization of knowledge and increased complexity of tasks in the workplace and other arenas, it is likely that the use of interdisciplinary teams will become more common (Schrage, 1990). Thus, it is important to describe, understand, and ultimately influence the communicative processes that influence productivity and the quality of outcomes from interdisciplinary collaboration. Journet (1993) observed that belonging to a specific discipline brings with it an allegiance to valuing specific types of events or data, concepts, problems, methodologies, and methods of argumentation and explanation. These allegiances can obstruct effective discourse and team performance. Understanding how people from different disciplines negotiate some shared understanding and common frame of reference within which effective action can occur is extremely important if interdisciplinary teams are not to become exercises in futility or exasperation.

Many groups in the workplace are dysfunctional and fail to take advantage of the resources available in the group, whereas other groups are very effective. How do effective and ineffective interdisciplinary teams differ in terms of the interactions of group members? In this chapter, we review the literature pertinent to the development of a research agenda related to the analysis of the cognitive processes that may occur in such groups. Because the functioning of groups is studied in a variety of disciplines, we draw on research conducted in social psychology, sociology, cognitive and educational psychology, communication, organizational and industrial psychology, business, decision making, and education to inform our analysis. Although the title of this chapter indicates that the focus is on the cognitive processes involved in interdisciplinary groups, we wish to note here that cognitive processes (e.g., problem representation) are deeply wedded to the social structures in which they occur, and as such, we inevitably discuss the social context within which groups operate. We return to this issue later in the chapter.

OVERVIEW

First, we provide a definition of interdisciplinarity. We then briefly review and synthesize the research literature on the influences on groups that contribute to their effectiveness. The purpose of this brief review is to highlight the main categories of variables thought to be important to the analysis of groups and their outcomes. We then distinguish between natural groups that occur in the world (e.g., surgical teams, management teams) and laboratory-based groups and examine influences on the operation and effectiveness of natural groups. We illustrate the influences on natural groups by examining a number of examples of natural groups. Interdisciplinary teams can be considered to be natural groups. Finally, we review a model of distributed cognition that may provide the basis for understanding the complexity of interaction in interdisciplinary teams. From this analysis, we hope to develop methods for analyzing the performance of interdisciplinary teams and provide the basis for improving the efficacy of such teams. However, the scope of this chapter does not permit us to do so here.

INTERDISCIPLINARITY

We were interested in examining the influences on interdisciplinary teams and how these are reflected in the cognitive/social processes in these teams. Many terms are used to describe the relationships between members of different disciplines as they work together on joint projects including interdisciplinary, multidisciplinary, transdisciplinary, and polydisciplinary. *Interdisciplinary* is defined most broadly by the Office of Economic Cooperation and Development (a subdivision of the United Nations Education, Social, and Cultural Organization as ranging from "the simple communication of ideas to the mutual integration of organizing concepts, methodology, procedures, epistemology, terminology, data, and the organization of research and education in a fairly large field" (as cited in Thompson-Klein, 1990, p. 63). According to Thompson-Klein (1990), a noted expert on interdisciplinarity, the most important difference in terminology is between multidisciplinary and interdisciplinary. The term *multidisciplinary* often refers to teams or activities and occurs when people in a group act from the perspective of their own discipline. Interdisciplinary groups are ones that consciously try to integrate knowledge from the different disciplines included. Few groups in the real world reach such goals, and the pragmatic concerns imposed by time demands and other variables make it difficult to achieve this integration. For example, the field of education is multidisciplinary in the sense that it includes many disciplines but has failed to achieve interdisciplinarity, which would require the development of common ground (Levin & O'Donnell, 1999). In this chapter, we use the term *interdisciplinary* to refer to groups that begin as multidisciplinary groups as defined by Thompson-Klein (1990) but may develop more integrative views in the process of interaction and begin to approximate interdisciplinary groups.

INFLUENCES ON GROUP EFFECTIVENESS

Social psychologists and others have built an extensive research base on small-group problem solving (e.g., Dillenbourg, 1999; Paulus, 1989). Until recently, the findings related to understanding group process were largely derived from the experimental analysis of small-group functioning in tightly controlled experiments. Variables of interest within this literature have included leadership style, communicative processes, conformity, group composition, group effectiveness, and a host of other variables. The focus of much of this work has been on the analysis of task-oriented groups that are brought together to solve specific tasks within a laboratory context. The results of the research conducted in this area has pointed to the importance of particular variables (e.g., group size) in the effective use of groups. More recent work on the analysis of group problem solving has taken place in classrooms (e.g., Cowie & van der Aalsvoort; O'Donnell & King, 1999). The variables of interest in laboratory-based social psychological experimentation are also of interest in classroom-based research.

Sociologists have also studied small-group functioning extensively. They examined the influence of power, status, and ability on the interactions among members of small groups (Berger, Conner, & Fisek, 1974; Robinson & Balkwell, 1995). Individuals' power, prestige, and status groups and who exerts influences affect who talks in a group and exerts influence on final decisions or conclusions. The influence of prestige, power, and status on decision making in small groups has also been studied by sociologists.

Sociologists have also studied the effects of expectations within small-group functioning (Berger et al., 1974). Within sociology, a major thrust of the work on small groups has been in the analysis of expectation states (Meeker, 1981) and their influence on interaction (number and frequency of engagements, predictions about influence). *Expectation states* are sets of expectancies about group members' competencies to contribute to successful task completion. Consider the following example of a typical scenario in efforts at school reform. A group of high school biology teachers and college faculty in biology meet to discuss what kind of preparation in biology might be expected of entering freshmen. College faculty and the high school teachers may differ in their expectations of who can contribute effectively to such a discussion. These expectations will influence the direction and flow of discourse in the group, often in a negative direction.

Sociologists have also examined the interactions of groups as reflective of efforts to maintain identity or develop identity (Robinson & Balkwell, 1995; Robinson & Smith-Lovin, 1992). If individuals in a group believe themselves to have low status in the group and lacking in the skills to contribute effectively to the group, this individual may interpret their own nonparticipation in the distribution of interaction as confirming this identity. In a sense, they engage in a sustaining expectation (Cooper & Good, 1983) in which they interpret the situation according to an existing identity. Alternatively, individuals in a group can use the group context to develop new identities (e.g., leader).

The basic research in social psychology and sociology can contribute to an understanding of the functioning of interdisciplinary groups. This research speaks to the broader question of how people from different backgrounds and experience can effectively interact with one another. The implication of this research literature is that the interactions of interdisciplinary interaction may be examined for the influences of power, influence, and leadership in groups and to identify the purposes served by such interaction (e.g., maintaining existing identities).

Research on group functioning in airline crews, surgical teams, health care teams, business, and education have drawn on the basic research literature in social psychology, communication, and sociology to explain the functioning of groups within these particular domains. Among the key concerns related to group functioning in applied settings are the identification of the components of effective teams, the analysis of key personality and cognitive variables important to group functioning, the identification of key variables in forming groups or maintaining groups, and the reward structures that influence group functioning. These concerns are also of importance to the understanding of interdisciplinary groups.

Effective groups share many characteristics (Hackman, 1990, 1995; Johnson & Johnson, 2002). Group members must constitute a "real team" (Hackman, 1995) in that members must be working on a task that requires all members to contribute skill. The task must be challenging. A curriculum development team, for example, might be faced with the task of reducing the number of topics covered in the ninth grade physics curriculum while amplifying the depth of coverage allocated to the topics retained. This type of task can be quite daunting, as people struggle to select necessary knowledge and ensure that this material is taught in a way that emphasizes critical skills. This kind of task will be much more readily accomplished if all team members believe they have something to contribute to the final product. Effective teams are also characterized by the interdependence of team members (Hackman, 1995) such that the outcomes from the group depend on all members. An easily understood example in this regard is a surgical team in which all team members (doctors and nurses) must support one another's activities in the interests of the patient's welfare.

Having teams in which all members believe they have something to contribute and whose outcomes are bound together implies that effective groups must be genuinely collaborative. In describing the distinctions that exist among various forms of peer interaction, "Genuine collaboration occurs when participants are equal in power and interchange among group members is characterized by mutuality (Damon & Phelps, 1989; De Lisi & Golbeck, 1999). In other words, all participants can be expected to contribute to the task, and each one's opinion is as valuable as any other's opinion. In such collaboration, there is an opportunity for introduction of dissonant information, the creation of disequilibrium or discomfort, and reequilibration (De Lisi & Golbeck, 1999). Most commonly, groups in the real world do not experience such mutuality and equality in interaction. The example given previously of the surgical team illustrates a context in which there is little equality among interactants. Status on a surgical team is important, and power is not shared. Although mutuality of influence and power might seem desirable, groups can nevertheless be effective even if they do not experience the types of collaboration delineated by Damon and Phelps (1989). Most natural groups (i.e., groups outside the laboratory) are not genuinely collaborative, as there is invariably a power/authority structure in place and differential accountability for outcomes.

Johnson and Johnson (2002) characterized an effective group as having three core activities: (a) accomplishing its goals, (b) maintaining good relations among its members, and (c) developing and adapting to changing conditions in ways that improve its effectiveness. Johnson and Johnson's (2002) delineation of effective groups is based on their extensive research and training efforts in schools and various kinds of work settings. Johnson and Johnson's 2002 book on group theory and performance is widely used in business, counseling, educational, and communication courses. Inherent in the characterization of effective groups as proposed by Johnson and Johnson (2002) is the notion that groups have multiple functions (McGrath, 1991). According to McGrath, groups function to produce something of worth, support group members, and promote group member well-being. The activities of accomplishing goals and adapting to changing conditions map directly onto the function of producing something of worth. The activities of maintaining good relations that will most likely involve adapting to changing conditions serve the function of supporting group members. When group members feel unsupported, a team or group may disintegrate into a cycle of frustration and avoidance (Eisenstat & Cohen, 1990). On occasion, the fulfillment of one function compromises the group's ability to serve another function. The outcomes from task-oriented groups are influenced by a variety of factors, and many frameworks exist to describe the factors that affect outcomes from such groups (e.g., Johnson & Johnson, 2002; McGrath, 1991). Generally, such frameworks address the influences of the task itself, input factors such as personality or cognitive profiles of the participating group members, and processing factors such as the tools available, time constraints, or incentive conditions under which the task is performed.

Tasks

The key variable in effective task-oriented group action is the task with which group members are engaged. The nature of the task makes particular cognitive and/or social processes salient, and these, in turn, can influence the functioning of the group. A group that is intended to be product oriented may concern itself with member support as a direct consequence of cognitive and affective responses to the task by group members. For example, a management team that is charged with "downsizing" and must come up with a plan for how to do so may get redirected from their task by the need to provide reassurance to various members of the team about their prospects for continued employment.

Tasks that are ambiguous can promote feelings of challenge and intrinsic motivation (Reeve, 1996). Nonroutine tasks are more interesting than routine tasks (Deci & Ryan, 1985). When individuals are interested in the task, they are more likely to persist with the task even in the face of apparent failure or frustration. Individual performance is best under conditions of optimal challenge (Deci & Ryan, 1985), but finding the group equivalent of optimal challenge is difficult (Dobos, 1996). When the consequences of task completion are very important or the group is working under significant time pressure, the information processing capabilities of group members can be significantly impaired, and selective attention to information can be enhanced (Frey, 1986). Under conditions of high stress, task ambiguity will significantly add to the stress experienced by group members and result in affective and cognitive conditions in a group that do not promote effective interaction (Doyle, 1983). The complexity of the task also influences the salience of specific cognitive, affective, or social processes. For example, if the task is a complex task, some members of a group may feel inhibited from participating effectively because of concerns about social comparison and maintaining status in the group. Alternatively, if a task is routine, the group members may fail to engage in appropriate levels of metacognitive activity. Hackman (1995) explained a major accident with one airline on this basis. The crew of the flight in question was inexperienced in flying in icy conditions. A normal part of their routine was to go through a series of checks prior to takeoff. One of these involved checking that the plane had been deiced. Although the crewmembers had a lot of flying experience, they were based in a warm climate, and their routine response to the deicing check had been negative. Failing to attend to the new conditions, they proceeded with other checks. In the particular circumstances, the failure of the group to engage in sufficient metacognitive activity with respect to the routine checks resulted in a major accident.

One of the clearest frameworks for characterizing tasks was produced by Steiner (1972). According to Steiner, most tasks fall into one of four different categories. First, there are additive tasks in which the contributions of each member of a group are combined to form a single group product. An example of an additive task is cleaning one's house. A number of people can work on this task simultaneously, and the results of their individual efforts are additive. Preparation of a group report can be divided among group members and the various parts added together to form the group report at the end. The second category of tasks is compensatory tasks in which the efforts of group members are averaged to produce a single group outcome. For example, a group of students may combine their efforts to produce one videotape for a class. Other group members who do have technical expertise may offset the lack of technical knowledge of some of the group members. The third category of tasks identified by Steiner (1972) is disjunctive tasks in which the group's product is determined by the performance of its best or most competent member. Team quiz shows often use this format. The success of the team requires that any member of the team can have the right answer, but only one person must have it. When the task is disjunctive, larger groups increase the probability that one of the members of the group will be competent to do the task. The final category of tasks is conjunctive tasks in which the group's final product is determined by the group's weakest link. A task such as a tug-of-war is a conjunctive task. Cooperative learning tasks in which all group members are expected to become accomplished at the target task, and any student's paper can be selected as representative of the group's accomplishments, can be considered a conjunctive task.

Some tasks, particularly more open-ended tasks such as those provided to policymaking groups, may be framed according to one of the four structures just described. The way in which the problem is framed may depend in part on the interaction patterns of the group. The nature of the task is important because it creates an arena for the social and cognitive processes brought to bear on the task by the members of the group. The pressures experienced by a group whose outcomes are determined by the weakest member of the group are markedly different than on tasks in which only one person need be competent. It is against this backdrop of task type that variables such as the entering characteristics of the group members (e.g., ability, personality) and the conditions under which the task is performed (highly contingent or not) must be understood.

Social and cognitive processes in group interaction are very tightly interwoven. Research on group interaction has often focused on either individual-level properties (e.g., leadership) or on system-level properties (e.g., norms of the group) but rarely on both (Hirokawa & Johnston, 1989). Individual-level and system-level properties intersect at the task and the interpretation of the task. Effective task performance requires that the optimal cognitive, affective, metacognitive, and social skills be available in the group (O'Donnell & Dansereau, 1992). If one participant engages excessively in metacognitive activity such as excessively editing the production of intermediate products (Flower & Hayes, 1981), other participants must place greater emphases on social and affective processes to maintain balance in the group.

The nature of the task (additive, conjunctive, disjunctive, compensatory) will contribute a great deal to the nature of the discourse among group members as they discover whose skills are needed for the satisfactory completion of the task. Interaction patterns in groups differed as a function of the type of task in which participants were engaged (Chinn, O'Donnell, & Jinks, 2000; Morris, 1996; Sorenson, 1971). Depending on the nature of the task, the relationships among group members, and the degree to which they experience interdependence (Deutsch, 1949; Johnson & Johnson, 1992), the group may suffer process losses in which group members are diverted from their purposes (Steiner, 1972) or may suffer reductions in productivity as a consequence of a lack of coordination. Process losses may occur as a consequence of social phenomena such as social loafing, free-rider, or sucker effects (Salomon & Globerson, 1989). Social loafing occurs when one member allows other people to do the work (Karau & Williams, 1995; Latane, Williams, & Harkins, 1979). These effects are often observed on additive tasks. When free riders or social loafers are detected in a group, individuals in the group may compensate for the lack of contribution by some members. In other cases, individuals in a group may react to the detection of free riders by actually reducing their own level of effort, a consequence known as the sucker effect. In circumstances in which some members of a group do not contribute, there are significant affective consequences to the group's functioning. The nature of the task can contribute a great deal to the possibility that social and affective processes will dominant group interaction or whether cognitive and metacognitive interchange do. One important consequence of task structure for group tasks is that it can signal to participants which members have status in the group and which members do not. The presence of such status differences (addressed in more detail following) can have important influences on the nature of the discourse among group members.

Social Influences

Although groups have the potential to be more productive than individualistic efforts for many tasks, the vulnerability of effective group processing to cognitive and social influences must be recognized. Much of the previous discussion on the effects of the task on group functioning is pertinent also to a discussion of the social influences on group performance. Reactions to the nature of the task can provoke affective or cognitive responses that may take center stage in the group interaction.

In a scathing critique of the emphasis on the use of teams in the workplace, Sinclair (1992) railed against "the tyranny of a team ideology" (p. 611) and argued that the emphasis on teams masked coercion and conflict in the workplace. If even some group members share such feelings, the possibility of productive interaction is in jeopardy. Effective leadership in a group can maintain the productive direction of a group, whereas poor leadership can amplify the effects of cognitive and social limitations.

Leadership in a group has a significant influence on group productivity. Leaders are often classified as people or task oriented (Fiedler, 1978; Fiedler & Garcia, 1987), although more recent studies of leadership behavior (e.g., Zaccoro, Foti, & Kenny, 1991) have suggested that effective leadership requires more adaptability on the part of the leader than is involved in trying to fit a person-oriented or task-oriented style to the task at hand. Leadership skills are important to the coordination of cognitive and social processing of the group members. The task for a leader in a group is to provide the necessary amount of structure and direction to the group discussion or performance without limiting other individuals' creativity and productivity. In analyzing differences between effective and ineffective teams in the real world, Hackman (1990) concluded that effective teams were generally characterized by effective leadership that set goals for the team but allowed team members autonomy in determining the means by which such goals might be accomplished. Leaders will also play different roles in the group. McGrath (1991) noted that group interaction can be dedicated to a variety of goals including production or maintaining good relationships. Successful leaders seem to fit their style to the necessary function.

Other limits on effective group processing include the pressure for consensus in groups (Sinclair, 1992) and the propensity for individuals to move too quickly toward consensus (Foushee, 1984; Janis, 1982; Nemeth & Staw, 1989). "Groupthink" (Janis, 1982) is characterized by a decrease in the exchange of discrepant or unsettling information. Generally, arguments against ideas are less commonly shared (Stasser, 1992). The joint influence of a press toward conformity and an unwillingness to share information are likely to lead to process losses.

Status

Status differences are often very salient in group interactions. Status, as used here, denotes expected competency for the performance of the task at hand. For example, a surgeon in a surgical team has high status associated with his or her experience, expected and previously demonstrated competence, and his or her centrality to the task at hand. A surgeon's status in relation to that of the attending anesthesiologist is less clear. Most health care teams have clear hierarchies of skill and influence. Even in groups in which such status differences are not immediately salient, status differentiation frequently emerges in groups with some members becoming identified as having competencies necessary to the task and other members being identified as having less competency for the task (Berger, Rosenholtz, & Zelditch, 1980; Cohen, Lotan, & Catanzarite, 1990).

High-status individuals take and receive more opportunities to influence the group process, talk more, answer questions, are accorded power in the group, and can ignore others in the group (Meeker, 1981). Low-status individuals in a group, in contrast, are often help seekers, are frequently ignored, say little, and when they do, their input is often ignored or devalued (Cohen, 1994; Cohen et al., 1990). Cicourel's

(1990) analysis of the interaction during the process of medical diagnosis demonstrates that status differences strongly influence the nature of the discourse. When attending physicians interact with medical residents, they are evaluating the resident's knowledge and making judgments of competence. The clear status differences in expertise and authority are manifest in the type of discourse in which the residents and attending physicians engage.

Groups with clear status boundaries often limit the exchange of information. In aircrews, pilots have high status in comparison to the first officer or other crew members. Subordinates in such crews often go along with the interpretation of events provided by the leader (in this case, the pilot) even when they experience doubt about the decision being made (Foushee, 1984; Foushee & Helmreich, 1988). When a subordinate does provide information that is contrary to the high-status individual's interpretation of events, the subordinate is often ignored. An airline accident in recent years illustrates this problem. The pilot of an aircraft was approaching the Dallas airport but was warned by his first officer that he might need to circle the airport one more time to improve the angle of descent. The first officer's warning was ignored by the pilot who attempted to land the plane and ran the plane off the runway. When status characteristics of group participants serve to limit disagreement and operate to increase the selectivity of information available to group members, the likelihood that the group will be ineffective is increased.

Status in a group serves both cognitive and social functions. It signals a stratification of worth, but it also influences the cognitive processing of the group in which some people are regarded as reliable sources of information (McClane, 1991), whereas others are not. Ability is the primary status characteristic (O'Donnell & O'Kelly, 1994). Other characteristics of individuals are often assumed, however, to be indicative of ability, although there is no necessary relation between such characteristics and performance. Salient characteristics of individuals that are used to infer the presence or absence of ability are called *diffuse status characteristics* (Meeker, 1981). Gender is an example of a diffuse status characteristic. Common stereotypes of women and men suggest that women will be less skilled at mathematics than men, whereas men will be less skilled at writing than women. When people enter into group interaction with expectancies that group members will have greater or less competence based on diffuse status characteristics such as race, ethnicity, or gender, group interaction will be influenced by those expectancies. Ideas expressed by some will be attended to, whereas those of others will be ignored. People with low status in a group often disengage from the group process. Their withdrawal of effort can be subsequently treated as social loafing, and a vicious cycle of low effort and recrimination can result.

LIMITS ON COGNITIVE PROCESSING

Even without status differences that limit exchange of and receptivity to information, there has been extensive documentation of the selective nature of information intake and retrieval (e.g., Anderson & Pichert, 1978; Donald, 1987; Frey, 1986). Effective processing of information available to a group is difficult even under optimal circumstances. People often pay attention to only that information that is schema relevant. As people do not necessarily always activate the same schemata (Anderson & Pichert, 1978), the information received by various members in a group can differ greatly. Information is often processed inaccurately as a consequence of such phenomena as the confirmatory bias (Crocker, 1981). Limitations on individual cognitive systems also constraint the efficacy of group processing. Leaders of groups are often cognitively overloaded (Ben-Bassat & Taylor, 1982), resulting in judgment errors.

To be productive, groups must have all the information necessary to make good choices. Influences such as the press for conformity or status differences may limit individual's willingness to contradict other group members' or may result in an individual devaluing the knowledge he or she has. Deficiencies in the information available to decision-making groups often result in poor decisions. Ball (1994) showed how such informational deficiencies permeated the decision making of President Kennedy and his advisors in considering their policies on the conduct of the Vietnam War. Foushee (1984), in an analysis of aircrew functioning, found that the unwillingness of subordinates to share information limited the effectiveness of teams.

Status is not the only reason why group members do not share information. A series of studies by Stasser and colleagues have suggested that pooling information sources may be the exception rather than the rule (Stasser, 1992; Stasser, Taylor, & Hanna, 1989; Stasser & Titus, 1985, 1987). Stasser and Titus (1985) proposed a model of group discussion called the *information sampling* model. This model suggests that information that is shared by members of a group is more likely to be mentioned in group discussion than information known only to a single individual. As a result, shared information is reiterated, and efforts to structure dialogue result in the known information being accurately retrieved. This seems to be a natural consequence of attempting to make oneself understood.

Under certain circumstances, individuals can be encouraged to discuss unshared information. In one experiment (Stasser & Stewart, 1992), half of the participants were informed that there was a correct solution to a problem, and their task was to find it. The remaining half of the participants were informed that it might be possible to solve the problem, but this was uncertain. Beliefs about the whether the problem could be solved or not strongly influenced the degree to which participants shared information not known to the entire group. Participants who were informed that the solution could be found actually shared more information that was not jointly held at the beginning than did members in the uncertain solution group. Apparently, when participants were encouraged to think that the problem was insoluble, they appeared to reduce their efforts to solve the problem and did not inspect the relevance of their privately held knowledge to the task at hand.

The literature available from basic research on group interaction has indicated that effective group or team functioning is very complex and subject to influence by a wide variety of variables. Answers to some key questions about groups (e.g., how do groups sustain effective functioning over time) cannot be answered in this literature, although it has provided some basic frameworks for interpreting results and framing new questions. In the end, the groups we wish to understand are the groups that exist outside the laboratory. Laboratory-based research has provided us with some insights into what might be important variables to consider when examining more natural groups.

NATURAL GROUPS

Overview

Despite the importance of groups and their prevalence, the research on small-group communications, interactions, and productivity suffers from a number of shortcomings (Frey, 1994). First, most of the research that has documented the processes and products of small-group interaction has been derived in laboratory-based research studies rather than natural groups (Scheerhorn, Geist, & Teboul, 1994). Examples of natural groups include management teams, aircrews, health care teams, and customer service teams. In a review of the literature published during the 1980s, Cragan and Wright (1990) observed that only 13% of the available studies in communication sampled groups in organizational or applied settings in contrast to a huge body of work related to groups in the laboratory. There seems to be no reason to believe that the picture emerging from other disciplines would be radically different. Although much valuable information can be gleaned from work in laboratories, the conditions present in laboratory-based versus natural groups are sufficiently different that one might reasonably assume that the findings from the laboratory cannot be imported directly into the analysis of the functioning of natural groups.

Additional criticisms of the research conducted on group processes has related to the types of tasks selected for study. Frey (1988) reviewed the group communication literature available from the 1980s and found that less than 5% of studies were concerned with tasks unrelated to decision making. Scheerhorn et al. (1994) argued for the investigation of group tasks other than decision making.

The tasks in which people engage in natural groups are radically different than those used in laboratories. Tasks used in laboratory groups and natural groups differ in terms of complexity, the commitment required of participants, the duration of the task involvement, task ambiguity and consequences, and the necessity for groups to attend to multiple functions such as productivity, member well-being, and member support (Hackman, 1990; Hirowaka, 1990). Natural groups work together over some extended time period. As described previously, these characteristics of the task in which people are engaged have implications for the kinds of social, affective, or cognitive processes brought to bear on the task. According to Hirokawa (1990), group performance will be largely dependent on input variables (e.g., personalities) when the task is simple. When tasks are complex or contingencies are high, process variables (such as resource allocation or coordinating activities) may become more important.

Many of the tasks in which people engage in the real world involve extensive involvement over time. Participants may enter into the group without knowledge of one another, but they soon develop such knowledge. The relationships that develop must be maintained over time. This is in contrast to laboratory-based research in which strangers engage in brief tasks that have little personal consequence for themselves. The duration of the members' involvement with one another requires more skill in coordinating activities, adjusting to changing environments, and coping with the dynamics of ongoing interaction. Natural groups may also be embedded within a larger organizational context, and thus, group development may also involve the relationships between specific groups and the larger groups in which they are embedded (Wheelan et al., 1994). Natural groups may have members leave and new members enter, which disrupts established patterns of interaction (Schopler & Galinsky, 1990).

In summary, natural groups differ from laboratory-based groups on a number of important variables that might be expected to influence the quality of interaction in groups and the products that emerge from them. They differ from laboratory-based groups in the types of tasks in which they engage, the duration of the group involvement, and the relationships among group members.

The research conducted on laboratory-based groups typically involves carefully conducted experimental research in which the researchers exercise a great deal of control over the variables in operation. Stringent control conditions are also invariably present so that appropriate inferences about causality of observed effects can be made. In general, studying natural groups in the real world does not permit this type of research. Instead, a variety of research methodologies are employed in attempting to understand how and why natural groups work and under what conditions they are ineffective. Examples of methodologies used in the study of natural groups include case studies involving interviews and detailed observations, protocol analysis, surveys, and on occasion, observed behavior is quantified and analyzed.

In the following pages, we describe groups in a number of natural settings. Many of the problems with groups that were identified in basic laboratory research can be readily identified within naturally occurring groups. Significant problems related to communication among group members, the operation of status in groups, and difficulties in coordinating group activities are commonly experienced problems. If anything, these problems are exacerbated in the natural context.

Natural Groups at Work

We present specific examples of natural groups such as health care teams and computer software design teams. The goal of the inclusion of these specific examples is to illustrate some of the general problems alluded to in the earlier part of this chapter.

Health care teams (HCTs) are made up of a variety of individuals from various occupations (e.g., nurse, pathologist) or disciplines (medicine, psychology, social work) for the purpose of providing health care services. In discussing the importance of communication in the field of health care, Kreps (1988) noted that a central concern in HCTs is that "health professionals are often ethnocentric about the legitimacy of their profession in comparison with other health care professional groups" (p. 253). Such beliefs hamper communication, and when there are real status differences among participants on an HCT related to medical expertise, communication among team members and coordination of activities can be impaired. Effective delivery of services requires that members of HCTs coordinate their activities in a manner that is logical, coherent, and efficient. Berteotti and Seibold (1994) illustrated many of the problems experienced in complex natural teams in their extensive case study of the operation of a hospice team. The hospice team consisted of volunteers and medical personnel. The volunteers spent a minimum of 4 hr a week working with patients or their families.

Two different strategies were used to coordinate team efforts. One strategy, known as the group method, used weekly team meetings to exchange information and plan for the work that needed to be accomplished. The second strategy, known as the feedback method, relied on written and oral communication of information. Although there were over 100 volunteers who worked at the hospice, they only had a single representative who attended the team meetings. This structural feature of the group composition led to the ineffective use of a feedback method of coordination. These meetings invariably resulted in hostility toward the director of the volunteers and the medical director. Participants in the meetings described the medical director as rude and unhelpful. The medical director was impervious to the other members' complaints. Complaints were also directed toward the director of volunteers. Dissatisfaction with these two key administrators resulted in poor coordination of efforts.

Medical personnel complained about the volunteers because they felt they were intruding on their own roles. Volunteers, in turn, felt unappreciated and felt that they received inadequate communication. The central problems in this particular health care team involved definition of roles and coordination of activities. Both of these required effective communication, which was apparently absent. The issue of role definitions is an important one, as it links to perceptions of status and legitimate authority. Medical personnel had expectations for the volunteers that excluded volunteers from adequate participation. They were viewed as transitory and not central to the mission of the team. Volunteers, however, wished to be involved and participate more and felt their roles to be central to the delivery of services to patients and their families. This conflict in role definition among the two major sets of constituents hampered the effective functioning of the team. Communication among team members either in the group meetings or via other mechanisms was poor.

Status is a particularly important variable in medical teams, as there is a clear hierarchy of skills, knowledge, and prestige. It is not comforting to realize that one arena in which status problems get played out is in the operating theater (Dennison & Sutton, 1990; Helmreich & Schaefer, 1994). Dennison and Sutton (1990) observed a team of operating room nurses for a week. The team consisted of 6 to 10 nurses who specialized in open heart and thoracic surgery, ear, nose, and throat operations. The use of teams of nurses represented a huge improvement in the efficient use of the nursing staff. Scheduling of operations was more efficient. All of the nurses on the team learned to be skillful in all the types of operations that were conducted. Compared to how nurses were treated prior to the use of teams, nurses in this team had quite a degree of autonomy. Despite the positive benefits of working in a team, nurses still felt very controlled in their work environment, and doctors usually exercised the control. In comparison to doctors and surgeons, nurses had very low status. In this instance, status and gender were utterly confounded. Nurses (all female) had a separate lounge from doctors (who were men). In the operating theater, doctors could joke, talk, and laugh but exercised tight control over when and if nurses could say anything. In the context of the operating theater, nurses were expected to do exactly what the doctors told them. Nurses rejected this notion of their role and were increasingly stressed as a consequence.

Surgical teams do not always have this clearly hierarchical status structure in which surgeons are in total control. In a study conducted in a European operating room (Helmreich & Schaefer, 1994), status conflicts emerged between the surgeon and the anesthesiologist and their associated support staffs. The central problem was that the anesthesiologist was unwilling to accept that the surgeon had higher status, and in turn, his support staff of nurses, orderlies, and technicians supported him. The support staff of the surgeon in like manner aligned themselves with their leader. Within this context of status competition, there was a breakdown in communication about the patient's needs. This particular study demonstrated that error in the operating theater can easily happen as a function of a breakdown in communication and coordination of activities. Some of this type of breakdown is associated with social processes related to status differentiation. In these examples from the medical field, one can see the problems that arise in natural groups that also arise in laboratory-based groups such as problems in communication as a result of status. One also sees problems that do not arise in laboratory studies such as those of role conflict and role definition problems over time as in the case of the hospice team. The relationships among participants who must continue to interact over time result in problems that occur in natural, ongoing groups.

Problems of coordination are also evident in other groups such as software design teams. Software design teams spent as little as 40% of their time on issues related to design but spent a great deal of time on monitoring progress and in orchestrating and sharing expertise (Olson, Olson, Carter, & Storrøsten, 1992). In designing a software product, the participants often start with a reasonably good representation of the desired end product or the functions that the software is intended to accomplish. Many other natural groups have less initial clarity about their goals and intended products. The 10 software teams studied by Olson et al. were surprisingly similar in the distribution of their cognitive activities within the context of their team discussions. The discourse in meetings contained large amounts of clarification. Thirty percent of the time in these groups was spent on summarizing progress, which seemed to serve a coordination role. Participants offered many design alternatives, but few of the offered alternatives received any evaluation. This may have been a function of having too much information simultaneously available.

Herbsleb et al. (1995) noted that software development places enormous cognitive, organizational, and management demands on participants in a software development team. Developing software to suit client needs requires that the client's domain knowledge and needs must be articulated and captured. The extraction of the client's goals can take a number of recursive episodes, and in the exchanges that go back and forth, much information may be forgotten (Walz, Elam, & Curtis, 1993). Object-oriented design or programming has been touted as a solution to many of the problems, particularly those associated with coordination and communication, that arise in traditional software design methods. One of the advantages of object-oriented design is that processing and data structures are contained in the same object. Herbsleb et al. (1995) examined the use of object-oriented programming in an industrial context and conducted interviews with experienced practitioners of object-oriented design. Based on Herbsleb et al.'s study, they concluded that teams using this approach spent less time in discussions of clarifications of design and spent more time on asking why questions. Communication and coordination seemed to have been facilitated in these design teams.

The problems of coordination that are endemic to these kinds of work teams have resulted in the development and evaluation of a variety of tools to support the cognitive processing of individuals working together on complex tasks. Simple tools for supporting interaction include whiteboards, flip charts, shared text editors, and recording devices such as audiotapes or videotapes. More complex computer-based tools for supporting group performance have also been developed. The impact of two broad technological support systems for group processes were evaluated by Kraemer and Pinsonneault (1990). These systems are known as Group Decision Support Systems (GDSS) and Group Communication Support Systems (GCSS). Kraemer and Pinsonneault defined the purpose of GDSS as systems that attempt to structure the group decision process through provision of decision models usually in some automated form. GCSS are systems primarily designed to support the communication process and typically include tools for regulating information such as storage and retrieval, representational tools such as large video displays, and support for collaboration. Kraemer and Pinsonneault reviewed the empirical research and findings on the impact of these systems and concluded that these tools increase the depth of analysis of the problems under discussion by group members and increase participation by members. Oddly enough, although GCSS has many positive benefits such as increased participation by group members, cooperation is decreased, and the quality of decision making actually decreases. This latter consequence may in part by due to a lack of trust occasioned by the decrease in cooperation.

In summary, natural groups suffer from many of the problems seen in small laboratory-based groups (e.g., status problems), but in addition, they are characterized by significant problems related to communication and coordination. Difficulties in communication and coordination are exacerbated in naturalistic settings by the time frame in which groups operate, changes in conditions under which they operate, and issues related to status and control that contaminate relationships. Recent efforts in the arena of software design and business have attempted to reduce some of these problems of communication and coordination through the use of various tools to support information processing and communication in groups. The use of these tools has resulted in positive outcomes on specific tasks.

INTERDISCIPLINARY TEAMS AS NATURAL GROUPS

What can one learn from the preceding discussion about the functioning of interdisciplinary teams? Interdisciplinary teams are more likely to share the characteristics of natural groups than groups in laboratory settings. These teams are very common in human service delivery and are becoming increasingly so (Schrage, 1990). Teams involving personnel from a variety of disciplines are also common in other work environments such as business, medicine, and education. We described some of the problems faced by such groups in our discussion of natural groups. Among these problems are difficulties with communication, managing group composition issues that may involve status differentiation, and difficulties in coordinating activities that include the appropriate exchange of information. Groups that represent different disciplines are often brought together for specific tasks such as developing policy or making decisions that affect a variety of people. Members of such teams may interact with one another over an extended period of time. Depending on the disciplines represented in such natural groups as task forces or other groups, seniority and status may be totally confounded with diffuse status characteristics such as gender or race. In a previous section, we examined some of the cognitive and social constraints on effective teamwork in groups in general and illustrated these problems by reference to particular examples of groups in natural settings. Interdisciplinary teams pose special risks in these regards.

Potential Difficulties in Interdisciplinary Teams

Each individual in an interdisciplinary team works from the perspective of his or her own discipline. Journet (1993) pointed to some of the problems posed to interdisciplinary interaction as a consequence of individuals' allegiance to a discipline. Disciplines differ in what events or data are interpretable, what methods they espouse, and what kinds of explanations are deemed satisfactory. Such differences can give rise to perceived status differences among representatives of different disciplines, and we have shown in preceding sections how such status differences can affect interaction. Journet further characterized the disciplines as differing in their contents, the persona of the investigator, and what constitutes a significant event. Journet proposed that efforts to integrate across disciplines requires the development of what she termed a "boundary rhetoric," one that negotiates the boundaries of the various disciplines involved. Journet described the difficulties involved clearly. As a consequence of these difficulties, most teams involving members from different disciplines never function as interdisciplinary, integrative teams as described by Thompson-Klein (1990) in her description of interdisciplinarity.

Interdisciplinary teams must solve what Krauss and Fussell (1990, 1991) referred to as the "mutual knowledge" problem. They must come to understand what their collaborators intend. Experts in a discipline share a common referential base, and communication among such experts is facilitated by the use of common referents. In interdisciplinary groups, such common ground must be developed (Clark & Brennan 1991). It may be that the need to develop such a common ground is not recognized by all members of the team, further adding to the difficulty of the interdisciplinary discourse. Team members must come to understand what their teammates know or do not know. This is not a simple task, as our earlier review of work on hidden knowledge suggests. In the work of Stasser and his colleagues (e.g., Stasser, 1992), people in groups often rehearsed already shared information and did not reveal less distributed knowledge. Within the context of interdisciplinary teams, such reluctance to share different knowledge would pose tremendous problems in negotiating a shared knowledge base.

Sjölander (1985, as cited in Thompson-Klein, 1990) described 10 stages or phases through which an interdisciplinary group or project pass through. Presumably, these stages are not fixed and serve to describe common patterns that might be observed during interdisciplinary interaction. Not all interdisciplinary teams might be expected to follow the pattern described by Sjölander. Interdisciplinary teams, however, that work together over extended periods of time may be at risk of experiencing many of the phases of interdisciplinary group development.

During the first phase of interdisciplinary group development, people in different disciplines spend time "singing the same old songs" (Thompson-Klein, 1990, p. 71). In this stage, participants adhere to their allegiances to their own disciplines (Journet, 1993). Each person represents his or her own discipline's viewpoint and interprets the group's common problem according to the cognitive and social schemata they understand in their own discipline (Thagard, 1994). Shifting perspective is often uncomfortable (Heider, 1958), and in the next stage, participants' adherence to their disciplinary views is almost counter-independent, as participants may regard others from different disciplines as having nothing to contribute. Attitudes such as these may result from an adherence to status hierarchies. As might be inferred from the material presented earlier on the influences on group learning, when members of a group assume such attitudes, negative reactions occur.

The third stage of Sjölander's development is that of retreating into abstractions. Discussion at the abstract level is safer and less likely to be challenged. When groups can discuss very general ideas that do not involve concrete actions, there is little disagreement. Few people would argue with the principle of improving safety on our streets. It is only when you get to the details of how and who pays for such improvements that we can observe disagreement. At Stage 4, participants attempt to develop a common language of discourse and may develop "definition sickness" (Thompson-Klein, 1990, p. 71). Participants engage in an effort to move groups forward or come to a common language by defining terms, limits of engagement, and so forth. Inevitably, too much time is spent on this activity. In curriculum meetings related to discussions of new standards in mathematics and science, participants often get trapped in esoteric discussions of the differences between *assessment, evaluation,* and *testing*. Too much of this kind of discussion can result in definition sickness, as the group may not be able to progress to general understandings of common concepts. Stage 5 marks the beginning of fruitful discussion that tends to skip from topic to topic in somewhat incoherent ways. Participants find points of connection in their various points of view and may skip around among subtopics. In Stage 6, the effort to develop a common jargon continues in what Sjölander referred to as the "glass bead game" (after a Herman Hesse novel). This continued effort to develop a common understanding through a common jargon is abandoned in Stage 7 in which participants, frustrated by wallowing in abstraction and constructing unfamiliar jargon, experience great failure. Group members who have not yet abdicated begin to appreciate other perspectives in Stage 8 but feel like traitors to their own disciplines. Sjölander referred to this stage as the "what is happening to me" stage. Stage 9 marks the continued engagement with the perspectives of people from different disciplines. This stage is referred to as "getting to know the enemy" and involves seeing others' perspectives on their own terms. Finally, in Stage 10, having negotiated a path through the borders of the various disciplines, participants in these teams are ready to really begin a discussion of the problem at hand.

Sjölander (1985, as cited in Thompson-Klein, 1990) clearly delineates the problems associated with mutual knowledge building. In expert groups (Krauss & Fussell, 1990), participants communicate almost telegraphically, using technical terms as referents for entire constellations of knowledge and strategies. When experts in a variety of fields are placed in interdisciplinary groups, their normal (within-discipline) discourse is altered, and they must work to derive new referents as facilitators of complex and rapid communication.

Based on the analysis presented previously, it seems most likely that the problems experienced by interdisciplinary groups will include the development of an appropriate task or problem representation and the development of shared knowledge. The social and affective consequences of these problems may be significant, returning the groups' purposes to functions other than productivity. The use of cognitive tools to facilitate the construction of mutual knowledge or communicate about what is known will be very important to the effective functioning of such groups. The use of such tools may also reduce the problems associated with affective responses.

A MODEL OF GROUP COGNITION

Theories of small-group performance have tended to focus on individual-level properties of the group (e.g., members' attitudes, personalities) or on system-level properties (e.g., group norms, decision rules) but not both, although effective group performance involves a complex interaction between individual- and system-level processes (Hirokawa & Johnston, 1989; Poole, Seibold, & McPhee, 1985). A broader view of group functioning is necessary to describe interdisciplinary group performance.

Many analyses of group functioning stem from an input-process-product model (e.g., Kraemer & Pinsonneault, 1990). Such analyses can rarely take into account in meaningful ways the interactivity among variables. Models of group functioning that revolve around a collection of individual information processors do not provide effective descriptions of complex group processing. Krauss and Fussell (1990) argued convincingly for the importance of the development of mutual knowledge to communicate effectively. We showed in some of the examples of natural groups how failures to communicate can negate the effectiveness of the group.

What does it mean to develop mutual knowledge and how can this occur in interdisciplinary teams whose members converse with the referential basis of their own disciplines (Krauss & Fussell, 1990) and with allegiance to their own disciplines (Journet, 1993; Shalinsky, 1989)?

One of the most influential models of human information processing was that proposed by Atkinson and Shiffrin (1968). Although major criticisms of this model have been made, and significant modifications to it have been advanced (Baddeley, 1990), the basic descriptions of the functions of human information processing described in that model are still valid. These functions include storage and retrieval of information in long-term memory, working memory, and executive control of the transfer of information from short- to long-term memory and vice versa. Criticisms of information-processing approaches (often referred to as symbol-processing approaches) have failed to dislodge the utility of these constructs.

In examining some of the difficulties experienced by natural groups in terms of communication and coordination, one can see that limits on individual cognitive processing exacerbate difficulties that might arise simply as a function of the social context in which people are working. The software design teams, without additional support, could not keep track of proposed alternative designs so that they could evaluate options. Communication among members of the hospice team was hampered by the reliance of the large number of volunteers on a single source of information, the volunteer director, for information related to the conditions at the hospice. The introduction of computer-based tools to support some aspects of individual functioning (e.g., representational capacity) in the software design teams improved the quality of decisions made and produced more equality in the distribution of participation. Earlier in the chapter, we also discussed the importance of informational deficiencies in a group. When important relevant information is not available to all members of the group, effective decision making or problem solving can be impaired.

Smith (1993), in describing what he termed "collective intelligence," began by delineating the types of knowledge that people exchange when in groups. First, there is the tangible knowledge in a group that includes the target products (e.g., policy statement) that provide evidence of the successful completion of the group's task. En route to the production of these targets, groups may produce instrumental products (accompanying graphs of supporting evidence) that support the work but are not part of the target product. A second kind of knowledge that is available is intangible and includes both public and private knowledge. Intangible knowledge is what is available in participants' heads. This knowledge may be public and shared or private. Ball (1994) provided some interesting insights on the costs of too much intangible knowledge remaining private. Groups also produce ephemeral products. These are transitory representations produced during collaboration to facilitate communication or other processes. They are typically destroyed and do not become part of the final product. According to Smith (1993), the task of collective intelligence in groups is to effectively manage the appropriate flow of information in groups and preserve the important features of accurate storage and retrieval of information, activation of appropriate knowledge, and control of the flow of information. In earlier parts of this chapter, we described some of the influences that can limit the intelligent performance of groups.

A model of "group intelligence" provides a novel way of considering the interactions within interdisciplinary groups. The research questions that stem from such a consideration include whether the flow of information in a group can be modeled in such a way as to allow us to pinpoint problem areas in the maintenance of an adequate flow of information. We need to examine the kinds of ephemeral products that are produced and understand the purposes they serve within a group.

REFERENCES

Anderson, R. C., & Pichert, J. (1978). Recall of previously unrecallable information following a shift in perspective. *Journal of Verbal Learning and Verbal Behavior, 17,* 1–12.

Atkinson, R. C., & Shiffrin, R. M. (1968). Human memory: A proposed system and its control processes. In K. W. Spence (Ed.), *The psychology of learning and motivation: Advances in research and theory* (Vol. 2, pp. 89–195). New York: Academic.

Baddeley, A. (1990). *Human memory: Theory and practice*. Needham Heights, MA: Allyn & Bacon.

Ball, M. A. (1994). Vacillating about Vietnam: Secrecy, duplicity, and confusion in the communication of President Kennedy and his advisors. In L. R. Frey (Ed.), *Group communication in context* (pp. 181–198). Hillsdale, NJ: Lawrence Erlbaum Associates.

Ben-Bassat, I., & Taylor, R. N. (1982). Behavioral aspects of information processing for the design of management information systems. *IEEE Transactions on Systems, Man, and Cybernetics, SMC, 12,* 439–450.

Berger, J., Conner, T. L., & Fisek, M. H. (1974). *Expectation states theory: A theoretical research program.* Cambridge, MA: Winthrop.

Berger, J., Rosenholtz, S. J., & Zelditch, M., Jr. (1980). Status organizing processes. *Annual Review of Sociology, 37,* 241–255.

Berteotti, C. R., & Seibold, D. R. (1994). Coordination and role definition problems in health care teams: A hospice case study. In L. R. Frey (Ed.), *Group communication in context: Studies of natural groups* (pp. 107–131). Hillsdale, NJ: Lawrence Erlbaum Associates.

Chinn, C. A., O'Donnell, A. M., & Jinks, T. S. (2000). The structure of discourse in collaborative learning. *Journal of Experimental Education, 69,* 77–97.

Cicourel, A. V. (1990). The integration of distributed knowledge in collaborative medical diagnosis. In J. Galegher, R. E. Kraut, & C. Egido (Eds.), *Intellectual teamwork: Social and technological foundations of cooperative work* (pp. 221–242). Hillsdale, NJ: Lawrence Erlbaum Associates.

Clark, H. H., & Brennan, S. E. (1991). Grounding in communication. In L. B. Resnick, J. M. Levine, & S. D. Teasley (Eds.), *Perspectives on socially shared cognition* (pp. 127–149). Washington, DC: American Psychological Association.

Cohen, E. (1994). Restructuring the classroom: Conditions for productive small groups. *Review of Educational Research, 64,* 1–35.

Cohen, E., Lotan, R., & Catanzarite, L. (1990). Treating status problems in the cooperative classroom. In S. Sharan (Ed.), *Cooperative learning: Theory and research* (pp. 203–229). New York: Praeger.

Cooper, H. M., & Good, T. (1983). *Pygmalion grows up: Studies in the expectation communication process.* New York: Longman.

Cowie, H., & van der Aalsvoort, G. (Eds.). (1999). *Social interaction in learning and instruction: The meaning of discourse for the construction of knowledge.* Amsterdam: Pergamon.

Cragan, J. F., & Wright, D. W. (1990). Small group communication research of the 1980's: A synthesis and critique. *Communication Studies, 41,* 212–236.

Crocker, J. (1981). Judgment of covariation by social perceivers. *Psychological Bulletin, 90,* 272–292.

Damon, W., & Phelps, E. (1989). Critical distinctions among three approaches to peer education. *International Journal of Educational Research, 13,* 9–20.

Deci, E. L., & Ryan, R. M. (1985). *Intrinsic motivation and self-determination in human behavior.* New York: Plenum.

De Lisi, R., & Golbeck, S. L. (1999). Implications of Piagetian theory for peer learning. In A. M. O'Connell & A. King (Eds.), *Cognitive perspectives on peer learning.* Mahwah, NJ: Lawrence Erlbaum Associates.

Dennison, D. R., & Sutton, R. I. (1990). Operating room nurses. In J. R. Hackman (Ed.), *Groups that work (and those that don't)* (pp. 293–308). San Francisco: Jossey-Bass.

Deutsch, M. (1949). A theory of cooperation and competition. *Human Relations, 2,* 129–152.

Dillenbourg, P. (1999). *Collaborative learning: Cognitive and computational approaches.* Amsterdam: Pergamon.

Dobos, J. A. (1996). Collaborative learning: Effects of student expectations and communication apprehension on student motivation. *Communication Education, 45,* 118–134.

Donald, J. G. (1987). Learning schemata: Methods of representing cognitive, content, and curriculum structures in higher education. *Instructional Science, 16,* 187–211.

Doyle, W. (1983). Academic work. *Review of Educational Research, 53,* 159–200.

Eisenstat, R. A., & Cohen, S. G. (1990). Summary: Top management groups. In J. R. Hackman (Ed.), *Groups that work (and those that don't)* (pp. 78–86). San Francisco: Jossey-Bass.

Fiedler, F. E. (1978). Contingency model and the leadership process. In L. Berkowitz (Ed.), *Advances in experimental social psychology* (Vol. 11). New York: Academic.

Fiedler, F. E. (1996). Research on leadership selection and training: One view of the future. *Administrative Science Quarterly, 41,* 241–250.

Fiedler, F. E., & Garcia, E. (1987). *New approaches to effective leadership: Cognitive resources and organizational performance*. New York: Wiley.

Flower, L., & Hayes, J. R. (1981). A cognitive process theory of writing. *College Composition and Communication, 32,* 365–387.

Foushee, H. C. (1984). Dyads and triads at 35,000 feet: Factors affecting group process and aircrew performance. *American Psychologist, 39,* 885–893.

Foushee, H. C., & Helmreich, R. L. (1988). Group interaction and flight crew performance. In E. L. Wiener & D. C. Nagel (Eds.), *Human factors in aviation*. San Diego, CA: Academic.

Frey, D. (1986). Recent research on selective exposure to information. *Advances in Experimental Social Psychology, 19,* 41–80.

Frey, L. R. (1988, November). *Meeting the challenges posed during the 70s: A critical review of small group communication research during the 80s*. Paper presented at the meeting of the Speech Communication Association, New Orleans, LA.

Frey, L. R. (1994). Introduction: The call of the field, studying communication in natural groups. In L. R. Frey (Ed.), *Group communication in context* (pp. ix–xiv). Hillsdale, NJ: Lawrence Erlbaum Associates.

Hackman, H. R. (Ed.). (1990). *Groups that work (and those that don't)*. San Francisco: Jossey-Bass.

Hackman, J. R. (1995, December). *Groups that fly*. Presentation to the Department of Psychology, Princeton University, Princeton, NJ.

Heider, F. (1958). *The psychology of interpersonal relations*. New York: Wiley.

Helmreich, R. L., & Schaefer, H. (1994). Team performance in the operating room. In M. S. Bogner (Ed.), *Human error in medicine* (pp. 225–253).

Herbsleb, J. D., Klein, H., Olson, G. M., Brunner, H., Olson, J. S., & Harding, J. (1995). Object-oriented analysis and design in software project teams. *Human–Computer Interaction, 10,* 249–292.

Hirokawa, R. Y. (1990). The role of communication in group decision-making efficacy: A task-contingency perspective. *Small Group Research, 21,* 190–204.

Hirokawa, R. Y., & Johnston, D. D. (1989). Towards a general theory of group decision making: Development of an integrated model. *Small Group Research, 20,* 500–523.

Janis, I. L. (1982). *Groupthink: Psychological studies of policy decisions and fiascoes* (2nd ed.). Boston: Houghton-Mifflin.

Johnson, D. W., & Johnson, F. (2002). *Joining together: Group theory and group skills* (4th ed.). Boston: Allyn & Bacon.

Johnson, D. W., & Johnson, R. (1992). Positive interdependence: Key to effective cooperation. In R. Hertz-Lazarowitz & N. Miller (Eds.), *Interaction in coopera-*

tive groups: The theoretical anatomy of group learning (pp. 174–199). New York: Cambridge University Press.

Journet, D. (1993). Interdisciplinary discourse and "boundary rhetoric." *Written Communication, 10,* 510–541.

Karau, S. J., & Williams, K. D. (1995). Social loafing: Research findings, implications, and future directions. *Current Directions in Psychological Science, 4,* 134–140.

Kraemer, K. I., & Pinsonneault, A. (1990). Technology and groups: Assessment of the empirical research. In J. Galegher, R. E. Kraut, & C. Egido (Eds.), *Intellectual teamwork: Social and technological foundations of cooperative work* (pp. 375–405). Hillsdale, NJ: Lawrence Erlbaum Associates.

Krauss, R. M., & Fussell, S. R. (1990). Mutual knowledge and communicative effectiveness. In J. Galegher, R. E. Kraut, & C. Egido (Eds.), *Intellectual teamwork: Social and technological foundations of cooperative work* (pp. 111–145). Hillsdale, NJ: Lawrence Erlbaum Associates.

Krauss, R. M., & Fussell, S. R. (1991). Constructing shared communicative environments. In L. B. Resnick, J. M. Levine, & S. D. Teasley (Eds.), *Perspectives on socially shared cognition* (pp. 172–200). Washington, DC: American Psychological Association.

Kreps, G. L. (1988). The pervasive role of information in health and health care: Implications for health communication policy. In J. Anderson (Ed.), *Communication Yearbook II* (pp. 238–276). Newbury Park, CA: Sage.

Latane, B., Williams, K., & Harkins, S. (1979). Many hands make light the work: The causes and consequences of social loafing. *Journal of Personality and Social Psychology, 37,* 822–832.

Levin, J. R., & O'Donnell, A. M. (1999). What to do with educational research's credibility gaps? *Issues in Education, 5,* 177–229.

McClane, W. E. (1991). Implications of member role differentiation. *Group and Organization Studies, 16,* 102–113.

McGrath, J. E. (1991). Time, interaction, and performance (TIP): A theory of groups. *Small Group Research, 22,* 147–174.

Meeker, B. F. (1981). Expectation states and interpersonal behavior. In M. Rosenberg & R. H. Turner (Eds.), *Social psychology: Sociological perspectives* (pp. 290–319). New York: Basic Books.

Morris, C. G. (1966). Task effects on group interaction. *Journal of Personality and Social Psychology, 4,* 545–554.

Nemeth, C. J., & Staw, B. M. (1989). The tradeoffs of social control and innovation in groups and organizations. *Advances in Experimental Social Psychology, 22,* 175–210.

O'Donnell, A. M., & Dansereau, D. F. (1992). Scripted cooperation in student dyads: A method for analyzing and enhancing academic learning and performance. In R. Hertz-Lazarowitz & N. Miller (Eds.), *Interaction in cooperative groups: The theoretical anatomy of group learning* (pp. 120–141). New York: Cambridge University Press.

O'Donnell, A. M., & King, A. (Eds.). (1999). *Cognitive perspectives on peer learning.* Mahwah, NJ: Lawrence Erlbaum Associates.

O'Donnell, A. M., & O'Kelly, J. (1994). Learning from peers: Beyond the rhetoric of positive results. *Educational Psychology Review, 6,* 321–349.

Olson, G. M., Olson, J. S., Carter, M. R., & Storrøsten, M. (1992). Small group design meetings: An analysis of collaboration. *Human–Computer Interaction, 7,* 347–374.

Paulus, P. B. (Ed.). (1989). *Psychology of influence* (2nd ed.). Hillsdale, NJ: Lawrence Erlbaum Associates.
Poole, M. S., Seibold, D. R., & McPhee, R. D. (1985). Group decision-making as a structurational process. *Quarterly Journal of Speech, 71,* 74–102.
Reeve, J. (1996). *Motivating others: Nurturing inner motivational resources.* Needham Heights, MA: Allyn & Bacon.
Robinson, D. T., & Balkwell, J. W. (1995). Density, transitivity, and diffuse status in task-oriented groups. *Social Psychology Quarterly, 58,* 241.
Robinson, D. T., & Smith-Lovin, L. (1992). Selective interaction as a strategy for identity maintenance: An affect-control model. *Social Psychology Quarterly, 55,* 12.
Salomon, G., & Globerson, T. (1989). When teams do not function the way they ought to. *International Journal of Educational Research, 13,* 89–99.
Scheerhorn, D., Geist, P., & Teboul, J. B. (1994). Beyond decision making in decision making groups: Implications for the study of group communications. In L. R. Frey (Ed.), *Group communication in context* (pp. 247–262). Hillsdale, NJ: Lawrence Erlbaum Associates.
Schopler, J. H., & Galinsky, M. J. (1990). Can open-ended groups move beyond beginnings. *Small Group Research, 21,* 435–449.
Schrage, M. (1990). *Shared minds.* New York: Random House.
Shalinsky, W. (1989). Polydisciplinary groups in the human services. *Small Group Behavior, 20,* 203–219.
Sinclair, A. (1992). The tyranny of a team ideology. *Organization Studies, 13,* 611–626.
Sjölander, S. (1985). Long-term and short-term interdisciplinary work: Difficulties, pitfalls, and built-in failures. In L. Levin & I. Lind (Eds.), *Interdisciplinarity revisited* (pp. 85–101). Stockholm: OECD, SNBUC, Linköing University.
Smith, J. B. (1993). *Collective intelligence in computer-based collaboration.* Hillsdale, NJ: Lawrence Erlbaum Associates.
Sorenson, J. R. (1971). Task demands, group interaction, and group performance. *Sociometry, 34,* 483–495.
Stasser, G. (1992). Pooling of unshared information during group discussion. In S. Worchel, W. Wood, & J. H. Simpson (Eds.), *Process and productivity* (pp. 48–67). Newbury Park, CA: Sage.
Stasser, G., & Stewart, D. (1992). Discovery of hidden profiles by decision-making groups: Solving a problem versus making a judgment. *Journal of Personality and Social Psychology, 63,* 426–434.
Stasser, G., Taylor, L. A., & Hanna, C. (1989). Information sampling in structured and unstructured discussions of three and six person groups. *Journal of Personality and Social Psychology, 57,* 67–78.
Stasser, G., & Titus, W. (1985). Pooling of unshared information in group decision making: Biased information sampled during discussion. *Journal of Personality and Social Psychology, 48,* 1467–1478.
Stasser, G., & Titus, W. (1987). Effects of information load and percentage of shared information on the dissemination of unshared information during group discussion. *Journal of Personality and Social Psychology, 55,* 81–93.
Steiner, I. D. (1972). *Process and productivity.* New York: Academic.
Thagard, P. (1994). Mind, society, and the growth of knowledge. *Philosophy of Science, 61,* 629–645.
Thompson-Klein, J. (1990). *Interdisciplinarity.* Detroit, MI: Wayne State University Press.

Walz, D. B., Elam, J. J., & Curtis, B. (1993). Inside a software design team: Knowledge acquisition, sharing, and integration. *Communications of the ACM, 36,* 62–77.

Wheelan, S. A., McKeage, R. L., Verdi, A. F., Abraham, M., Krasick, C., & Johnston, F. (1994). Communication and developmental patterns in a system of interacting groups. In L. R. Frey (Ed.), *Group communication in context: Studies of natural groups* (pp. 153–178). Hillsdale, NJ: Lawrence Erlbaum Associates.

Zaccoro, S. J., Foti, R. J., & Kenny, D. A. (1991). Self-monitoring and trait-based variance in leadership: An investigation of leader flexibility across multiple group situations. *Journal of Applied Psychology, 77,* 525–535.

II

Studies of Interdisciplinarity in the Wild

4

Seeing in Depth*

Charles Goodwin
University of California Los Angeles

The ship is the heterotopia par excellence.
—Foucault (1986, p. 27)

In so far as scientific knowledge does not float free in some abstract, context-free domain but is instead situated, a key question arises: How can one describe the way in which the concrete place where scientific work is done has consequences for the knowledge produced there (Ophir & Shapin, 1991; Shapin, 1988)? The description of such a space raises a host of questions. Thus, Lynch (1991) noted that "the place of laboratory work is not a locale within a unitary physical space, since it is constituted by the actions that *dwell* grammatically within it" (p. 53). From such a perspective, relevant spaces are reflexively constituted through the organization of the actions that simultaneously make use

*This chapter reprinted by permission of Sage Publications Ltd. from "Seeing in Depth" by Charles Goodwin, *Social Studies of Science, 25,* 237–279, Sage Publications, 1995.

of the structure(s) provided by particular places while articulating and shaping them as meaningful entities appropriate to the activity in progress. In this chapter I describe the interdigitation into a common course of action of a diverse patchwork of different kinds of spaces and representational technologies by differently positioned actors working together to take samples on an oceanographic research vessel.[1]

A research ship constitutes a bounded, tool-saturated environment for the doing of a range of different kinds of science. A most salient characteristic of oceanographic ships as work sites is their heterogeneity. Because of the cost of chartering a ship, scientists from quite different disciplines, each pursuing his or her own research project, are forced not only to cooperate in a common endeavor but literally to set up their laboratories next to each other. This creates unique possibilities for communication across disciplines. Unlike the publication and discussion of findings that occur at a conference or through journals, scientists on a research ship are directly exposed not only to the ideas but also to the tools and work practices of their colleagues as they bump elbows while trying to pursue their separate projects in the limited space available. Moreover, to get their science done, they must work closely not only with other scientists but also with sailors. This collection of participants from diverse disciplines and occupations, with separate tool kits making possible different kinds of research endeavors, is strongly segregated from ordinary social life on land as it sails alone through the sea.

The primary object of study for scientists on the ship is the sea, and they spend a great deal of effort and money to position themselves precisely at specific points within it. However, many of the spaces that are most important to them are found not outside the ship but within its laboratories. Representations provided by various kinds of printed and electronic documents are the objects of intense scrutiny.[2] Rather than constituting collections of information in the abstract, such inscriptions are themselves spatial arenas for the organization and production of meaningful action. An analytical framework is thus needed that can en-

[1]Very relevant analysis of how oceanography has developed as a discipline within the political and economic structure of world capitalism, and the United States in particular can be found in Mukerji (1989). In this chapter, with its focus on in situ work practices within the laboratories of such a ship, I provide a perspective that complements Mukerji's.

[2]How representations are used in the organization of scientific practice has been the topic of much insightful analysis (e.g., Knorr-Cetina & Amann, 1990; Lynch, 1988; Latour & Woolgar, 1979, 1988). For a comparison of the practices used by archaeologists to make maps with those used by lawyers to shape and contest perception of graphic images at a trial, see Goodwin (1994).

compass at least (a) the spatial organization of the laboratory, (b) the visible orientation frameworks created by the positioning of the human bodies that inhabit the laboratory, (c) the frameworks for perception and action situated within the documents being attended to, and (d) the spaces and phenomena that those inscriptions represent—for example, features in the sea. Foucault used the term *heterotopia* to mark "a relatively segregated place in which several spatial settings coexist, each being both concrete and symbolically loaded" (Ophir & Shapin, 1991, p. 13).[3] Ophir and Shapin proposed that in the modern West, the sites where science is done are fundamentally heterotopic spaces. Central to such spaces is a segregated world inhabited by a restricted range of social actors, which contains within it a second space where the phenomena that animate the discourse of a particular scientific discipline are made visible. A major function of such places is to force "the invisible to manifest itself, to leave traces, to betray a hidden presence. Yet the invisible appears only to the eyes of those authorized to observe it. The heterotopic site is at one and the same time a mechanism of social exclusion and a means of epistemically constituting conditions of visibility" (Ophir & Shapin, 1991, pp. 13–14).

Among the heterotopic places described by Foucault (1986) are theaters and the cinema, "a very odd rectangular room, at the end of which, on a two-dimensional screen, one sees the projection of a three-dimensional space" (p. 25). A laboratory on the ship contains a conjunction of spaces that is both analogous and, in relevant ways, quite different. In the following, two scientific technicians, Phyllis and George, stare intently at a pair of two-dimensional inscriptions provided by two different tools, a computer and a Precision Depth Recorder (PDR), which contain representations of the sea they are probing (see Fig. 4.1).

Like the screen in a cinema, these inscriptions are the focus of intense, engrossing scrutiny. Indeed they are the place in this laboratory where phenomena in the world the scientists are trying to study, the sea under their ship, are made visible. Moreover, like an unfolding movie, these inscription surfaces are not static but instead show a spectacle of relevant, meaningful events that are constantly changing in significant ways. However, unlike the story in a cinema, the drama that these screens contain is available to few, if any, members of the larger society on shore. Indeed, even within the tiny group visible here, the ability to see what these images have to offer is unevenly distributed; the man on

[3]See also Foucault (1986).

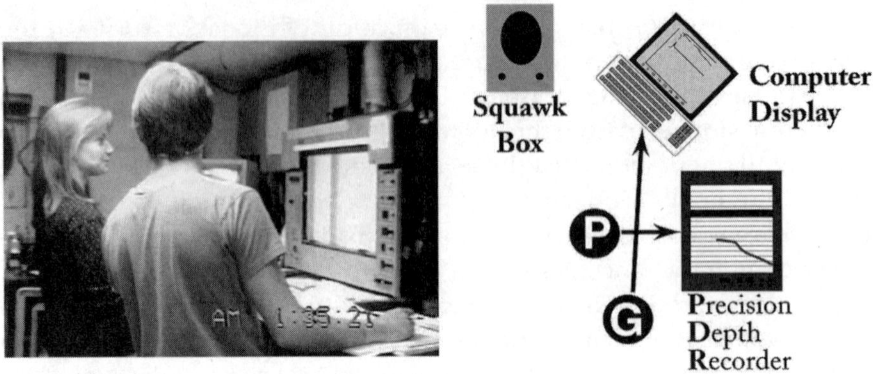

FIG. 4.1. Two scientists positioned in ship's laboratory, heterotopic space.

the right does not know how to competently read the complicated squiggles on the PDR that is the focus of his coworker's attention. In that two different screens are being intently scrutinized, the space is more like a multiplex than a single cinema, but a strange one in which the audience is positioned to watch, simultaneously, two different shows. However, unlike the discontinuous stories told in separate films, these two separate screens each provide very different representations of exactly the same place. Moreover, their carefully chosen audience does not sit passively until the images on the screen come to an end but instead uses them to perform consequential actions while the events they display are still unfolding. The screens provide not just a window into the sea but the resources required to move other inscription devices within it including some of the machines that are producing these very representations. The audience for these images is simultaneously the crew that produces them, a crew that reaches through the images to move things in the world they represent.

The flow of images is sporadically accompanied by a relevant soundtrack. However, rather than capturing the noises that are occurring at the place being looked at, it takes the form of talk over a squawk box from a third member of their team who is working in a different place. Most of his talk consists of reports about where he thinks they are at the moment: for example, "11 meters." The two scientists are thus attending not only to visual representations of the place they are investigating but also to spoken ones. Unlike the self-contained world of the cinema, the multiple spaces they must attend to to do their work encompass a patchwork of mutually relevant but discontinuous places, including not only the sea they are sampling and the laboratory where they are working with its disparate spatial representations but also other places on the ship to which they have highly structured but very limited access.

Positioning in space(s) is central to the work that is occurring here. Understanding that work requires both ethnographic analysis and detailed examination of the specific activity in progress.[4]

SAMPLING GRID

These two people are part of a large team of scientists on the AmasSeds project investigating what happens when the Amazon River meets the Atlantic Ocean. The Amazon is far and away the largest river in the world (one of the islands in its mouth is the size of Switzerland). As it hits the Atlantic ocean, a range of very complex processes occur. The scientific project is especially interested in tracking the way in which the river and the sediments it carries mix with the waters of the ocean and deposit sediments on the sea floor. Figure 4.2A is an example of one of the products they are trying to produce. It is a graph that uses salinity differences to trace where the fresh river water goes as it moves into the sea.

How is such a product made? First, a sampling grid is imposed on the area of interest (Fig. 4.2B). The decision as to precisely where samples are to be taken is the outcome of an intense political process both among the scientists whose different research agendas require different kinds of data (e.g., some are particularly interested in sediments near the shore, whereas others want to study phenomena that are best observed further out to sea) and between the scientists and the Brazilian government, which was unwilling to allow an American ship loaded with equipment for collecting vast amounts of data to probe too closely in its territorial waters (an observer from the Brazilian navy was present on the ship at all times). The grid was thus shaped not only by the competing theoretical interests of different disciplines but also by Brazil's reaction to America's history of imperialism in South America. Finally, the characteristics of the tools being used also constrained where samples could be taken. The draught of the ship, how deeply it sank into the water, limited quite forcefully its ability to sail into shallow water with-

[4]Close attention to the situated details through which courses of practical action are accomplished in endogenous settings is central to ethnomethodologically informed studies of practice in science and the workplace (e.g., Button, 1992; Garfinkel, 1986; Garfinkel, Lynch, & Livingston, 1981; Heath & Luff, 1996, 2000; Heath & Nichols, 1977; Lynch, 1985, 1993; Sharrock & Anderson, 1994; Sharrock & Button, 1994; Suchman, 1987, 1992).

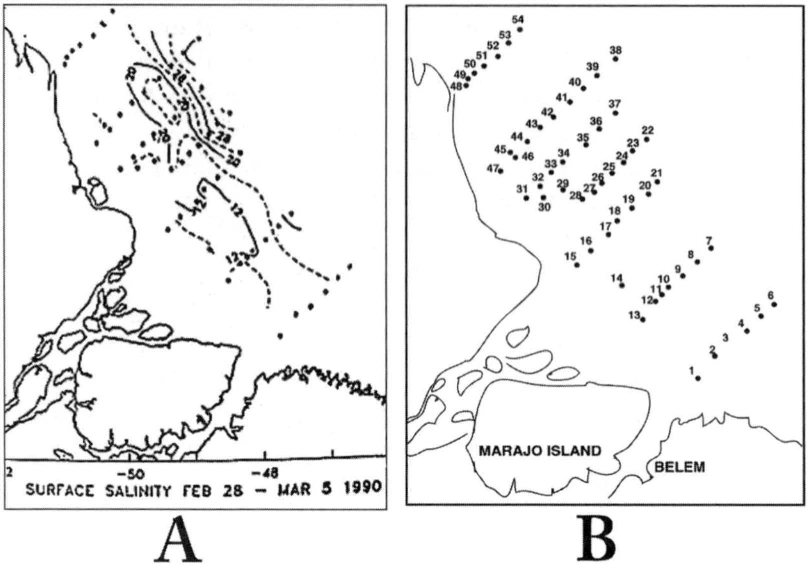

FIG. 4.2. Surface salinity graph (A) and sampling grid (B).

out running aground, and this was complicated by the fact that the charts being used were known not to be accurate. Many of the scientists had a very strong interest in gathering data much closer to shore but were unable to do so because of the sampling grid that emerged from the sum of these political negotiations and technical constraints.[5] The social and political processes required for an American research ship to do field research in the territory of another country also had the effect of constituting the project as one of international collaboration in which Brazilian as well as American scientists were very active participants.

For the actual collection of data, an oceanographic research ship is hired at considerable expense. The ship chosen for the Amazon study was 170 ft (51.8 meters) long, had a gross tonnage of 281 tons (255 metric tons), and a draught of 10.5 ft (3.2 meters). It was staffed by six officers and six crew members, all of whom were men. It sailed from Florida (where some teams of scientists loaded their equipment) to the port of Belém on the Amazon River near its mouth. From there it made a series of cruises into the area defined by the sampling grid, most of which lasted approximately 10 days to 2 weeks. Typically, scientific teams

[5]On-shore samples were taken in other phases of the project by other teams of scientists.

would change during the 1- to 2-day layover between cruises in Belém, although some teams might sail on successive legs of the study. One of the goals of the project was to sample the same place at different points during the year (when seasonal changes in the river would lead to sharp differences in the amount of water being discharged, the corresponding sediment load, and other relevant phenomena). Thus, over the course of the project, the same scientific teams would make multiple cruises.

At sea, the ship sails to each point or station on the grid. Once there, it stops, and samples are taken. Intense activity occurs "on station," as people from different scientific teams move about the ship to collect data. Frequently, teams converge at specific places (such as the quarter-deck where instruments are lowered into the sea), and there, members of one team will sometimes help others move their equipment and fill sample bottles. As soon as samples have been collected, a process that typically took less than 2 hr, the ship sets sail for the next station, usually reaching it within another 2 hr. While in transit, each team retreats back into its own laboratory where they process the samples that have been collected at previous stations and prepare for the next station.

In that the sampling grid is written on a piece of paper, a map, one might be tempted to analyze it primarily as an inscription, an immutable mobile (Latour & Woolgar, 1979) that allows scientists sitting in their offices in North America to plan where they can best test their theories in the seas off another continent. Although entirely valid, such an analysis ignores the way in which the sampling grid structures the lifeworld of those on the ship. The major factor governing the distribution of their activities is the distinction between being on station and "in transit." Inhabitants of the ship do different things and frequently work in different places at each of these times. Like the seasons in an agricultural community, the sampling grid establishes the basic rhythms that structure the life of scientists working on the ship.

Because ship time is so expensive, there is a strong emphasis on collecting as much data as possible. The ship moves quickly from station to station, pausing to allow those on board time to rest is a luxury that simply cannot be afforded. If, as frequently happens, a work crew does not have enough personnel to organize two shifts, it is not at all unusual for scientists and technicians to work for 36 hr or 48 hr without stopping to sleep and to continue day after day at a pace in which they might average only four hours of sleep a day. Night and day lose their meaning as frameworks for the organization of work. Instead of taking time to go to their bunks, technicians sometimes drop to the floor next to their laboratory benches to try and catch a quick nap before the ship reaches its next station. Rather than existing only in a conceptual space defined by the scientific theories it is probing or as geographical coordinates on a map, the sampling grid as an inhabited space structures the work, movement, and lived experience of those caught within it as inexorably as do the clock hands on the assembly line that Charlie Chaplin (1936) tried to follow with his own hands in *Modern Times*.

Because of its position in their world, scientists on the ship can find events in the sampling pattern that would be completely invisible to the reader seeing it in a journal report. For example, the points on the grid are arrayed in sets of lines extending out to sea, with closely spaced stations within each line but much larger gaps between each line. The ship will take much longer to traverse these gaps (e.g., to go from Station 6 at the end of the first line in the south to Station 7 at the beginning of the next), and those on the ship see in these places times when rest might be possible.

In contrast, the ship's crew worked a regular schedule consisting of four hours on duty followed by eight hours off. In a variety of ways, their lives were segregated (although not entirely) from those of the scientists. Most of them worked most of the time in different areas of the ship (e.g., the wheelhouse, engine room, and kitchen), and they had their own sleeping quarters. Everyone ate at the same times in the ship's single dining room, but scientists and crew were assigned to separate tables. Many of the tasks performed on station, such as lowering instruments into the sea, required close collaboration between crew and scientists, but as I show in more detail later in this chapter, actors from these separate occupations were usually stationed at different places (e.g., a crewman on the bridge would handle the ship's winches for scientists working on deck or in a laboratory). In brief, the very small space provided by the ship contained two distinct communities, each drawn from different social backgrounds and possessing separate sets of skills. Despite their very close proximity in space and activity and the fact that sailors and scientists would sometimes visit with each other in off hours, these communities lived and worked within quite different temporal and spatial lifeworlds.

Wittgenstein (1958) argued that the meaning of a representation is not its bearer (for instance, the territory marked by the sampling grid superimposed on a standard map) but is rather the grammatical processes used to articulate the representation within a relevant language game. The sample grid is embedded successively within a variety of different, although related, activities. Months before the ship leaves port, the grid is built through an intense political process involving scientists with different agendas and whole nations. At sea, positioning the ship at the points specified by the map is an intricate, artful, ongoing situated accomplishment. Global satellite positioning systems are used to place the ship as close as possible to a point defined simultaneously as (a) a patch of actual ocean water that can both float the ship and provide a water column to study; (b) a position defined within a global system of latitude and longitude; and (c) a station in the sampling plan—that is, a point constituted through its embeddedness within a larger structural system, a network of other places that will give this one contrastive meaning within the research plan. As noted previously, the tasks of carrying out the structure specified by the sampling grid build a lived temporal and spatial lifeworld for those charged with accomplishing its regularities.[6] Moreover, the two communities on the ship have different relations to the symbolically loaded spaces constituted by these different, although interrelated, systems of meaning. The ship's crew is responsible for placing the ship at the latitude and longitude defined for each station (aforementioned b), something that requires competent deployment of a range of skilled practices and tools. However, they have no professional interest in the larger research processes within which work at the station (aforementioned c) is embedded. The crew's interest in the station ends when the ship is properly positioned. For the scientists, the work done at a particular station is just a beginning, a small part of a larger, still unfolding process, one that will involve considerable work not only at other points at sea but also back at their laboratories on the mainland, at conferences, in the pages of journals, and so on.

After the voyage, as the grid is moved to other activity systems, it becomes a different kind of object: a way of coordinating measurements made at different times in a single space, a field for visibly showing patterns, a boundary object that can be used to compare the findings of different scientific disciplines (Star & Griesemer, 1989), and the like. Both the larger goals of the research project (for instance, measuring properties of the interface between river and ocean including their change over time) and invariant, portable features of the grid itself provide

[6]For analysis of the situated work required to produce such regularities as a schedule, map, or record of changes that occurred in a sample medium on successive days, see Lynch (1985, 1988), and Suchman (1992).

deep continuity between these separate activities. However, at each stage in this process, different grammars instantiated within locally relevant activity systems, each with its own constellation of tools and practices, are deployed properly to use the grid and to "see" quite different things within the phenomenal field it provides.

CONVERGENT DIVERSITY

At each station, separate teams of scientists, each pursuing their own research agenda, set out to work (see Fig. 4.3). In that different kinds of scientists share the ship on each cruise, what will be done at each station will vary considerably from cruise to cruise. At a typical station from the cruise I investigate here, two teams of physical oceanographers drop separate instrument packages into the sea. One team of geochemists collects water samples at different depths and at some stations uses a box core to obtain a column of mud from the sea floor. The water and mud are used as sources of data by biologists as well as

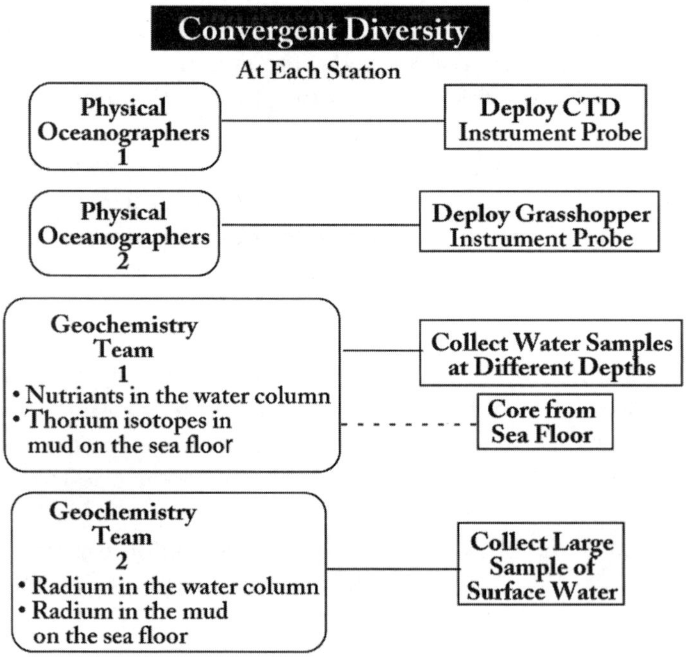

FIG. 4.3. Activities of different research teams aboard ship illustrate case of convergent diversity.

geochemists. A second team of geochemists collects a very large sample of surface water to track the source and distribution of different components of the water.

Although all of these scientists want to look at what happens at this particular place, the things they are interested in are in fact quite diverse. Phenomena of interest to one field might be quite irrelevant to another. In a very real sense, although all of these groups are in precisely the same place and probing exactly the same patch of sea, each will in fact see something quite different there. However, as I will show in more detail, the interests and findings of the separate groups of scientists are not incommensurate with each other. Although one reason that they share the ship is to distribute the considerable expense of running it, another is that the findings of these separate disciplines complement each other. Having geochemists, physical oceanographers, and biologists take samples at precisely the same spot provides a perspective on the processes they are investigating that would be impossible for a single discipline. Despite their disciplinary boundaries, they attend conferences together where they exchange and compare findings. The sea they are all investigating thus constitutes a clear example of a boundary object that facilitates collaboration across disciplines while being constituted differently within each.[7]

In addition to shaping the objects around which collaboration is structured, such phenomena also have consequences for the organization of activity in particular kinds of places. Thus, the activity that occurs on deck at each station as different groups take samples provides an example of *convergent diversity*. By this, I mean a place where separate individuals, groups, or teams converge; however, when they converge, they do not all work with each other in the pursuit of a single plan of action but instead follow rather separate agendas, which may interlock at points with the agendas of others. Points of convergent diversity are thus characterized by interrelated heterogeneity. Although most research in the social sciences (recent investigations of science being an exception) has concerned itself with single activities shaped by coordinated action around a common focus,[8] points of convergent diversity appear to be both com-

[7]The sea floor itself provided a particularly clear example of a boundary object. Different research projects defined where it occurred in different ways (e.g., how much sediment had to be present before a sample stopped being muddy water and started to be the mud of the sea floor), and talks were initiated across disciplines to try to create a common definition.

[8]Note, for example, Goffman's (1964) classic definition of an *encounter* as "a single, albeit moving, focus of visual and cognitive attention" (p. 135).

mon and important. Indeed, they provide a prototypical example of the kind of multiactivity settings that were investigated by the "Workplace Project" (Brun-Cottan et al., 1991; Suchman, 1992).

TOOLS

Convergent diversity is instantiated concretely in some of the tools used on the ship. The heterogeneous organization of such tools has strong consequences for the way in which phenomena of interest to the scientists are perceived, manipulated, sampled, and studied and for the organization of interaction within work practices.

One of the most important tools used by physical oceanographers is the "CTD." This is an instrument probe that is lowered into the sea on a cable where it makes a range of measurements about the physical properties of the water it passes through, including its conductivity, temperature (or more precisely resistance across a set of platinum sensors that is translated into a temperature measurement), and pressure (translated through appropriate equations to an extremely accurate measurement of depth). Measurements made by these instruments are sent back to the ship on an electrical cable where they serve as input to a computer. The computer both translates sensor readings into measurements of temperature, depth, and so on and uses those figures to graph changes in the water column as the CTD moves through it. The CTD is thus a complicated tool, one that brings together both precise and expensive instruments that are sent to the bottom of the sea and a relevant body of theory that is reified as a set of equations and algorithms in a computer on the ship. The most important component of the CTD is not the probe itself but the equations used to translate the measurements obtained from it into accurate, meaningful data. The development of this tool has a complex history that weaves through a number of different disciplines. It began not as a tool for physical oceanographers but as a chemical instrument. The crucial equations governing (for instance) the translation of conductivity to salinity were formulated by physical chemists. This instrument was then appropriated by physical oceanographers for their work—for example, as a way of getting information about temperature and salinity that could be used to measure the density of water and thus to investigate issues such as how buoyant some parts of the ocean are relative to others. Perhaps because of the central importance of the information it makes visible (e.g., accurate measurements of the precise depth at which significant features of the water column change), the CTD is a tool that has a history of being appropriated by one discipline after another. Indeed, several different research teams had CTDs of their own on the cruise I investigate here. One, owned by a team of physical oceanographers, was particularly important. Lowering it to the sea floor was a central component of the work done at every single station.

For their work, one team of geochemists needed samples of actual water collected at different depths. To obtain these samples, they used a "Niskin bottle"—essentially, a long tube with stoppers at each end that are closed when a signal is sent from the ship, thus trapping the water in them at that particular depth. How do the geochemists get their Niskin bottles to the depths where they want to take samples? The physical oceanographers already have a platform descending to the bottom, the CTD. The geochemists attach their instruments to that platform producing a heterogeneous tool in which the CTD of the physical oceanographers is surrounded by a ring of the geochemists' Niskin bottles (see Fig. 4.4). What results is a complex tool that ties together two different scientific disciplines and two different communities of practice that have a common interest in studying the sea. Like the sea itself, the "CTD rosette" is a boundary object (see Fig. 4.4).

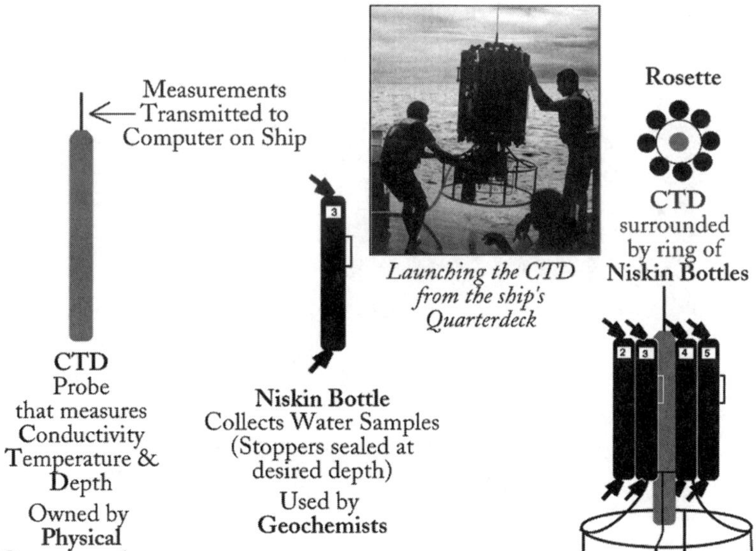

FIG. 4.4. Convergent diversity instantiated in a heterogeneous tool (CTE rosette), a boundary object.

CTD as a Tool for Perception

What the scientists want to study is the distribution of different kinds of phenomena in the water column. As river water pushes out on the surface, sea water moves back beneath it, producing a water column of considerable complexity. The sea under the ship is thus not homogeneous but instead consists of a patchwork of different kinds of water and sediments (greatly oversimplified in Fig. 4.5). The scientists are particularly interested in properties and distribution of these bodies of water and in the location of "fronts," places where two different kinds of water meet.[9]

How can these underwater features be seen so that they can be sampled and studied? The CTD sends measurements it is making back to the ship as it moves through the water. The computer graphs made from these data provide a continuously changing picture of relevant features in the water column. For example, as the CTD passes through a front, salinity and temperature change. Computer software that displays such changes graphically provides a way of "seeing" these fronts and other features that are of interest to the scientists. The CTD thus provides the

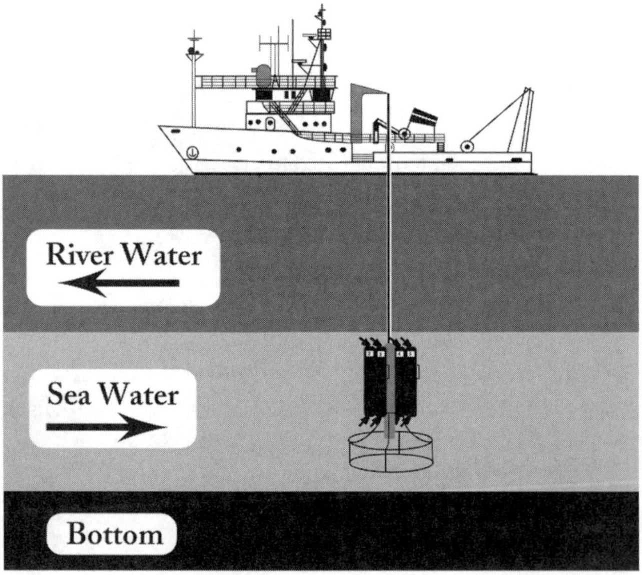

FIG. 4.5. Simplified illustration of the water column being sampled.

[9]See Friedman (1989) for a historical study of the development of the conception of polar front in meteorology.

scientists on the ship with perceptual access to the world they are sampling while simultaneously shaping what they are able to see there (e.g., just those properties of the environment that their sensors capture and software can make visible as organized patterns). This is true not only in the narrow sense of "new hardware" that makes it possible to see things such as salinity differences as graphs on computer screens, but more crucially in the "theoretical" sense that both the objects made visible (such as fronts and gradients), and the current interest of the scientists in such objects are provided by the historical development of a particular theoretical field.

The history of the tools being used is not, however, confined to a single discipline. Equations developed by physical chemists to describe the relation between conductivity and salinity provide another discipline, physical oceanography, with tools for probing phenomena of interest to it. Appropriation across disciplines is central to this process. Where does this occur, and how might it be organized? Aspects of this process I now investigate in more detail.

I noted previously that on research ships, scientists from different disciplines are required to do laboratory work in close proximity to each other. The two people gazing intently at separate screens we examined earlier are scientists working in the ship's CTD laboratory. Phyllis (P) is a physical oceanographer, and George (G) is a geochemist (see Fig. 4.6). How does the fact that the CTD with its ring of Niskin bottles brings together the tools of their two separate disciplines organize their work? An initial possibility might be that they operate side-by-side in parallel but independently of each other. This turns out not to be the case. The geochemist is staring intently at the CTD display (see Fig. 4.7). Although this CTD is owned by and gathering data for another discipline, physical oceanography, George can use the picture of the water column it makes available to determine where to take his own samples. The juxtaposition of tools thus produces a creative synergy, as a tool embedded within the work practices of one discipline provides new resources and opportunities to view phenomena for another.

MULTIPLE PERCEPTUAL FRAMEWORKS

The physical oceanographer (Phyllis) is intently scrutinizing not the CTD display but instead a PDR. For simplicity, I treat this as a form of sonar, which records echoes of phenomena in the water (including the bottom) on a moving paper chart, and I refer to it as a "sonar chart" in

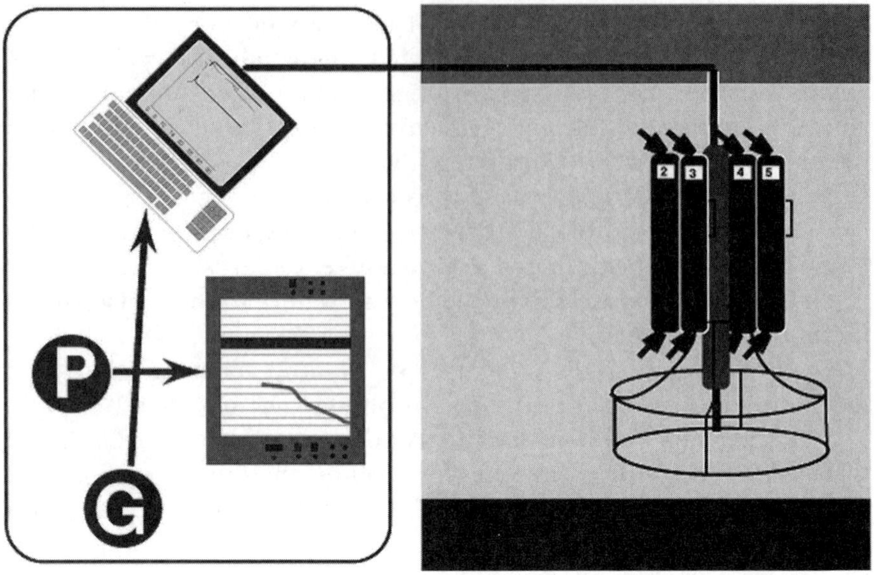

FIG. 4.6. A hybrid tool produces separate screens observed by two scientists.

FIG. 4.7. Tool juxtaposition produces creative synergy between scientists from different disciplines.

this chapter (see Fig. 4.8). The CTD is one of the most important tools of Phyllis' profession, an instrument that sits at the cutting edge of current technology in physical oceanography. The pressure gauge it carries provides a far more accurate measure of depth than the complicated, confusing image provided by the sonar chart. Why then isn't she staring at the CTD display as intently as the geochemist?

To answer this question, it is necessary to look more closely at the work involved in taking samples with this CTD. The device, with the equipment attached to it, costs approximately $25,000. If she slams it into the bottom, there is a real chance that she will cut the cable and lose the instrument (anticipating such a possibility, the ship carried a second CTD as backup). However, both she and the geochemist want to get as close to the bottom as possible, for it is there that some of the phenomena of most interest to them are to be found. She must thus walk a very fine line between two conflicting constraints: (a) getting as close to the bottom as possible (b) without actually hitting it. To do this, she begins a CTD cast by lowering the instrument almost to the bottom but with an adequate safety margin to prevent actually hitting it. During the descent, data are collected and displayed, and it is these graphs that the geochemist uses to help him decide where to take samples as the instru-

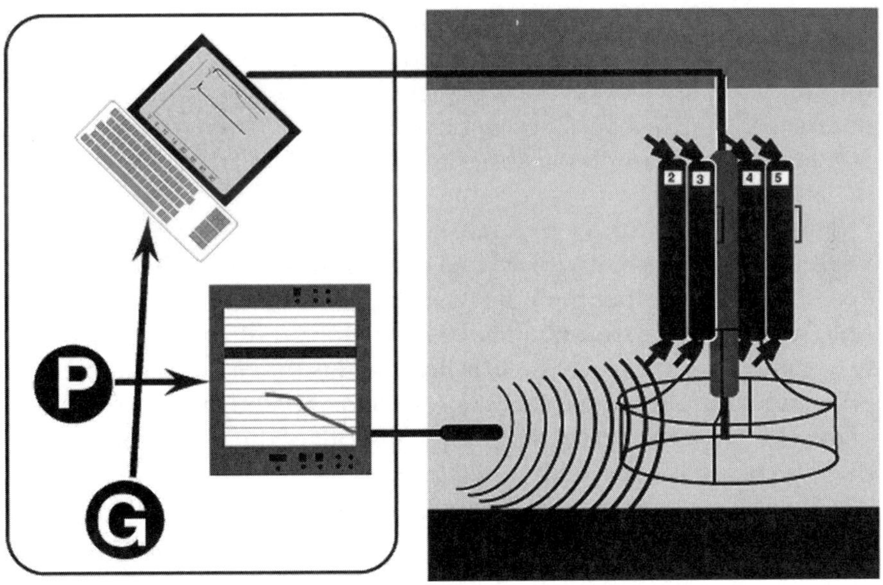

FIG. 4.8. Physical oceanographer prefers geochemist's "sonar" to judge position of CTD relative to bottom.

ment platform ascends. Once the CTD reaches its safe point near the bottom, the physical oceanographer reanalyzes the situation in the light of the new information provided by the descent and decides exactly how much deeper she can safely go.

The most accurate measure of depth available to her is provided by the pressure gauge on the CTD. However, what is crucial for her current work is not accuracy in the abstract but instead a measurement of depth that is relevant to the tasks she is currently engaged in—that is, the position of the CTD relative to the bottom. The tool that best makes visible this relation is not the pressure gauge (which reports only the position of the CTD while being oblivious to the bottom) but the sonar (PDR) that juxtaposes the CTD to the bottom while giving a less accurate measure of the absolute depth of the CTD.

Depth is thus dealt with by these scientists not as an abstract, context-free measurement[10] but instead as something to be defined indexically—that is to say, with reference to something else. What that something else is is defined by the specific activity that the act of measuring is helping to accomplish. Thus, for the geochemist, the relevant position of the CTD is constituted by its relation to features of the water column that he wants to sample. For the physical oceanographer in her capacity as "driver" of the CTD (but not necessarily in her capacity as research scientist), relevant depth is defined in terms of the relation between the CTD and the bottom. Each activity requires a different view of the environment they are working in together. Therefore, each attends to a different tool, which shapes perception of that environment in a different way but one relevant to the specific tasks that the party doing the looking is engaged in.

There is yet a third, very relevant participant in the CTD cast—Warren, the winch operator who actually lowers the CTD through the ocean. The winch operator is not a scientist but a sailor. His task and the skills he brings to it are embedded within a long tradition of seamanship (e.g., lifting heavy objects such as fishing nets into a boat). Indeed, the perceptual requirements for his task are instantiated in the architecture of the ship. In addition to the window in front of the wheelhouse, a second window has been built in the back so that the winch operator can see the objects he is manipulating. By using the tools available to him, he is thus building on the work of his predecessors in this job who have

---

[10]For analysis of measurement as an indexical, situated process, see Lynch (1991), Sacks (1989), and Sacks (1992).

developed solutions to the systematic problems posed by such tasks and embodied these solutions in concrete tools (Leont'ev, 1981). What is found here is quite literally a historically constituted *architecture for perception*, a history that is instantiated not in the texts that report earlier political events but rather in the tools built by anonymous ancestors that shape in quite detailed ways the life and activity of their successors.

The window overlooking the quarterdeck where the CTD will be lifted back on to the ship has close task-relevant parallels to the sonar display that the physical oceanographer uses. Just as she was most concerned about not having the CTD hit the bottom, Warren risks losing it if he slams it into the top of his crane by reeling in too much cable. Just as the sonar display enabled Phyllis to see the relation between the CTD and the bottom, the wheelhouse window provides Warren with a view of the crane that will lift the CTD out of the water, the quarterdeck where it will be placed, and the sea it will emerge from.

It is absolutely crucial that the winch operator know when the instrument package is approaching the surface. However, he is in a different location from the two scientists and cannot see the images visible on their screens. The only way that he can measure the position of the CTD is by the amount of cable he has played out. It is known by everyone that currents can pull the CTD so that the length of cable deployed can give a very inaccurate reading of absolute depth. Despite this, cable length is the depth measurement provided by the tools that the winch operator is working with, and, if he zeros it correctly when he launches the CTD, it will accurately tell him when the probe is returning to the ship. Its measurements work for the particular tasks he is charged with accomplishing.

Central to the activity of deploying the CTD is the task of positioning it in appropriate places. However, within this activity there are in fact a number of different, task-relevant views of where the CTD is (see Fig. 4.9). The activity of deploying the CTD thus involves the articulation in real time of multiple views of how the tool being worked with is positioned within its relevant environment. Although three parties are collaborating in the activity of moving the same tool through its environment, each has different perceptual access to that environment, that access being shaped by the tools that each is using and these tools being selected in terms of the specific tasks that each is facing. What is involved in this activity is not simply a division of labor but a division of perception.

It is frequently argued in anthropology that the analyst must work to get the participant's perspective. However, there is no single partici-

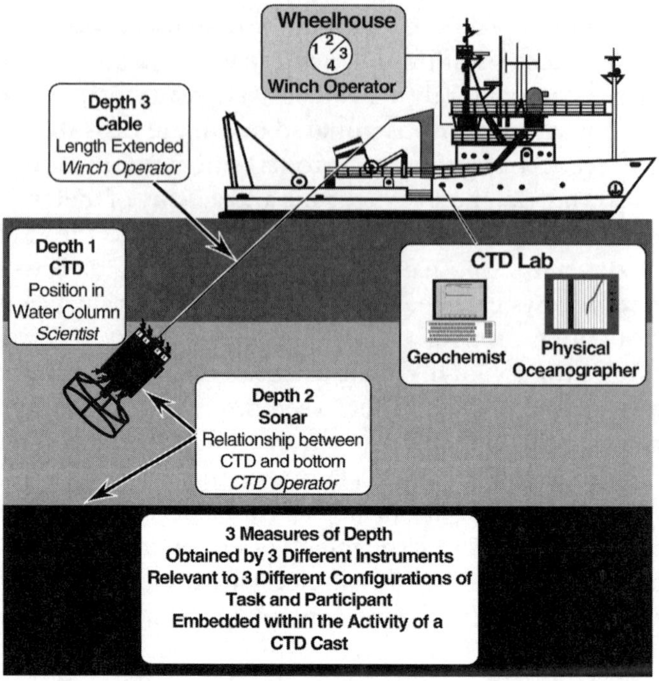

FIG. 4.9. Division of perception as a form of social organization.

pant's perspective but instead multiple perspectives (Haraway, 1988). Moreover, these alternative views on what is to be seen are not random, idiosyncratic, or haphazard but instead are systematic products of the organization of the endogenous activities in progress.

What is at issue here are processes of perception. The organization of this perception is not, however, located in the psychology of the individual brain and its associated cognitive processes but is instead lodged within and constituted through situated endogenous social practices. Such perception is a form of social organization in its own right.

A clear demonstration of the situated nature of the perceptual processes being examined here is found in the way in which they require for their accomplishment the tools that build the setting that makes the activity possible in the first place—for example, the sonar charts, computer graphs, and the like that constitute a CTD laboratory. These tools shape perception through the way in which they construct representations. The structure of representations in scientific practice has been the topic of insightful analysis.[11] In this chapter, I complement this work by looking at how representations are articulated by differentiated participants to accomplish something within temporally unfolding sequences of action. Perception and action are inextricably linked. Moreover, the activity of these scientists and sailors is structured not only by their tools but also by the perceptual frameworks provided by their disciplines and the routine work practices that have developed for the accomplishment of this task. Consistent with Hutchins' (1990) analysis of distributed cognition, the "knowledge' " required to perform a CTD cast is not lodged within any single individual but is instead distributed throughout a system that includes not only human actors of very different types but nonhuman actors as well (Latour, 1987). Perception is something that is instantiated in situated social practices rather than in the individual brain.

ARTICULATING THE DOCUMENT SURFACE

In that the separate perceptual frameworks of each participant must be integrated into a common task (for instance, putting the CTD in a particular position), the task of translating the view from one perspective into the frame of reference of another is posed. Investigation of how this is done provides the opportunity to look in more detail at how two- dimensional inscriptions, such as documents and images on computer screens, are (a) organized as conjunctions of diverse spaces with heterogeneous properties and (b) articulated as frameworks for the production of meaning and action.

The CTD display on the computer in the ship's laboratory provides a graph of the water column that the geochemist can use to guide his sampling (see Fig. 4.10). However, to use this graph to collect samples, the geochemist must be able to tell the winch operator where to go. To do this, he has to translate the information on the graph into a statement that can be expressed in meters, the only measurement system available to the winch operator who must reel in a specific amount of cable. The CTD display provides the resources for doing this—a scale at the bottom of the screen (as well as a range of other scales that I ignore in this discussion). Like many documents, the display is a complex heterogeneous surface bringing together on a single flat plane structurally different types of information (e.g., a representation of the environment that

[11]See Footnote 2.

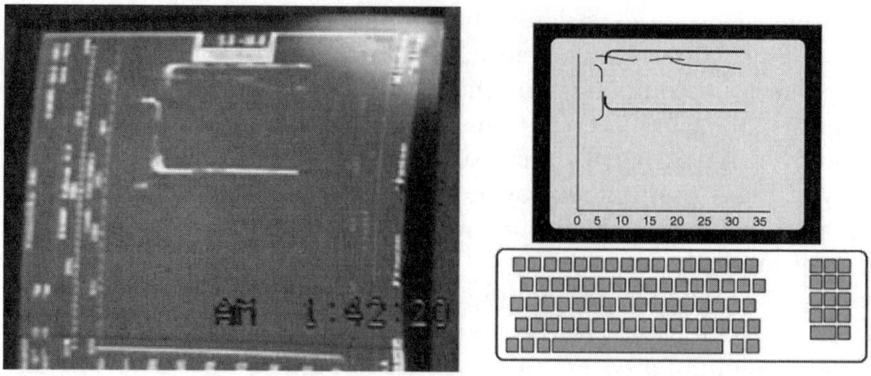

FIG. 4.10. CTD graphic display brings together structurally different representations.

the scientists are sampling and tools that can be used to work with that representation). However, the ruler at the bottom of the screen is distant from the graph of the structures being sampled. The geochemist is thus faced with the task of bringing together two points of information that are distributed spatially on the document surface he is working with. He does this by using another tool. The pencil he has been using to make entries in the log book is now placed on top of the screen so that information provided by the scale can be juxtaposed to the graph (see Fig. 4.11). The task of reading the screen in a work-relevant way thus leads to a situated improvisation, as implements designed for other purposes are tailored to local projects. Within this process, the object in George's hand becomes two different tools when embedded within alternative activities—in this case, a writing instrument when log entries are to be made and a straight edge when distant points have to be juxtaposed on the computer screen.[12] For its part, the screen is not simply a flat inscription, a place where information is to be apprehended through vision alone, but the base of a three-dimensional work area, something that can be touched and manipulated to shape the material it provides into the phenomenal objects required for the tasks of the moment.[13] Reciprocally, the marks on the screen as instantiations of the fea-

[12]The possibilities for such mutation are not unlimited; crucial to the use of the pencil as a way of measuring events on the screen in the present case is its size, straightness, and the fact that it is readily at hand.

[13]The transformation of events visible in the pixels of a screen into new discursive objects is by no means a neutral process. See Goodwin (1994) for analysis of how the Los Angeles policemen who beat Rodney King worked with the video image of his body writhing under their blows to "demonstrate" that he was in fact the aggressor, struggling to rise and attack them.

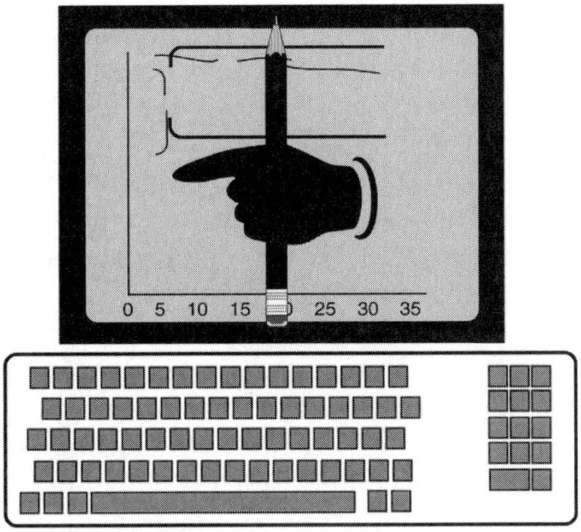

FIG. 4.11. A pencil is a tool for juxtaposing scale and graph.

tures being investigated by the scientists provide a framework of intelligibility that constitutes what George's hand is doing as meaningful action rather than aimless movement. Like the mirror in Cocteau's *Orphee* (Paulve, Film du Palais-Royal, & Cocteau, 1950) the graph on the screen is not merely something to be looked at but instead an open gateway to a world where the human body can move and act within new frameworks of meaning.[14]

Like a playing field that builds a landscape within which certain moves, such as a "goal," become both possible and visible, the graph on the computer screen creates an arena for the perception and constitution of relevant action. Consider, for example, the access that the scientists in the laboratory have to the actions of their coworker in the wheelhouse. As the sampling run unfolds, the scientists will tell the winch operator to move to a place, specified as a depth, where they want to take samples. In that the CTD will provide a more precise and up-to-date picture of the sample place as it gets close to it, instructions for further movement may be given to the winch operator before samples are actually taken. Relevant instructions and acknowledgments are given over the ship's squawk box: the physical oceanographer tells the winch operator to move to a particular

[14]For analysis of a physicist transporting himself from one kind of space to another as his hand moves over the surface of a graph, see Ochs, Jacoby, and Gonzales (1994).

depth (e.g., "Warren, bring the CTD up to 11 meters please"). When the winch operator arrives at what he thinks is the correct spot, he tells the scientists he has completed the task given him by announcing the new location of the CTD ("11 meters").

The work-relevant activities of the winch operator are available not only in the reports he makes over the intercom but also through the way in which he moves the CTD, a process that is visible to the scientists in the laboratory as changes in the graphs they are looking at. These graphs provide mediated access to not only the sea they are studying but also their coworker. As representations of the lived activity of another human being, the squiggles traced on the graph are quite different from the talk heard over the squawk box. Almost everything that one thinks of as constituting the embodied performance of a human actor has been stripped away: the visible body itself, language, the features of a human voice that allow participants in interaction to recognize a specific individual, his affect, the stance he is taking toward the activity in progress, and the like. However, although the graph offers an extraordinarily attenuated vision of the winch operator as an embodied coparticipant, it provides the scientists with their best record of precisely those features of his action that are most relevant to the task at hand—in this case, where he has placed the CTD. To determine what to say next to the winch operator, the scientists will in fact pay more attention to his actions as visible on their graphs than they will to what he says. Goffman defined the primordial site of human interaction, the *social situation*, as "an environment of mutual monitoring possibilities, anywhere within which an individual will find himself accessible to the naked senses of all others who are 'present,' and similarly find them accessible to him" (Goffman, 1964, p. 135). Here, rather than Goffman's simple but clear case of immediate embodied presence to the naked senses of others, what one finds is a complex texture of mediated access as the scientists attend to multiple representations of the winch operator's action (both his talk and the traces of his activity on their graphs) available through media with quite disparate properties. Although the graphs lose most of what Goffman defined as crucial for the organization of interaction, they in fact provide the most pertinent arena for the perception of the winch operator's action, situating what he is doing as moves on the very playing field that is structuring the activity in progress—that is, as movements occurring within a particular landscape: the features of the water column that the scientists are trying to sample, which are instantiated in the graphs.

Crucial to the status of the marks on the graph as interactive events that help structure the future course of the activity in progress is their unfolding temporal organization within a project that has not yet come to completion. By monitoring through their changing graphs both the actions of the winch operator and the features that his probe has revealed, the scientists decide what to do next.[15]

This temporal horizon is lost (or at least radically transformed)[16] when the CTD cast is completed, and these same marks become records of a past event rather than resources for the shaping of future action. The graphs visible on the scientists' displays during the CTD cast are thus not flat, timeless, two-dimensional inscriptions but instead constitute inhabited spaces that provide an architecture for the perception, monitoring, and production of relevant action as the night's work unfolds in lived time.

Similar arguments can be made about the language that occurs here. Consider, for example, the winch operator reporting that he has reached a requested depth by saying "11 meters." Like the graph, these words provide a representation of an event in the sea under the ship that is relevant to the activity in progress: a statement about the current depth of the CTD. Many approaches to the philosophy of language would treat a statement such as this as a proposition about some possible state of affairs in the world it describes. A central game played with propositions, one that links language to the world, is evaluating their

---

[15]It is by no means unusual for temporal processes of human interaction to occur entirely within the space constituted by an electronic document. At one of the airlines studied by the Xerox PARC Workplace Project (e.g., Brun-Cottan et al., 1991), flap settings for planes taking off all around the country were computed at a single control room in Texas. These settings can only be computed after all of the passengers, fuel, and baggage have been loaded on the plane. Between the moment that a plane left the gate and the time it reached the end of the runway, workers in the local Ops Room in California would be monitoring an electronic document on their computers to check on whether Texas had put the appropriate number in a particular place on the computer form. Although their entire interaction is filtered through the keyhole of these numbers in one cell of a document on a computer screen, the workers in Texas and those in California work together under very tight time constraints to get planes safely off the ground. In even more attenuated fashion than the situation of the ship (where sequences of talk are also exchanged), the space constituted through an electronic document provides an arena for the production and monitoring of meaningful action within temporal sequences of interaction.

[16]As data inscriptions back on the mainland, the graphs will of course have a new projective horizon in terms of the analysis they help develop. However, the prospective horizon that is structuring action here—the problem of where to go to take the next sample—will no longer be available. The loss of the possibility of agency within a temporally unfolding situation that occurs when the graphs become records is a central component of the process through which the embodied work being investigated here is erased within subsequent reports that will use the products of this night's work as data.

truthfulness. Within such a framework, the winch operator's statement is accurate and truthful if the CTD is actually 11 meters under the sea and false if it is not.

Is this in fact the way that the scientists listening to the report evaluate it? Note that they are in an especially strong place for making such a evaluation. Not only do they have an interest as scientists in measuring this depth, but the pressure gauge on the CTD, when interpreted by the equations in the computer producing the display that the scientists are looking at, gives the most accurate reading of depth possible with the tools currently available to science. Despite this, those in the laboratory do not reply to the winch operator by calling him a liar because their instruments show the probe to be actually 10.25 meters deep or gloat over the superior knowledge provided by the expensive tools of modern science when compared with the crude instruments of the working sailor. Instead of using the framework provided by a correspondence theory of truth to interpret "11 meters," they hear that utterance as an appropriate sequential move within a relevant language game[17]—that is, as a report that the last instruction given the winch operator ("Bring the CTD up to 11 meters") has been accomplished so that the sample run can now move to its next stage. Indeed, differences in the perspectival frameworks provided by the tool kits being used by alternatively positioned actors are one of the things that make such a move necessary. In that everyone knows that the measurements of the winch operator are at best approximations of absolute depth (e.g., currents can pull the probe horizontally), the scientists cannot simply look at their graphs to see when the CTD reaches the requested depth but must instead get a report from the perspective of the winch operator, situated within the phenomenal world provided by his tools, to know that he is finished. The winch operator's report is properly heard not by looking in the abstract to the world it describes but instead by embedding it within a relevant language game. Investigating the endogenous organization of situated activities makes it possible to develop a framework for the study of representations that does not create an arbitrary division between language or "mental" phenomena and material objects such as maps, graphs, and other inscriptions but instead analyzes the

---

[17]Analysis of how utterances are organized as actions within the unfolding sequential structure of talk-in-interaction lies at the heart of the approach to the analysis of conversation initiated by Sacks in collaboration with Schegloff and Jefferson (e.g., Jefferson, 1973, 1987; Sacks, 1963, 1974, 1992; Sacks, Schegloff, & Jefferson, 1974; Schegloff, Jefferson, & Sacks, 1977; Schegloff & Sacks, 1973). For examples of how such a framework can elucidate the organization of talk and action in work settings, see Drew and Heritage (1992).

meaningfulness of any representation by describing the grammar for articulating it—that is, how to use it to make an appropriate move within a relevant activity system.

SEEING IN COMMON

To place the pencil on the screen, George walks right in front of Phyllis (see Fig. 4.12). In that his movements occur within a setting for human interaction, they can be seen and interpreted as meaningful action by others. Phyllis comments on what he is noticing (see Fig. 4.13). Note that Phyllis does not inquire about what George is looking at. Instead, by treating what he is noticing as already visible to her, she demonstrates her ability not only competently to read a relevant display but independently to determine what in it he would find interesting.

What does it take to be able independently to notice a "nice feature"? Standing there, I did not see anything like a nice feature. The ability to see such an event is embedded within an endogenous community of practitioners, the work of which provides a guide for seeing—interpretative structures that locate particular phenomena as relevant and interesting—and the tools and intellectual frameworks that make such phenomena visible in the first place (Goodwin, 1994). Such seeing constitutes an instantiation of culture as practice. Note also that in seeing this event, Phyllis is integrating analysis of two very different kinds of spaces: (a) the events instantiated on the surface of the document being examined and (b) movement through the laboratory itself by an actor performing an activity that becomes recognizable by attending to both his actions and the tool he is working with.

Phyllis embeds what is being seen here within a temporal horizon as well. With her "again" in "That nice feature again," she links the phenomena now visible on the screen to events seen earlier in other spaces that the ship has sampled.

Why is it relevant for one coworker to be able to see what another is doing? One very basic answer to this question is provided by the task(s) of producing collaborative action. Phyllis and George will be working together to take samples as the CTD is lifted to the surface. Three and half min later, after taking a set of samples at 22 meters, Phyllis asks George where the next sample will be taken (see Fig. 4.14). In his reply in line 5, George does not provide a precise number but a visibly approximate one: "About eleven meters." Moreover, his answer is noticeably delayed by two long pauses (lines 2 and 4) and an "Uh:m"

FIG. 4.12. George's move in front of Phyllis is interpreted in context.

Phyllis: I looked at that.
It was ni:ce.
(0.2)
Yeah.
(1.6)
That ni:ce feature ag[ain.
George: [Yeah.

FIG. 4.13. Phyllis requires no verbal explanation to interpret George's noticing.

accompanied by a hand gesture displaying uncertainty. For an addressee accustomed to the unambiguous instructions that typically occur in such sequences, such displays can raise the question of why this particular number has such a penumbra around it. At the beginning of the pause in line 6, Phyllis turns and starts moving toward the intercom she uses to give depth instructions to the winch operator. However, instead of completing this action (e.g., by telling the winch operator to go to 11 meters, as in line 11) she turns back to George with the query in line 7: "Wanna try en hit that?" This utterance ends with an indexical *that* accompanied by a hand gesture pointing to the computer display. What she appears to be referencing is the "nice feature" that she had earlier commented on, and that George's activity of measuring on the screen

1 Phyllis: Where next.
2 (0.8)
3 George: Uh:m.
4 (1.0) 
5 About eleven meters.
6 (1.1)
7 Phyllis: Wanna try en hit that? 
8 George: Yeah.
9 (2.2)
10 Phyllis: Warren.
11 Bring the CTD up to eleven meters please.

FIG. 4.14. Producing collaborative action involves talk within locally relevant spaces.

had made so salient. Her question thus offers an hypothesis that would account for the approximate character of the number she has just been given—namely, that George is interested not in a particular depth per se but in the feature they've observed. Indeed, when they get to 11 meters, they do not take a sample there but instead begin an elaborate chase as they work to position the CTD precisely at the feature so that they can sample it. The talk that occurs here is thus tied retrospectively to prior action (for instance, to the earlier noticing and measuring of the feature), whereas prospectively it sets the agenda for what Phyllis and George will do together in the future. Making sense out of that talk, such that Phyllis knows what George is trying to do when deciding where to take samples (e.g., to "hit the feature"), requires not only competence in the larger work practices that constitute their domains of professional responsibility but also close attention to both the meaningful articulation of a range of different kinds of locally relevant spaces and the details of how talk is produced within sequences of human interaction.

Access to such phenomena is restricted to those in the CTD laboratory, the only people positioned to see both each other and the representations that are guiding the sampling. Thus, although the winch operator is an important coparticipant in the activity, he knows at this point only that the CTD is to move to a particular depth, not that a feature is being hunted—and indeed, the study of features is not part of his work. Once again, alternatively positioned actors have quite different access to what it is that they are doing together.

HYBRID SPACES: SPACE AS LOCALLY ORGANIZED, HISTORICALLY SITUATED PRACTICE

Central to the work that these scientists are doing is their placement within and organization of spaces of many different kinds. Lynch (1991) drew attention to the problem of locating where the action is occurring in scientific work in terms of the distributed *activity field* implicated by a specific course of action. However, most analysis of the human use of space has focused on the organization of more bounded, self-contained, and internally consistent types of space—for example, spatial frameworks that provide organization for human interaction,[18]

[18]See, for example, Kendon (1990) for analysis of how the organization of human bodies in space provides frameworks for the organization of their interaction with each other and Duranti (1992) for analysis of how the movement of bodies through spaces constituted socially with specific cultures provides interpretive frameworks for the organization of meaningful action.

the way in which different languages encode spatial relationships (Hanks, 1990; Levinson, 1992), the organization of graphic displays by scientists (Lynch & Woolgar, 1988), and so on. When the organization of endogenous *activities* is taken as the relevant unit of analysis (as it is here), this conceptually neat and clean division of phenomena into isolated, self-contained systems becomes inadequate. The activities of the scientists in the CTD laboratory repetitively and systematically cut across the ways in which space has been partitioned by social scientists for systematic study. The spaces they inhabit and articulate to get their work done are "hybrid spaces." Thus, the spatial organization of events on the computer screen is consequential for and intimately tied to the interactive organization of human bodies in the space of the CTD laboratory. The participation frameworks being sustained by these bodies include orientation not only to other human beings but also to tools and documents of various types. In attending to these documents, the scientists are organizing their actions with reference to spaces that extend far beyond the laboratory itself. In brief, to do their night's work, the scientists on the ship must juxtapose a heterogeneous collection of very different kinds of spaces. To study this process, one must look not just at each of these separate orders of space (which do require systematic analysis as coherent phenomena in their own right) but also at the activity that the scientists are engaged in as an unfolding course of practical action within which particular kinds of space emerge as relevant at particular moments and are then articulated with reference to each other. For example, manipulating the surface of the computer screen to position a probe at a particular place in the sea below the ship requires noticeable movement through interactive space that makes visible to others present an orientation to particular theoretical events instantiated on the screen and so on. Although those working in the laboratory attend seamlessly to various orders of space as interconnected components of coherent courses of action, it is useful to distinguish at least provisionally and perhaps inaccurately some of the structure and complexity of the different kinds of space they so easily move through. In the 10-sec sequence I have been examining, the spaces being attended to by these scientists include at least the following (see Fig. 4.15).

1. The environment they are studying: the sea outside the ship.
2. A representation of that environment displayed by tools that make visible structures of interest to the scientists.

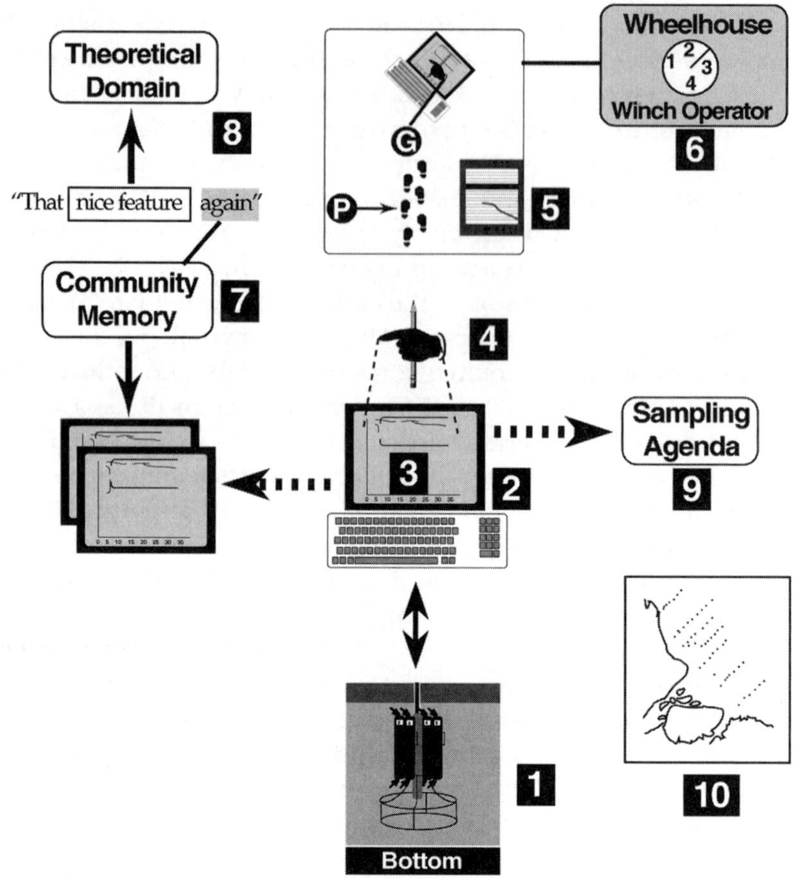

FIG. 4.15. A conceptualization of multiple spaces being attended to by scientists.

3. The spatial organization of the screen displaying that representation.

4. Transformation of this two-dimensional screen into a three-dimensional locus for visible activity as the document surface is articulated in a work-relevant way.

5. The work setting itself, the CTD laboratory, which includes both tools distributed in space and orientation frameworks being sustained by its inhabitants. Thus, to get to the screen where he wants to take measurements, the geochemist must traverse the orientation space constituted through the physical oceanographer's gaze toward the sonar screen she is working with.

6. This setting is linked to other work settings also implicated in the organization of the activity such as the wheelhouse where the winch operator is positioned. A *culture* sustained by this community of practitioners provides for the appropriate seeing of both the representations provided by their tools and the work that the participants are doing—for example, the transformation of squiggles on a screen into an independently seeable "nice feature."

7. This community is also able to link what is currently being seen to other spaces seen in the past.

8. A conceptual, theory-defined world (built in part through graphs and other spatial artifacts), which the present instance helps further to constitute.

9. The collaborative seeing that occurs here has a prospective orientation as well, as it sets the agenda for future work sampling the feature identified here.

10. The ship on which the scientists are working has been positioned at a place defined by the sampling grid being used to organize data collection on this cruise. This point is constituted through a conjunction of space constituted by a particular theoretical agenda (shaped by a variety of political processes), a global framework of latitude and longitude, and an actual physical spot in the ocean. Locating this spot requires another intricate conjuncture of space and activity (navigation satellites, maps, the work of other crew members, and the like), whereas the spot as a sample point links the products of this night's work to a larger scientific project.

To take their samples, these scientists must navigate through an array of different kinds of space, articulating one with reference to another and improvizing within them to perform an ordinary night's work, another station. The mundane nature of this work rests on an infrastructure of historically sedimented practice that is mobilized as a situated, temporally unfolding process to accomplish the work at hand.

Such processes are quite relevant to the question of how human cognition is to be analyzed. The analysis of spatial organization has recently become a major focus of research within cognitive science. However, within such research, the organization of space is conceptualized as a mental entity, divorced from practical action in endogenous settings. By way of contrast, the analysis developed here focuses on human cognition as a historically constituted, socially distributed process encompassing tools as well as multiple human beings situated in structurally different positions.[19] Restricting the analysis of cognition to processes located within the brain (including the sedimentation there of processes that have a larger social life, e.g., de Saussure's, 1966, *langue*) gives a very inadequate view of human cognition. As has long been recognized by Vygotsky (1962) and his followers, crucial to the development of human cognition is the ability of our species to secrete cognitive artifacts (including but not restricted to language) into the external world where they can shape not only our own actions but also those of our colleagues and successors. Such an expanded view of cognition seems especially important for the analysis of space in that human beings perceive space from within socially organized settings and conceptualize, articulate, and traverse space through a rich collection of tools that have been appropriated from the cognitive activities of our predecessors (maps, graphs, ships, etc.). Central to the organization of space are local activities and processes of human interaction within which different orders of space are tied together into the structures necessary for the accomplishment of relevant action. It is only within endogenous activities in actual settings, with their constellations of relevant tasks and tools, that the full richness and complexity of human spatial cognition becomes visible.

ACKNOWLEDGMENTS

I am deeply indebted to Willard S. Moore for inviting an anthropologist to observe oceanographers doing their work at sea and for obtaining grant money that would allow me to sail as one of his technicians. Without his help and support, the research reported in this chapter would never have come about. I am also very deeply indebted to the scientists and crew on the ship for allowing me not only to observe but also to videotape the work that they were doing. Very special thanks to Heather Astwood, Ruth Gorski, Steve Harden, Holly Kelly, Dick Limeburner, and Robin Pope.

---

[19]This work is thus quite consistent with the approach to cognition that has emerged within studies that have investigated the sociology of scientific knowledge (e.g., Collins, 1985; Latour, 1987; Pickering, 1992; Pinch, 1988, and many others). It is also consistent with work on situated action in the workplace (Brun-Cottan et al., 1991; Suchman, 1987), and on activity theory (see Cole, 1985; Engeström, 1987; Leont'ev, 1981; Vygotsky, 1962). Recent developments in cognitive and linguistic anthropology that focus on cognition as a distributed process (see Hutchins, 1993) and that make use of the spatial organization of endogenous settings (Duranti, 1992) are also consistent and relevant.

Earlier versions of this analysis were presented at the conference on Rediscovering Skill in Science, Technology and Medicine at the Science Studies Centre, University of Bath (September 1990), The Second International Congress for Research on Activity (Lahti, Finland, May 1990), the 89th Annual Meeting of the American Anthropological Association (New Orleans, December 1990), and at a colloquium at Xerox PARC (April 1991). The current chapter is an edited version that originally appeared in *Social Studies of Science*, Vol. 25, No. 2 (May, 1995), 237–274, and is based on an invited address presented at the 20th Annual Conference of the Cognitive Science Society. I am deeply indebted to Francoise BrunCottan, Paul Drew, Candy Goodwin, Gitti Jordan, Mike Lynch, Adam Kendon, Billy Moore, Lucy Suchman and four anonymous reviewers for *Social Studies of Science* for helpful, insightful comments on earlier versions of this analysis.

REFERENCES

Brun-Cottan, F., Forbes, K., Goodwin, C., Goodwin, M. H., Jordan, B., Suchman, L., & Trigg, R. (1991). *The Workplace Project: Designing for diversity and change* [Video]. Xerox Palo Alto Research Center (Producer). Available from Xerox Parc, 333 Coyote Hill Rd., Palo Alto, CA 94304.

Button, G. (Ed.). (1992). *Technology in working order: Studies of work interaction and technology*. London: Routledge.

Chaplin, C. (Writer & Director). *Modern times* [Motion picture]. United States: Warner Home Video.

Cole, M. (1985). The zone of proximal development: Where culture and cognition create each other. In J. Wertsch (Ed.), *Culture, communication, and cognition: Vygotskian perspectives* (pp. 146–161). Cambridge, England: Cambridge University Press.

Collins, H. M. (1985). *Changing order: Replication and induction in scientific practice*. London: Sage.

de Saussure, F. (1966). *Course in general linguistics*. (W. Baskin, Trans.). New York: McGraw Hill. (Original work published 1916)

Drew, P., & Heritage, J. (Eds.). (1992). *Talk at work: Interaction in institutional settings*. Cambridge, England: Cambridge University Press.

Duranti, A. (1992). Language and bodies in social space: Samoan ceremonial greetings. *American Anthropologist, 94,* 657–691.

Engeström, Y. (1987). *Learning by expanding: An activity-theoretical approach to developmental research*. Helsinki, Finland: Orienta-Konsultit Oy.

Foucault, M. (1986). Of other spaces. *Diacritics, 16,* 22–27.

Friedman, R. M. (1989). *Appropriating the weather: Vilhelm Bjerknes and the construction of modern meteorology*. Ithaca, NY: Cornell University Press.

Garfinkel, H. (Ed.). (1986). *Ethnomethodological studies of work*. London: Routledge.

Garfinkel, H., Lynch, M., & Livingston, E. (1981). The work of a discovering science construed with materials from the optically discovered pulsar. *Philosophy of the Social Sciences, 11,* 131–158

Goffman, E. (1964). The neglected situation. *American Anthropologist, 66*(6), 133–136.
Goodwin, C. (1994). Professional vision. *American Anthropologist, 96,* 606–633.
Hanks, W. F. (1990). *Referential practice: Language and lived space among the Maya.* Chicago: The University of Chicago Press.
Haraway, D. (1988). Situated knowledges: The science question in feminism and the privilege of partial perspective. *Feminist Studies, 14,* 575–599.
Heath, C., & Luff, P. (2000). *Technology in action (Learning in doing: Social, cognitive & computational perspectives).* Cambridge, England: Cambridge University Press.
Heath, C., & Nichols, G. (1997). Animated texts: Selective renditions of news stories. In L. B. Resnick, R. Säljö, & C. Pontecorvo (Eds.), *Discourse, tools and reasoning: Essays on situated cognition* (pp. 63–86). New York: Springer-Verlag.
Heath, C., & Luff, P. (1996). Convergent activities: Line control and passenger information on London underground. In Y. Engeström & D. Middleton (Eds.), *Cognition and communication at work* (pp. 96–129). Cambridge, England: Cambridge University Press.
Hutchins, E. (1990). The technology of team navigation. In J. Galegher, R. E. Kraut, & C. Egido (Eds.), *Intellectual teamwork: Social and technological foundations of cooperative work* (pp. 22–51). Hillsdale, NJ: Lawrence Erlbaum Associates.
Hutchins, E. (1993). Learning to navigate. In S. Chaiklin & J. Lave (Eds.), *Understanding practice: Perspectives on activity and context* (pp. 35–63). Cambridge, England: Cambridge University Press.
Jefferson, G. (1973). A case of precision timing in ordinary conversation: Overlapped tag-positioned address terms in closing sequences. *Semiotica, 9,* 47–96.
Jefferson, G. (1987). Exposed and embedded corrections. In G. Button & J. R. E. Lee (Eds.), *Talk and social organisation* (pp. 86–100). Clevedon, England: Multilingual Matters.
Kendon, A. (1990). *Conducting interaction: Patterns of behaviour in focused encounters.* Cambridge, England: Cambridge University Press.
Knorr-Cetina, K., & Amann, K. (1990). Image dissection in natural scientific inquiry. *Science, Technology, & Human Values, 15,* 259–283.
Latour, B. (1987). *Science in action.* Cambridge, MA: Harvard University Press.
Latour, B., & Woolgar, S. (1979). *Laboratory life: The social construction of scientific facts.* London: Sage.
Leont'ev, A. N. (1981). *Problems of the development of the mind.* Moscow: Progress Publishers.
Levinson, S. C. (1992). Primer for the field investigation of spatial description and conception. *Pragmatics, 2,* 5–47.
Lynch, M. (1985). *Art and artefact in laboratory science.* London: Routledge & Kegan Paul.
Lynch, M. (1988). The externalized retina: Selection and mathematization in the visual documentation of objects in the life sciences. *Human Studies, 11,* 201–204.
Lynch, M. (1991). Laboratory space and the technological complex: An investigation of topical contextures. *Science in Context, 4,* 51–78.
Lynch, M. (1993). *Scientific practice and ordinary action: Ethnomethodology and social studies of science.* Cambridge, England: Cambridge University Press.
Lynch, M., & Woolgar, S. (Eds.). (1988). *Representation in scientific practice.* Cambridge, MA: MIT Press.

Mukerji, C. (1989). *A fragile power: Scientists and the state*. Princeton, NJ: Princeton University Press.

Ochs, E., Jacoby, S., & Gonzales, P. (1994). Interpretive journeys: How physicists talk and travel through graphic space. *Configurations, 2,* 151–171.

Ophir, A., & Shapin, S. (1991). The place of knowledge: A methodological survey. *Science in Context, 4,* 3–21.

Paulve, A., Films du Palais-Royal (Producers), & Cocteau, J. (Writer/Director). (1950). *Orphée* [Motion picture]. United States: Discina International

Pickering A. (Ed.). (1992). *Science as practice and culture*. Chicago: The University of Chicago Press.

Pinch, T. (1988). Understanding technology: Some possible implications of work in the sociology of science. In B. Elliott (Ed.), *Technology and social process* (pp. 70–83). Edinburgh, Scotland: Edinburgh University Press.

Sacks, H. (1963). Sociological description. *Berkeley Journal of Sociology, 8,* 1–16.

Sacks, H. (1974). An analysis of the course of a joke's telling in conversation. In R. Bauman & J. Sherzer (Eds.), *Explorations in the ethnography of speaking* (pp. 337–353). Cambridge, England: Cambridge University Press.

Sacks, H. (1992). *Lectures on conversation: Volume I.* Oxford, England: Blackwell. Gail Jefferson (Ed.).

Sacks, H. (1989). On members' measurement systems. *Research on Language and Social Interaction, 22,* 45–60.

Sacks, H., Schegloff, E., & Jefferson, G. (1974). A simplest systematics for the organization of turn-taking for conversation. *Language, 50,* 696–735.

Schegloff, E., Jefferson, G., & Sacks, H. (1977). The preference for self-correction in the organization of repair in conversation. *Language, 53,* 361–382.

Schegloff, E., & Sacks, H. (1973). Opening up closings. *Semiotica, 8,* 289–327.

Shapin, S. (1988). The house of experiment in seventeenth-century England. *Isis, 79,* 373–404.

Sharrock, W., & Anderson, B. (1994). The user as a scenic feature of the design space. *Design Studies, 15,* 5–18.

Sharrock, W., & Button, G. (1994). Occasioned practices in the work of software engineers. In M. Jirotka & J. Goguen (Eds.), *Requirements engineering: Social and technical issues* (pp. 217–240). New York: Academic.

Star, S. L., & Griesemer, J. R. (1989). Institutional ecology, "translations" and boundary objects: Amateurs and professionals in Berkeley's Museum of Vertebrate Zoology, 1907–1939. *Social Studies of Science, 19,* 387–420.

Suchman, L. A. (1987). *Plans and situated actions: The problem of human machine communication*. Cambridge, England: Cambridge University Press.

Suchman, L. A. (1992). Technologies of accountability: Of lizards and airplanes. In G. Button (Ed.), *Technology in working order: Studies of work interaction and technology* (pp. 113–126). London: Routledge.

Vygotsky, L. S. (1962). *Thought and language*. (E. Hanfmann & G. Vaker, Trans.). Cambridge, MA: MIT Press. (Original work published 1934)

Wittgenstein, L. (1958). *Philosophical investigations* (2nd ed.). G. E. M. Anscombe (Trans.) & R. Rhees (Eds.). Oxford, England: Blackwell. (Original work published 1953)

5

Disrupting Representational Infrastructure in Conversations Across Disciplines

Rogers Hall
Vanderbilt University

Reed Stevens
University of Washington

Tony Torralba
University of California, Berkeley

In this chapter, we analyze conversations in consulting meetings where people work across disciplines to design things. We focus on interactional processes through which people disrupt and attempt to change representational technologies for scientific and technical classification. Our case material is drawn from ethnographic and cognitive studies of work in field entomology and architectural design. In both cases, we found common structures of interaction when people work across disciplines. These include selective use of talk, embodied action, and inscription to animate representational states that make up design alternatives. Participants from different disciplines animate situations in strikingly different ways, but these differences can either go unremarked or be put into coordinated use without explicit, shared understandings. Differences become remarkable either when a design proposal runs counter to deeply held disciplinary objectives or threatens to destabilize a wider network of representational technologies. These kinds of disruptions and their consequences for representational infrastructure are a central problem for research on distributed cognition.

Working conversations between disciplinary specialists are places in which differences in perspective lead people to see and act on represented objects in different ways. Under what conditions do participants in these conversations notice and use disciplinary differences to change existing systems of representation?

We compare cases of meetings in which people consult across disciplines on design problems. In the first case, entomologists solicit advice from a statistician to develop a new device for classifying insects. They hope this device will help to stabilize a complex network of representational technologies they use to study insect behavior and taxonomy. In the second case, architects, structural engineers, and historical preservationists meet to work on a design scheme for remodeling a historic public library. They (along with their clients) hope to strengthen the building so that it can be classified as safe for public use. Both cases are part of a larger, comparative, ethnographic and cognitive study of how mathematics is used in design-oriented work (Hall, 1998; Hall & Stevens, 1995, 1996).

Representational change is a central problem for theories of distributed cognition. Understanding, skill, inference, memory, or learning are distributed in ways that allow for coordinated and effective activity, and studying these activities requires that researchers "move the boundaries of the cognitive unit of analysis out beyond the skin of the individual person" (Hutchins, 1995, p. xiv). Theories of distributed cognition need to address the development of representational infrastructures for constructing, propagating, and interpreting representational states. Here, we analyze what happens in interaction when infrastructure breaks down, is resisted by participants in activity, or is purposely changed. We call these *disruptions to representational infrastructure*.

Our approach to this question looks for phenomena within the details of talk-in-interaction as people work together (Schegloff, 1992). We also pay close attention to how talk, embodied action, and physical inscription (e.g., writing, drawing, computing) are assembled to construct, juxtapose, and evaluate representational states (Goodwin, 1994; Hall, 1996; Hutchins, 1995; Stevens & Hall, 1998). Our analysis focuses specifically on interactional processes of animation through which participants in conversation position themselves and others with respect to relevant events (Goodwin, 1990; Ochs, Jacoby, & Gonzales, 1994). Processes of animation are used both (a) to assemble representational states during consulting meetings and (b) to examine critically these very processes of assembly (i.e., to disrupt existing infrastructure). Disrupting representational infrastructure opens up or challenges historical arrangements for making design worlds visible and sensible (Bucciarelli, 1994; Suchman, 1995). We thus look closely at how participants make historical material relevant in their interaction (Engestrom, 1999a, 1999b; Hall, 1999). This analysis of history made present is supported by ethnographic studies in the settings from which these cases were drawn.

In the remainder of this chapter, we discuss assumptions that are important for our analysis. These concern the distributed and historical character of representational infrastructures that implement scientific or technical classification, the role that people play in making and using this infrastructure, and the settings and conditions under which infrastructure can be disrupted and changed. We then present two case studies and analyze structures of interaction when consulting work turns to changing the way representational states are assembled. We turn finally to a comparative analysis across these cases of interdisciplinary consulting, treated as a site for disrupting and changing representational infrastructures that support classification.

REPRESENTATIONAL INFRASTRUCTURE SUPPORTS SCIENTIFIC AND TECHNICAL CLASSIFICATION

Categories and classification have been extensively studied as phenomena of individual cognition (Lakoff, 1987; Neisser, 1987; Smith & Medin, 1981). In these studies, the cultural or social basis of how people sort out objects and events in their lived worlds has usually been treated as secondary to or arising from the architecture of the individual mind (Anderson, 1990; Sperber, 1994). Our approach to classification reverses this assumption, drawing from an alternative body of work in cultural and historical psychology (Cole, 1996), conversation analysis and ethno-methodology (Goodwin, 1997; Lynch, 1991; Suchman, 1987), the sociology of science and technology (Bowker & Star, 1999; Latour,

1987; Star, 1996), and theories of distributed cognition (Hutchins, 1995). Here, systems of classification are socially constructed, maintained, refined or reorganized, and retired—practical accomplishments always distributed over technologies and people.

Within this broad area, two lines of research are important for our interest in how people disrupt or change representational infrastructures that support classification. The first is studies of information infrastructure done by Star, Bowker (Bowker & Star, 1994) and their colleagues across settings as diverse as the history of international classification of diseases, the development of a collaboratory for biologists working on the genetics of worms (Star & Ruhleder, 1996), and the development of a classification system for nursing (Bowker, Timmermans, & Star, 1995). These studies have documented how infrastructure is historically sedimented and critical for coordinating work across distinct communities. Under development, infrastructures for classification are deeply contested because they serve different and sometimes conflicting community interests. For example, new categories for classifying nurses' work create a two-edged visibility for practitioners, on one hand allowing them to claim distinct medical services as the province of their professional work but on the other hand opening up their work to scrutiny by doctors, hospital administrators, and insurance companies. New infrastructures stabilize and become useful to the extent that they provide for flexible, local use while still maintaining enough coherence to enable coordination across local sites and different communities (Star & Griesemer, 1989). Finally, infrastructure is difficult to study because it is embedded in stable work practices and so transparent in use, only becoming visible when work within or across communities breaks down (Star & Ruhleder, 1996).

The notion that representational infrastructure becomes visible during breakdowns has been taken up in a second line of research on the historical and distributed nature of "cognition in the wild" (Cole, Engestrom, & Vasquez, 1997; Hutchins, 1995). Engestrom, Brown, Christopher, and Gregory (1991/1997) studied learning at work, based on the idea that "disturbances" (p. 374) create a zone of proximal development for learning and organizational change. In a study of sidebar conversations between a judge and lawyers trying a civil court case, Engestrom et al. (1991/1997) found that talk sometimes shifted from "authoritative silencing" in which the judge decides in favor of one or the other lawyer into forms of "cooperation" or "reflective communication" about changes in court procedure. Different understandings by the judge and trial lawyers still blocked effective organizational change, leading Engestrom et al. (1991/1997) to call disturbances to infrastructure an "invisible battleground" (p. 384; Engestrom, 1999b).

Hutchins (1995) has also studied processes of adaptation during disruptions to representational infrastructure, analyzing how power loss to steering and navigational instruments creates a crisis for a team piloting a U.S. Navy helicopter carrier into harbor. Piloting is a distributed cognitive task accomplished by a team of navigators using a network of historically entrenched representational devices (e.g., a fathomoter, telescopic sighting devices, navigational charts showing relevant geographic features, a gyrocompass, and work routines that coordinate among these). Under normal conditions, the team finds the ship's position on the chart in an iterative "fix cycle" (Hutchins, 1995, pp. 26–29). A series of representational states can be juxtaposed to make decisions about the anticipated location of the ship. When this distributed network runs smoothly, the navigational system is robust: There are "ad hoc divisions of labor" (Hutchins, 1995, p. 220) in which team members complement each others' actions with devices, forms of "mutual monitoring" (Hutchins, 1995, p. 221) that provide redundant assessments of the ship's progress, and uses of public talk that support processes of evaluation and repair at a distance (e.g., announcing what a less experienced navigator ought to see over the intercom).

When power fails, this infrastructure is disrupted. Some devices become inoperable (e.g., the gyrocompass), and people on the navigation team need to reorganize their activities to meet the demands of a fast-moving fix cycle. Hutchins (1995) analyzed closely (pp. 321–345) how two team members reorganize the fix cycle—using different representational devices to regroup calculations and minimize effort at the chart table. Once the ship rests safe at anchor, what develops from the response to this disruption is a short-term change in the "cognitive ecology" (Hutchins, 1995, p. 346) of navigation aboard this ship. No explicit representation of the reorganized system is saved, the participants move on to other jobs, and conventional navigational practices resume once power (and layers of infrastructure above it) is restored.

For our analysis, Star's work (Bowker & Star, 1999; Star, 1996; Star & Ruhleder, 1996) provides an account of representational infrastructure, how it involves historically sedimented layers of technology and human practice that become visible during disruptions, and how changes in infrastructure cut across the interests and practices of multiple communities. Engestrom's studies (1999a, 1999b; Engestrom et al., 1991/1997) foreground disruptions (or disturbances) as a methodological strategy, following conflicting perspectives into episodes of local interaction in which people reflectively set out to change work organization. Finally, Hutchins' (1995; Hutchins & Klausen, 1996) work focuses directly on the link between infrastructure and ongoing activity, opening up the local, interactional work people do as they attempt to deal with an unexpected disruption. When a stable infrastructure breaks down unexpectedly or when participants are actively engaged in bringing the disruption about, how do people actively redistribute cognition?

SETTINGS DIFFER IN RESOURCES AVAILABLE FOR CHANGING REPRESENTATIONAL INFRASTRUCTURE

Ship or airline navigation is a particular kind of activity, perhaps typical of "high reliability" organizations (Roberts, 1993; Rochlin, La Porte, & Roberts, 1987). However, across a broader collection of organizational sites—civil courts, medical centers, scientific collaboratories, information system design, field entomology, and architectural design—what kinds of resources are available for changing representational infrastructure?

In our settings, people focus intensely on the classification of scientific or technical objects using layers of representational technology. Among the entomologists, work is dominated by the need to construct and stabilize new technologies for classifying insects, which are used to make novel claims about a wide variety of instances. The ontological gradient is reversed for architectural designers whose work is dominated by transforming a single instance (e.g., a building) so that it can be successfully classified by a wide variety of existing technical categories (e.g., state and municipal codes governing seismic stability for public buildings).

In both settings, difficulties arise that lead people to challenge or attempt to reorganize technologies for classification. In the case of field entomology, a consulting meeting has been called to find a new, more powerful representational device for classifying samples of insects—the disruption is already underway. In the case of architectural design, different specialists meet to find a design scheme that will preserve and strengthen a historically significant public building. As they consider different schemes, a conflict across disciplines leads to an extended challenge against existing technical categories.

Both cases involve disruptions to infrastructure, the first a matter of displacing one device with another and the second a matter of dissent from existing means of classification. By displacing one device with another, the entomologists can make their scientific claims about new or reorganized classes of insects stronger. By dissenting against existing means of classifying buildings, a design specialist can argue for a remodeling scheme that better preserves a historical building. Displacement and dissent are both means of disrupting representational infrastructure. However, they run time in different directions along a developmental axis for scientific and technical classification (Latour, 1987). Displacement anticipates dissent by making representational infrastructure stronger; dissent unravels and challenges the history of existing infrastructure, proposing alternate means and outcomes for classification.

MAKING A NEW DEVICE TO SORT TERMITES INTO SPECIES AND COLONIES

The *Bughouse* is our pseudonym for a group of entomologists working at a federal experiment station in Northern California. They do a combination of basic and applied research on "the chemical ecology of forest insects" (unless otherwise noted, quoted terms are used by participants; all names are pseudonyms). The leader of the BugHouse group, Mark, built a research program around the study of subterranean termites. These are "cryptic" species in the sense that they are not visible or even easily found in forested areas, yet they play a major role in forest ecosystems. For example, pine-boring beetles might kill a tree, and when the tree falls, termites and ants break up and consume the downed wood, eventually turning these materials back into the soil. Applied research at the BugHouse involves finding new technologies for detecting, monitoring, and "baiting" (or poisoning) termite colonies that have infested residential or commercial buildings.

To find and follow termites, researchers at the BugHouse superimpose two-dimensional "plots" of sampling stations over forested or residential areas, and then they use bundles of wooden slats in these stations to collect samples of foraging termites over regular intervals of time. Termite samples are classified by species and different colonies within a species, whereas wood bundles are analyzed for wood lost to termite feeding. Traditionally, entomologists have classified termite species based on morphological characters, such as the length of a mature worker's head, as well as on tests of their fighting behavior (i.e., termites from different species and even different colonies within a species will fight to the death). Researchers at the BugHouse are developing new methods of "chemical taxonomy" that make species and colony classifications by comparing chemical residues found on the exoskeletons of insects using gas chromatography and mass spectroscopy. They call these new means of classification "chemical fingerprints," and their claims for these methods, as well as new termite species they have discovered, are controversial in the field of entomology.

BugHouse members are quite skilled at comparing graphical profiles by visually picking out peaks that represent differences in the kinds of chemicals termites can secrete (see Fig. 5.1). Yet when they communicate the results of these visual comparisons in journal articles, their representational formats are cumbersome (e.g., tables comparing samples cover multiple journal pages). Finding a more efficient (and statistically powerful) way to compare chemical profiles would advance the group's efforts to stabilize their findings about termites and to make claims for methods of chemical taxonomy.

In the consulting meeting we analyze, the BugHouse team asked to meet with a statistician who also works in the experiment station with the explicit purpose of replacing one technology (visual comparison and multiple tests of significance) for implementing a classification with another. The BugHouse members at this meeting include Mark, the group leader and a senior scientist, Gary, another senior scientist interested in evolutionary problems, and Leah, rated as a "biological technician" but the group's resident expert on chemistry (i.e., she does all their gas chromatography work). The statistician, Browning (we use his last name as do members of the BugHouse team), has been invited to help develop new ways to compare chemical profiles and classify termite species or colonies.

Animating the Foraging Study Inscription Device (Turn 1)

As the meeting gets started, Mark (lead entomologist) proposes they focus on problems of community ecology in the team's study of termite foraging behavior. They have found three termite species, and he wants to use methods of chemical taxonomy to find and follow interactions between different termite colonies within these species. Mark's extended first turn animates the locations of samples within a study plot, the activities of termites and field workers in this plot, and the substantive link between fieldwork and laboratory analysis. As he puts

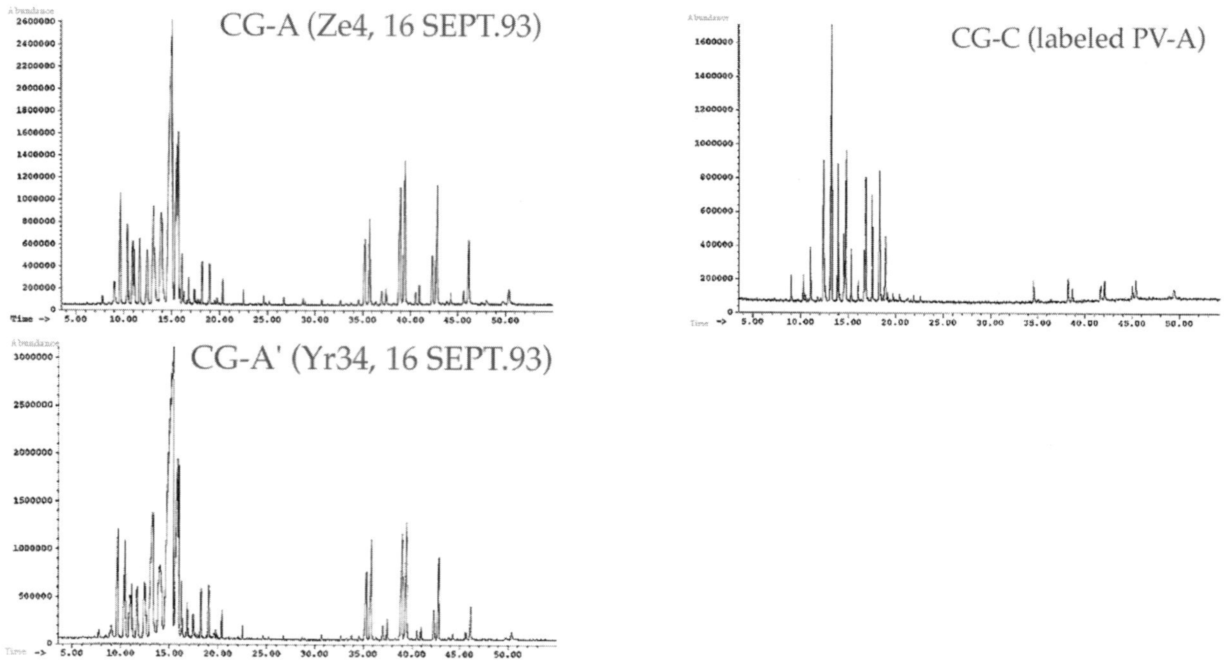

FIG. 5.1. Three chromatograms (CGs) produced from a waxy residue on the exoskeletons of termites. These graphs show the relative abundance (y-axis) of specific hydrocarbons that are identified by their retention time (x-axis) in a gas chromatography machine. CG–A and CG–A' are from the same termite species but different colonies because the graphs show "quantitative differences" in the abundance of hydrocarbon but no "qualitative difference" in which hydrocarbons are present. CG–C is from a different termite species than the others because the graph shows qualitative differences in which hydrocarbons are present.

it, species differences are "easy to find" as a "drastic qualitative change" in chemical profiles, but colony differences are more difficult to classify.

We divide Mark's extended turn into four parts (listed 1a to 1d), each accomplishing a different purpose as the meeting gets started. When transcribed closely (see Appendix for transcript conventions), Mark produces 10 visually prominent action sequences within this talk. To save space and foreground phenomena that are important for our analysis, we list only 6 of these following the utterance in which they appear (Actions 3 through 8, numbered within the turn in bold; CG = chromatogram):

1a Mark:	See my question … is around population or community ecology and not necessary … evolutionary relationships. Which is where this PAUP analysis that Gary talks about, uh, is involved.
1b	I just want to know … (**3**) if we have three species, how many different COLONIES of each species might we have (**4**) in this plot, and how might they interact. *(3) L hand open over table, beats/points at different locations on "three species" then beats over these positions on "COLONIES"* *(4) hands bound region at center on "plot," then fingers of each hand come together and fan open on "interact"*
1c	Eventually we'll want to know if, if they shift? (**5**)Like if this station, one day has species A and the next day has species B, that should be (**6**)easy to find. You know (**7**)when, when we sample these things monthly, (**8**)we'll know if we get a drastic qualitative change in the chromatogram. ***(5)** hands form a cylinder, pushed down on table and held at "station"* *(6) L points to stack of 3 CGs used earlier to show different species, including CG–C in Fig. 5.1* *(7) L circles twice, above position of cylinder formed at (**5**), as if pulling termites out of the ground* *(8) L points to stack of CGs, same area as (6)*
1d	What if we get a quantitative? You know, all of a sudden it looks different. And Leah's about to …

Mark's animation makes a complex chain of inscription at the BugHouse visible to the statistician and to us. In the first part (turn 1b), Mark shows how colonies are found inside the plots they have constructed over the forest floor. He does this by pointing to distinct locations on the table surface between he and Browning (turn 1b, Action 3, henceforth designated 1b(3)), and all participants in the meeting direct their gaze to this space. Then at 1b(4), Mark uses his hands to form a plot, held as a spatial enclosure over the separate locations he has just shown (see Fig. 5.2[a]). From the boundary of this plot, his fingers ("they") move to the center and then fan out as if depicting the interactions of individual insects. By coordinating talk with manual depiction, Mark animates important aspects of geography in their foraging study (i.e., termite samples are taken from widely dispersed locations in their

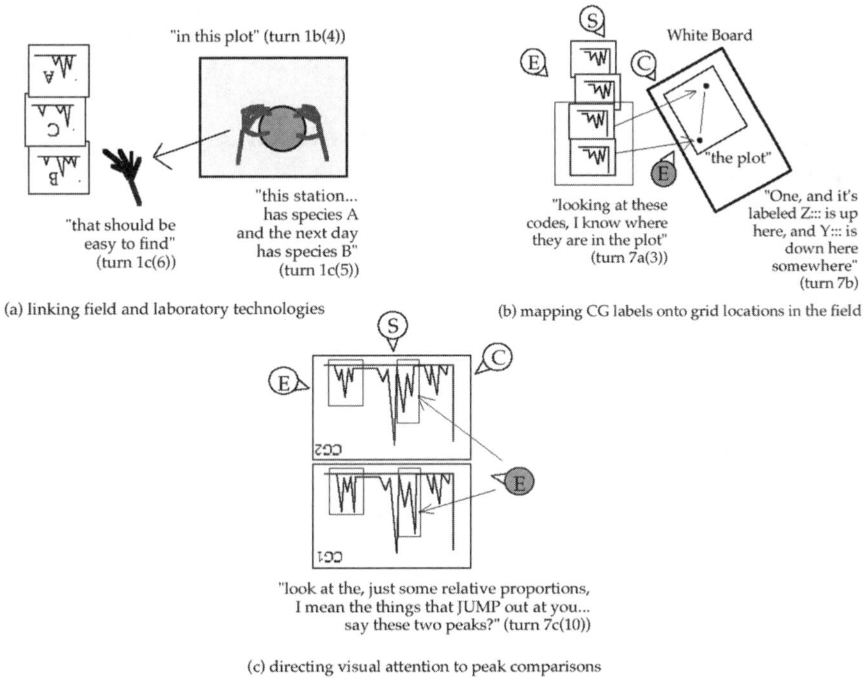

FIG. 5.2. Three aspects of coordination in Mark's animation of field and laboratory work at the BugHouse. In (a), turn 1, Mark animates a conceptual link between field and laboratory technologies for classifying species (qualitative differences) and colonies within species (quantitative differences). In (b), turn 7, he maps symbolic labels from chromatograms (CGs) onto a "high-tech drawing" of the distance between samples in their field plot. In (c), turns 7 and 9, Mark directs visual attention to the relative abundance of specific hydrocarbon peaks in two CGs selected from a stack of Species A.

two-dimensional plot) as well as the activities of termites within that geography (i.e., foraging interactions of individual termites/fingers).

In the second part of this animation (turn 1c), Mark describes how they follow the movement of termites over time, and he illustrates a critical relation between field and laboratory activities. At 1c(5), Mark forms and places a sampling "station" inside what he earlier called a plot. As we show in Fig. 5.2(a), this sampling station is held fixed as he describes finding different species on different days. At 1c(6), Mark points to a stack of CGs that he and Leah had assembled earlier to show different termite species (one of these is CG–C, in Fig. 5.1). Across these two utterances (1c(5) and 1c(6)), Mark shifts in talk and action from the field (i.e., this station formed with his hands) to their chemical analysis in the laboratory (i.e., a stack of CGs showing distinctly different chemical profiles). Different termite species would be easy to find because they show qualitative chemical differences in the laboratory. Mark repeats this animation of the relation between field and laboratory work at 1c(7) and 1c(8).

The Entomologist and a Chemist Assemble Instances for A "Good Comparing" (Turns 7 to 9)

Having shown the statistician how they collect and analyze termite samples, Mark and Leah arrange a stack of four CGs in front of Browning (turns 2–6, not shown). In the transcript following, these are labeled CG1 to CG4 (top to bottom) from the statistician's perspective as a conventional reader of graphs (i.e., his position in the interaction as constructed by Mark and Leah). In turns 7 to 9, Mark uses this stack to assemble two pairs of CGs that are ideal instances for the problem of classifying termite colonies. This is a complex undertaking; therefore, we give an overview of Mark's work before showing the transcript.

As Mark carefully explains, these pairs of CGs are ideal for developing a new classification device because they combine two sources of evidence. First, they show clearly visible differences or similarities in chemical abundance; second, they come from geographic locations that exceed or fall within the maximum size of colonies for this termite species (i.e., as reported in the literature, these termites can forage over areas of up to 100 m). Mark selects two pairs of CGs for Browning to consider. The first pair (CG1, CG2; these are CG–A and CG-A' in Fig. 5.1) is a good comparing for classifying different colonies because the samples are separated by over 200 m (i.e., no colony could be this large), there are no qualitative differences in the chemicals present, yet Mark finds quantitative differences "that JUMP out at you" (7c). He underscores the importance of distance by drawing their plot and sample locations at the whiteboard (see Fig. 5.2[b]). The second pair (CG2, CG3) probably comes from the same colony (9b) because there are no quantitative differences and the samples were taken only 30 m apart (i.e., colonies are known to forage over areas this large). By the end of turn 9, Mark has presented Browning with instances (i.e., pairs of CGs) that clearly fall to either side of the classification they want his help in making (i.e., different vs. same colony within a termite species):

7a Mark:	[This one (**1**)here is not pretty similar. And (**2**)this is a a good comparing right here, as a matter of fact, (**3**)beCAUS::E, looking at these codes I know where they are in the plot. *(1) R point to CG1, at top of the vertical column* *(2) shifts R point to CG2 while L holds on CG1; then R opens and holds over CG1 and CG2 on "comparing"* *(3) L point to plot codes on CG1*
7b	So if I were to, if I could just, let me just draw it real quick. This is one of the high tech drawings that I do. They have a square plot. One, and its labeled Z::: is up here, and Y::: is down here somewhere. So that distance is probably two hundred meters or greater, where those two occur.
7c	And if you look at the, just some relative proportions, (**10**)I mean the things that JUMP out at you … say these two peaks? *(10) R points at peaks 13 and 14 in CG1*
8 Browning:	[Yep.
9a Mark:	[(**1**)They're pretty radically different. Then(**2**) the arrangement of these … peaks here? (2 sec) *(1) R pen points at comparable peaks in CG2* *(2) R pen circles over peaks at tail of CG1 on "these …", then matching peaks in CG2 on "peaks here"*
9b	As opPOS::ED to something that might be … Yr34 and Yr19:: These might be from the same colony. (4 sec) Y'know just looking at em, they don't look that DIFFerent. (4 sec)
9c	What this nineteen, the Yr, Y is the plot, R is a row:::, it's given a letter number, and nineteen is the position

> on that row. So these are::: thirty four, fifteen, [thirty meters apart.

Mark's earlier animation of field and laboratory practices is expanded in two ways during turns 7 to 9, and this expansion is again produced for the consulting statistician. First, Mark clearly illustrates the evidential relation between chemical differences and geographic distance when classifying termite colonies, moving between symbolic labels on CGs and what he calls a "high tech drawing" of their field plot (see Fig. 5.2[b]). The location of samples in their plot, which he depicted earlier with his hands over the surface of the table, he now depicts as a relation between written codes and premeasured distances in the field (i.e., these symbolic codes index measurements used to construct the plot). Second, after establishing distance as one line of evidence, Mark sits down and effectively zooms into the selected CGs with his talk and hands to find chemical differences that "JUMP out at you" (turn 7c, see Fig. 5.2[c]). This combination of chemical differences and geographic distance is produced again for Mark's second instance, this time leading to a visual comparison in which "they [CGs] don't look that DIFFerent" (turn 9b and 9c).

Mark's coordinated use of talk, action, and inscription is a form of "highlighting" (Goodwin, 1994), a discipline-specific system of activity that simultaneously produces the chemical differences Mark is searching for and directs the attention of his recipients in conversation to both these differences and his practical methods for finding them. By *coordination* we mean the articulated use of evaluative and technical terms (e.g., "these peaks" as being "pretty radically different" amplifies chemical differences), repeated deictic points to select and juxtapose graphical elements (i.e., peaks within and across CGs), and the deliberate physical arrangement of inscriptions to facilitate specific visual operations (i.e., aligning CG pairs by their *x*-axis to compare chemical structure and abundance).

Making/Contesting a Statistical "Distance"

As the conversation continues (turns 10–22, not shown), Mark, Gary, and Leah provide still other examples of how they find quantitative differences between CGs. During this elaboration of their existing approach to classification, they arrange two CGs (CG1 and CG4) directly in front of Browning, noting that these show only quantitative differences and so should come from different termite colonies. With these instances at the center of shared visual space, the consulting statistician takes the conversational floor. Browning has literally been set up to talk about a new piece of representational technology. What he proposes, a distance between CGs that is computational rather than visual, may provide an alternative means of classifying termites. However, as we document following, his proposal meets with considerable resistance over how it will fit into existing work practices at the BugHouse.

The Statistician Contrasts Visual and Computational Means of Finding Differences (Turns 23–26). Browning prefaces his contribution as the joint work of "how *we* go about defining a distance," something that stands in contrast to what the BugHouse group currently does ("*You*'re trying to use *your* eyes …"; turn 23, italics added). However, when Browning mentions the "differences" that these entomologists have just been demonstrating, Mark leans back in his seat and looks up at the visiting statistician. Browning begins to pause, and Mark interrupts with an emphatic contrast between "DIFFERENCES" and "distances" (turn 24). As Mark interrupts, Browning quickly retracts his hand from the stack of CGs at 25(1), then he agrees to and repeats what Mark proposes as the preferred, technical term ("differences").

23 Browning: Ok let's, part, part of the thing is, is how you define,
again how we go about defining a distance (of it).
(2)You're trying to use your eyes to, to uh, to to match
these two up and see where there are differences and
[where there aren't … but]
(2) gaze to CGs, R point traces up and down vertical
on "your eyes" then continues vertical tracing across
horizontal axis of CG1 and CG4
24 Mark: [(1)DIFFERENCES], [not distances, differences]
(1) leans back, raises head and eyebrows, gaze to
Browning; Browning begins to pause
25 Browning: [(1)] Y, yeh, [(2)differences]
(1) quickly retracts R hand, held over CGs
(2) flat R hand beats over CGs
26 Mark: [Yeh]

Several things are important in this opening exchange. First, the visiting statistician contrasts what BugHouse team members currently do with their eyes (i.e., a collective "you") with what "we" as participants in this consultation across disciplines will do by means of calculation. This initial utterance explicitly marks changes in time (i.e., existing vs. proposed activity) and in disciplinary perspective (i.e., what "you" entomologists do versus what "we" will do together). Second, and in an apparent response to Mark's emphatic contrast, Browning momentarily retracts his hands from a pair of CGs that provides an instance of what Mark earlier called a good comparing (i.e., different colonies, same species). What gets interrupted is not just Browning's utterance but also the terms he uses and the way in which he uses his hands to coordinate language with parts of these graphical displays. Third, what Browning calls "a distance" (turn 23) and what Mark calls "distances" (turn 24) are not equivalent, and further evidence for quite different disciplinary meanings comes over the next several turns at talk.

The Statistician Reformulates His Proposal for a New "Distance" (Turns 27–29a, Not Shown). While Browning momentarily drops "distance" in favor of "differences," he quickly reformulates his proposed contribution in terms of the presumed needs of the entomological research community. As he puts it, they are working together ("*we're* trying"; italics added) to find a replicable or "more standard" way to convert multiple differences into a single number. Mark emphatically agrees with this reformulation of what they are doing ("I'd LOVE, I'd love that"), and the conversation continues. However, what the entomologists mean by "distances" (turn 24 and earlier) is still out of alignment with "the distance" that Browning proposes will take multiple differences into a single number.

The Statistician Animates Operations That Could Implement a New "Distance" (Turn 29b). Having regained the conversational floor, Browning next proposes a distance that involves a sequence of statistical operations. He animates these operations in layers over the surface of two CGs on the table surface between he and Mark. In this more detailed proposal, calculating a distance is manually illustrated as statistical operations over entire graphical profiles. These are quite different from the peak-wise comparisons of entomologists (turns 7–9).

29b Browning: Because again we may only want to take (3)the peaks
that are above a ... certain level. And that's per:fec:tly
legitimate to do. (4)That we want to get rid of these

smaller peaks and not let them be a nuisance, at all. That it's(**5**) only the higher peaks that we wanna(**6**) compare the distances from.
(3) R held flat, cuts horizontal through CG4 then holds as gaze rises to Mark
(4) both hands point to rightmost peaks in CG4, gaze from Mark to Leah
(5) R held flat, repeats horizontal cut of CG4
(6) R thumb and index finger line up large peaks across CG4 and CG1

29c Uh, we talked about, (**7**)sort of superimposing these things before and (**8**)then looking at the areas in between. Which is another way to characterize the difference. Um ...
(7) R index sweeps up from CG4 and traces across CG1
(8) R index and thumb form an interval, which traces across CG1 then holds and beats on "difference"

29d which means down here with the smaller peaks, it means this isn't gonna influence things very much. Which sort of says that the, um ... If one has a little peak here, but this one doesn't have any, that's still a little difference. Even though, qualitatively one has it and the other doesn't

Rather than picking chemicals out of each profile for comparison, Browning's first suggestion is to set a threshold on abundance (i.e., the vertical axis in each CG). Below this threshold, peaks (i.e., chemicals) that might contribute relevant biological differences would not be considered at all (see Fig. 5.3[a]). Rather than physically arranging CGs for visual inspection, Browning's second suggestion is to superimpose one CG onto another and then to calculate the difference in area between them (turn 29c; Fig. 5.3[b]). He illustrates this with a gesture (29c(8)) that sweeps across the proposed superimposition, collecting differences into a single quantity with his fingers. Whereas the entomologists handle CGs as collections of distinct chemical structures (i.e., the peak-wise comparisons animated by Mark in turns 7 and 9), the visiting statistician treats CGs as unit inputs to a complex calculation. What Mark called "distances" in his earlier, emphatic contrast with "DIFFERENCES" (turn 24) are clearly not what Browning means by "a distance" (turn 23), “one number,” or “how you characterize the distance” (turns 27 and 29a, not shown).

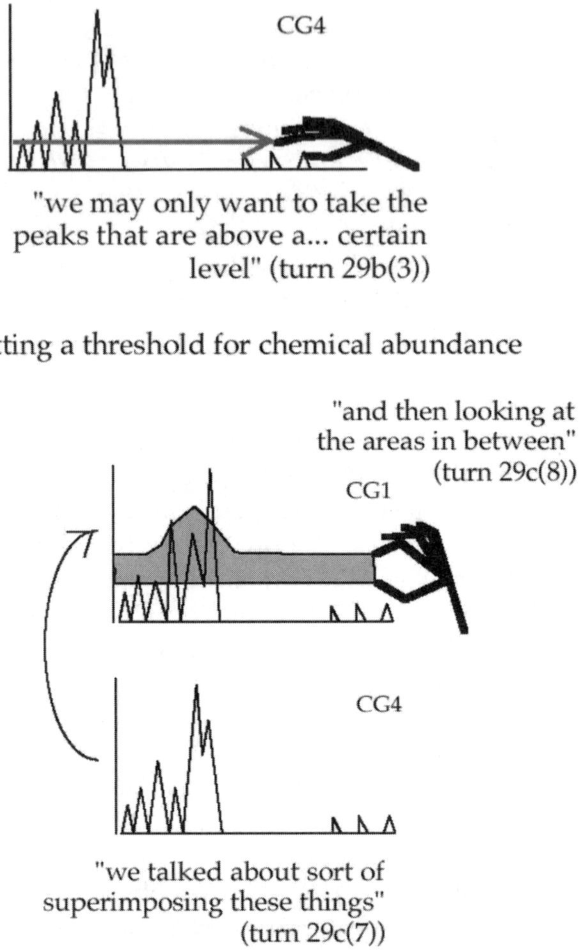

(a) setting a threshold for chemical abundance

(b) superimposing then accumulating differences in abundance between a pair of CG's

FIG. 5.3. The statistician (Browning) animates a layer of calculation over two chromatograms (CGs). In (a), he sets a threshold below which differences in chemical abundance would be ignored. In (b), he sweeps across the *x*-axis of two thresholded CGs that have been superimposed using an interval held by his fingers to accumulate differences in abundance as the “area” between graphs.

The Lead Entomologist Defends a Stable, Biological Meaning of Distance (Turns 32–36, Not Shown). Mark, who has been listening quietly since his earlier interruption, returns to the contested meaning of distance in their conversation. He again challenges Browning's language, this time explicitly proposing they "get rid of the distance" in Browning's earlier proposal and pointing out they will need to preserve their technical arrangements for mapping termites ("When I'm talking about *distance*, I'm talking about *geography*"; italics added). If distance is going to change meaning in a way that is sensible to entomologists (in and out of the BugHouse team), then the role of geography in holding together their work will need to be carefully preserved. This time Browning interrupts, summarizing his own proposal for a "measure" that takes a pair of CGs as discrete inputs and produces a clearly defined output.

Aligning Old and New Technologies to Classify Differences

We chose this consulting meeting because it had been arranged to solicit advice from a statistician (Browning) about how to improve existing representational technologies for comparing chemical profiles. The BugHouse team hoped to find a way that was simpler and possibly more powerful for detecting biologically relevant differences between termite samples. In our analysis, the interaction up to this point has produced three critical events:

Turns 1 to 9: Entomologists have animated a detailed account of work practices and representational technologies that span field and laboratory sites (i.e., the gridded plot, sampling schedules, geographic distance, and visual comparison of coded graphical profiles), culminating in the selection of instances that are ideal for classifying same/different termite colonies.

Turns 23 to 29: The statistician has proposed replacing their existing visual practices with a new representational technology, animated as a series of statistical operations that take multiple, peak-wise differences into a single number or measure (a "distance").

Turns 24 and 32 to 36: A conflict has emerged over distinctly different work practices and meanings associated with "distance"—what entomologists take as an independent source of evidence (geography) that can be combined with chemistry, the statistician treats as the outcome of a computational process.

As the conversation continues (turns 37–39, not shown), Browning consolidates aspects of his earlier proposals into a statistical process that takes "ALL of this information from all of these peaks" into a single number. This is a new, distinctly nongeographical meaning for distance. If a pair of CGs are "far enough away" from each other in this new terrain, judged against a distance that is established for "sampling error," then they can be classified as "definitely DIFFerent."

The Entomologists Provisionally Take Up a New, Statistical Approach to Comparing Instances (Turns 40–43). Although the details of Browning's distance measure still need to be worked out, senior members of the BugHouse team provisionally take up and use the new piece of representational technology. Gary (turn 40, following) appears to start a positive assessment of the consolidated proposal, and Mark (turn 41) interrupts to give a provisional but fully elaborated account of how the BugHouse might use the new classification device:

40 Gary:	Ah::: hah. [So
41a Mark:	[So if we came up with that statistic. If we … agreed on A statistic or a set of statistics, then we could look at injection error, we could look at aliquots for the same trap at the same time.
41b	Then see what those, (3)what those distances are between those chromatograms. And then maybe (4)set that as a STANDARD? *(3) hands, open and vertical, sweep back and forth over different sized horizontal intervals* *(4) fixed interval held between hands, beats on "STANDARD"*
41c	[Like if its], if its (5)this small or smaller, then there's (6)no difference. Ok. *(5) held interval is fixed at "this small" then scaled down for "smaller"* *(6) hands open out from held interval as if to offer on "no differences"*
42 Browning:	[Yeh, and, and] (5 sec) And there may be ways to sort of VALIDATE, whether or not those are useful measures by, again, the fighting behavior, that you did.
43 Mark:	Right.

Unlike his earlier, sustained efforts to block or edit Browning's proposal, Mark now explores how they could adopt a new, statistical meaning for distance. First, Mark's uptake of the new technology is strongly provisional, described as a hypothetical course of action (e.g., "So *if* we came up with" and "*If* we ... agreed on A statistic"; turn 41a, italics added) they might take in the future (e.g., "we *could* look at" and "*maybe* set that as"; turn 41a, italics added). This is in keeping with his earlier defense of a well-established, geographic meaning of distance that makes it possible for the BugHouse to coordinate practices for sampling, coding, and making inferences about termites (turns 7, 24, and 34).

Second, Mark rapidly describes a study they could do to calibrate the new technology in terms of existing practices, proposing that they systematically identify variation in laboratory techniques (i.e., "injection error" and preparation of "aliquots" or samples; turn 41a). He then describes these variations as "distances" between CGs (41b(3)) using his hands to show how they might calculate or "set ... a STANDARD" within this measured scale (41b(4) and Fig. 5.4[a]). Finally, and in sharp contrast to Browning's animation of the statistical operations that make up

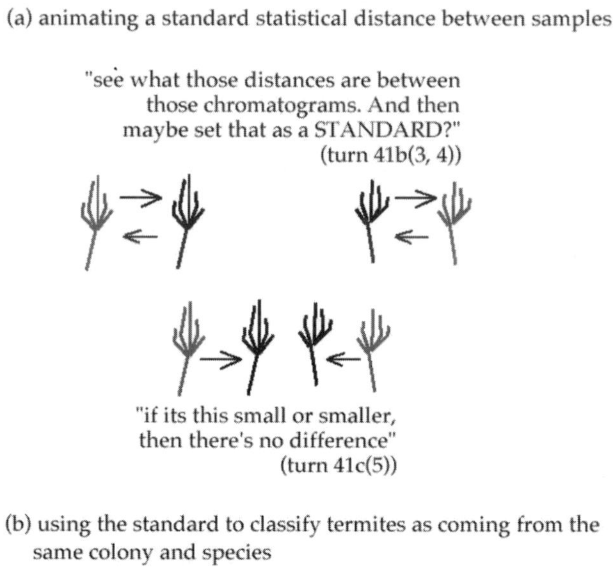

FIG. 5.4. An entomologist (Mark) provisionally adopts a new meaning for "distance" animated from the perspective of a user. In (a), Mark animates a study to construct a "STANDARD" statistical distance between samples; in (b), he demonstrates using this standard to classify samples as showing "no difference" (i.e., same species and colony).

the internal workings of this new technology, Mark animates a distinctly different perspective on its use. Building on his gestural depiction of a "STANDARD" distance associated with errors, he holds an interval "this small" at turn 41c(5) then slides his hands closer together to illustrate finding "no difference" between CGs at turn 41c(6) (see Fig. 5.4[b]).

Now 41 turns into this episode from a consulting meeting, the senior entomologist uses a new piece of technology to classify a hypothetical pair of chemical profiles (CGs) as coming from the same colony and species of termites. Presumably, if this new distance were to fall just outside the "STANDARD," the BugHouse team would classify samples as coming from different colonies of the same species (i.e., the situation of a "good comparing" in turn 7a). And if the distance were to fall well outside the standard, they would classify samples as coming from different species (i.e., a "qualitative" difference in chemical profiles that Mark called "easy to find" in turn 1c). In turn 42, Browning links the new device to another, existing technology for classification of termites (i.e., "fighting behavior"), and Mark agrees (turn 43).

USING EXISTING DEVICES TO SORT SAFE FROM UNSAFE PUBLIC BUILDINGS

JCArchitects is our pseudonym for a midsized architectural firm specializing in remodeling and restoring public buildings in Northern California. The firm consists of two principal architects (Charles and Jackie) as well as a changing staff of junior architects and architecture students. The work of this firm ranges from individual residential projects to large-scale public space (e.g., during our ethnographic study, they won a contract to design a multistructure community center for a local city). The projects we studied most closely were efforts to remodel two public libraries (Stevens, 1999), and each required design proposals to create access for people with disabilities and to strengthen the buildings against earthquakes. One of these libraries (Taraval, also a pseudonym) was located within 2 miles (3.22 kilometers) of a major geological fault line and constructed out of "unreinforced masonry" (URM, in this case bricks and mortar).

In the event of a large earthquake, which seismologists say is increasingly likely in this region of California, URM buildings pose substantial dangers to their occupants—the libraries could collapse altogether, crushing or trapping library staff and patrons. As a result, state and municipal governments have adopted a set of ordinances that require owners either to list URM buildings as public hazards (e.g., both libraries we studied were listed) or to retrofit these buildings so that they meet various structural code standards. Because these libraries were Carnegie-era buildings (i.e., historically significant structures), multiple code standards were applicable, creating options for classifying the buildings that ranged from very stringent (i.e., retrofitting a building to make it conform with criteria for new construction) to much less stringent (i.e., simply listing a building as a historic landmark). In direct relation to these code thresholds, the architects could choose from a range of conventional retrofitting schemes that were more or less invasive. As shown in Fig. 5.5(a), for example, meeting the Uniform Building Code for new construction (most stringent code) would require the most invasive design scheme, replacing entire walls in each library with structures made of hardened concrete (i.e., "gunnite"). As a design problem, then, architects need to attach particular instances (buildings) to a variety of general criteria (codes) to classify the building as safe/unsafe for public use.

To bid for and then complete design projects of this sort, JCArchitects typically assembled teams of consultants from several professional disciplines. For the Taraval library, this team was led by Charles (a principal architect; Jackie, his partner also participated) and included specialists in historical preservation (Doug and Pat in the transcripts that follow), structural engineering (Gary and Cal), heating and mechanical systems, and cost estimation. Our ethnographic study started just after JCArchitects' bid for the Taraval library was accepted by the city. At this stage in the project, a precise physical characterization of the library was still underway, and no design scheme had yet been proposed for attaching the library to any particular code. Reflecting this early stage of design, the architects and their consultants were actively considering multiple, alternative schemes.

The open-ended character of this design problem is illustrated in Fig. 5.5(b), which shows how the design team made a strategic decision to change their description of the existing building. Because the lower floor of this "two-story library" was partly below the ground, they decided to strengthen these walls, call them a "foundation," and then describe the building as a "one-story library." As part of their design scheme, changing the formal description of the library (as well as physical properties of its below-ground walls) allowed them to meet standard code thresholds using design schemes that were less invasive.

We characterize this as changing the instance (the Taraval library) to arrive at a more desirable way of attaching it to representational tech-

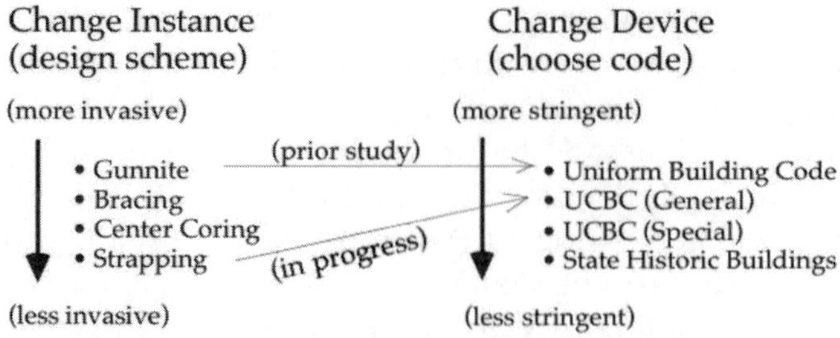

(a) attaching an instance (Taraval library) to different classification devices

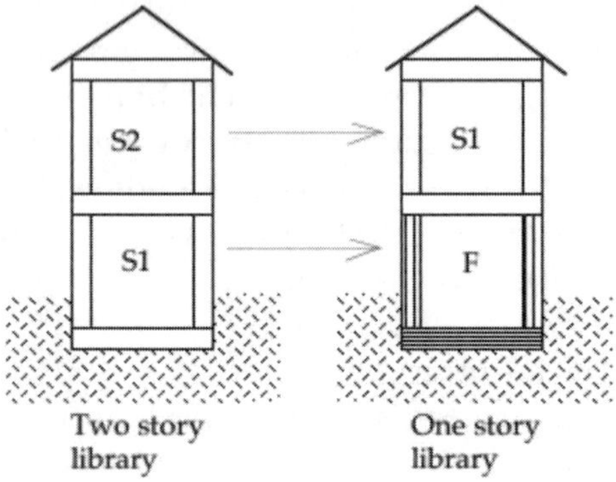

(b) changing the instance (Taraval library) for optimal attachment

FIG. 5.5. Working through design schemes that attach public libraries (instances) to existing code documents (a class). In part (a), the JCArchitect team's in-progress design scheme is less invasive and nearly as stringent as a prior feasibility study; in part (b), this in-progress scheme is enabled by changing the Taraval library from a two-story to a one-story building.

nologies that implement the classification (code thresholds defining safe/unsafe buildings). With these activities as background, the work in progress for the JCArchitects team is choosing a minimally invasive design scheme (e.g., strapping existing walls and ceilings) that will allow them to attach this building to an adequately stringent code (e.g., the "Uniform Code for Building Conservation" (UCBC) general procedure, as shown in Fig. 5.5[a]). Based on our observations and interviews, it was clear that interests varied across members of the JCArchitects team: Historical preservationists would be happy with a scheme that saved historical fabric (i.e., less invasive retrofitting); structural engineers would be happy with a scheme that made the building stronger (i.e., more invasive retrofitting); architects would be happy with a scheme that was as stringent as possible while still meeting constraints of budget, time, and the client's program of use.

Classifying Buildings and Anticipating Trouble

We present two episodes from this consulting meeting. In the first, a structural engineer (Gary) reports on his ongoing analysis of two public libraries. The team is in "good shape" on one library, but they are over code thresholds on the other (the Taraval library). When Gary proposes that they might be able to "live with" this excess, he is challenged by the lead architect (Charles), and a sustained disagreement begins between historical preservationists and structural engineers. One of the historical preservationists (Doug) recalls the engineering studies that were used to set code thresholds, arguing that the code thresholds adopted by state and municipal governments added in "arbitrary" margins of safety. When he proposes that the JCArchitects team convince the city and public to accept test values that exceed code thresholds, the structural engineers vigorously resist. This proposed challenge to an existing classification system (code) is resolved in the second episode, about 13 min after the first. The structural engineer (Gary) outlines a scheme to add seismic resisting elements to the library, distributing forces in a way that should allow the design team to live with a wall that exceeds code thresholds.

A Structural Engineer Attaches Libraries (Instances) to Seismic Thresholds (Classes; Episode 1, Turns 1–5). As this early design meeting gets underway, Gary (structural engineer) and Charles (lead architect) have been talking about which code guidelines they should use in a retrofitting scheme for the Taraval library. Gary reminds Charles that a prior, feasibility study used the Uniform Building Code for new construction (UBC in Fig. 5.5[a]) and as a result recommended using gunnite (hardened concrete) to replace some library walls. This design scheme would be maximally invasive and very costly. Charles has agreed that they should use the less stringent UCBC code guidelines, possibly avoiding gunnite altogether. However, he also anticipates the concerns of library patrons at an upcoming public meeting who will want to make the building safe for use by children's programs and other public functions. As Gary takes the conversational floor to describe his ongoing structural analysis, this complex set of trade-offs (i.e., historical fabric, cost, strength, and public reaction) is already under discussion:

1 Gary: And um ... um I find that on Portal, uh ... we're in, we're in very good shape in terms of h over t. The analysis shows that its ok?
2 Charles: Um [hm.
3 Gary: [On::: uh, Taraval, then ... we're not. We're at, we're uh, at one point one one, demand capacity ratio. So we're uh, and those (**1**)numbers are, you know are all plus or minus ten percent [this time, ok?
(1) L hand raises, opens, and rocks between little finger and thumb
4 Charles: [Um hm. Um hm.
5 Gary: Um. So um ... if we can::: if we can LIVE with a (**1**)bit of excess h over t on (**2**)the Taraval, then we can say that that's not that's not an issue.
(1) L hand raises, index finger traces vertical then horizontal strokes
(2) L point sweeps over plan for Taraval

There is trouble with the Taraval library, which is still in the process of being attached to code guidelines despite choosing a less stringent code and redescribing the building as a one-story library (Fig. 5.5). Although Gary shows that his calculations are only approximate (3(1)), one of the quantities he needs to produce for the UCBC code ("demand capacity ratio") already exceeds a tabled threshold. As he reports/shows in turn 5(1), the trouble stems from a height-to-thickness calculation ("h over t" animated with vertical and horizontal strokes; see Fig. 5.6[a]) for a particular wall in the library, but he thinks they may be able to "LIVE with" this excess.

The Lead Architect Allocates Seismic Trouble Across Disciplines (Episode 1, Turns 6–9). As someone anticipating what could be a hostile reception at a public meeting over these libraries, Charles (the leader of this design team) takes the next turn at talk. Whereas Gary reported that "*we're* not" in good shape on the Taraval library and wondered "if *we* can LIVE with excess" (turn 3, italics added), Charles puts his question directly to Gary and the discipline of structural engineering:

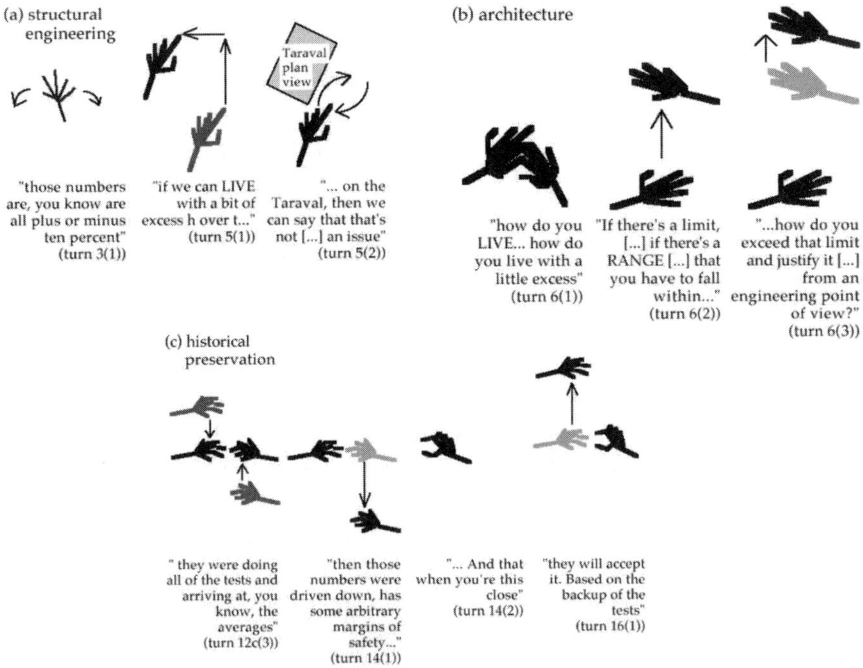

FIG. 5.6. Animating discipline-specific relations to technical concepts. In (a), a structural engineer (Gary) uses his hands to depict margins of error and the dimensional structure of calculations involved with finding h/t (height-to-thickness calculation) and demand capacity ratios; in (b), an architect (Charles) forms a range and excess for a code threshold they are attempting to meet; in (c), a historical preservationist (Doug) shows the calculation of averages, the adoption of "arbitrary" margins of safety, and their current test values in comparison to these "backup" margins.

6 Charles: How do you, how do you **(1)**LIVE … how do you live with a little excess h over t? If there's a **(2)**limit, you know, if there's a RANGE that you can, um … that you have to fall within, how do you, how do you **(3)**exceed that limit and justify it from a, from an engineering point of view?

(1) R hand beats on open L palm as if checking in a book
(2) hands form and sweep through a large vertical interval then hold and beat on "RANGE"
(3) R hand rises above top of held interval

7 Cal: Its tough, I think its tough. I think, I guess that its, maybe the actual idea is, that's what you're saying, we have to talk with the city, I guess, right?

8 Gary: Yes.

9 Cal: Yeh, SOMEtimes they overlook little things. But I think this is like a high, I think in a way, it's kind of a high profile job? I think they want to, I think at least the city wants to go by the book. They say if you, if you [go over it then (inaudible)]

Trouble with the library is a moving target both across disciplines and along different trajectories into the future. First, Charles' question allocates seismic trouble directly to "an *engineering* point of view" (turn 6, italics added), explicitly picking Gary and Cal out from the design team and providing a sharp contrast with Gary's use of a collective "we" in turns 1, 3, and 5. Consistent with this allocation of trouble, it is Cal, a co-owner of the structural engineering firm that employs Gary, who begins to answer Charles' question. As Jackie (co-principal architect) suggests later in this meeting, these engineers are "on the line" for matters of structural integrity. Second, and also an integral part of turn 6, Charles animates a decision process that will attach the Taraval library to code documents (see Fig. 5.6[b]). As he asks about how to "live with a little excess," his hands form an interval along a vertical scale that he calls a "RANGE" (turn 6(2)), then his right hand rises above the upper end of this interval to show a value that exceeds this "limit" (turn 6(3)). This animation from the perspective of a user of structural analysis makes their current trouble with attaching the Taraval library to code starkly visible. Third, Cal's answer is also a complex form of animation, starting with the idea that they will need to "talk with the city" (turn 7, confirmed by Gary in turn 8) and progressing into a future-time narrative in which Cal actually speaks for the city in what he projects to be a negative assessment (i.e., "They say if you, if you go over"; turn 9).

Challenging Code and Justifying Alternative Classifications

While the Taraval retrofitting scheme is still being developed, Cal has transported the team (figuratively) into a direct confrontation with the city architect's office. However, there are other design options to consider, and the historical preservationists "step in on" the conversation.

A Historical Preservationist Challenges Assumptions Behind the Existing Code (Episode 1, Turns 10–16). If the ongoing design scheme cannot attach the Taraval library to structural code guidelines, then the team will need to consider more invasive (and costly) retrofitting schemes. Not surprisingly, the historical preservationists (Pat and Doug) are the next people to speak on this problem, and they mount a sustained, past-time challenge to the studies and assumptions used in developing code guidelines. Faced with a troublesome instance, their design solution is to dissent from the representational technologies implementing the class:

10 Doug:	[Ah::=
11 Pat:	=No.=
12a Doug:	=No::: I'm gonna, I'm gonna step in on that. First,] first of all, um, I've got boxes of the test data that provided (**1**)these numbers. That was done by ((*firm*)) and company during the, I think the eighties. *(1) R pencil points at UCBC code*
12b	Now in what, what they're telling me is that(**2**) the h over t values for masonry under their testings was, was in excess of twenty. *(2) R pencil points at Taraval plans*
12c	And that after the test, and there was, (**3**)they were doing all of the tests and arriving at, you know, the averages and the things that they do? *(3) flat hands raise and lower then form a horizontal threshold at eye level and hold on "do"*
13 Charles:	Yeh.
14 Doug:	And (**1**)then those numbers were driven down, has ... some arbitrary margins of safety. And that when you're this (**2**)close ... *(1) R hand rises above then drops below threshold still held by L hand* *(2) R index finger and thumb form a small interval raised to eye level and held*
15 (Gary):	Um hm.
16 Doug:	That in many communities they will accept it. (**1**)Based on the backup of the tests that have been [performed.

(1) hands again form horizontal threshold at eye level, then R holds while L rapidly sweeps out a vertical interval above

While Cal's response to Charles' question took the JCArchitects team forward in time into a confrontation with the city, Doug's challenge takes them all backward in time to a set of studies involving test structures, simulation of seismic events, and assumptions about margins of safety. We have obtained these studies and they contain the kinds of test structures, experiments, and statistical analysis that Doug mentions here and in other parts of the consulting meeting. A prestigious engineering firm with federal support conducted these studies, and state and municipal regulators then selectively adopted their recommendations. What Doug calls "arbitrary margins of safety" were introduced during this adoption process.

This is again a complex form of animation. First, Doug and Pat baldly interrupt Cal's ongoing narrative about talking to the city, taking the conversational floor to propose a potentially stronger position in the pending argument. Their proposal reverses time in an effort to destabilize existing practices of classification. Second, Doug's past-time account of studies yielding "boxes of the test data" (turn 12) is produced with a gestural depiction that reproduces Charles' earlier animation of a "RANGE," "limit," and "excess." This is a complex assembly with several components (see Fig. 5.6[c]). At 12c(3), Doug depicts a process of calculating "averages and the things that they do" by moving his hands along a vertical range. As he speaks, his hands converge on a threshold that he holds with both hands inside the vertical range (i.e., similar to Charles' "limit," at 6(2)). Then at 14(1) Doug uses his right hand to illustrate how this threshold, still held by his left hand, was "driven down" by what he calls "arbitrary margins of safety" to establish an even lower threshold, which he shows with the resting position of his right hand. According to Doug, it is this second, lower threshold that they are now just failing to meet. At 14(2), he shows this comparatively small excess (i.e., "you're this close") with his right hand, then he goes on to argue that "in many communities they will accept it" once they have been informed about the original tests (i.e., "the backup of tests," again animated at 16(1)).

Anticipating Different Contexts of Justification (Episode 1, Turns 17–22). The JCArchitects design team and the Taraval library have again been brought forward into a hypothetical meeting with the city. Although the library just fails to meet structural code criteria (i.e., the existing representational infrastructure would classify it as an "unsafe" public building), the team is considering an argument that these criteria (and the class of "safe" buildings they can be used to identify) are based on arbitrary historical judgments. Rather than consider more invasive retrofitting procedures, Doug (a historical preservationist) has proposed a radical alternative. They can simply tell the city that the classification system is no good. As with his response to Gary's proposal that they might "LIVE with" a code violation, Charles (lead architect) quickly takes the conversational floor:

17 Charles:	[Ok, so how do you, how do you document that? I mean how do [you arrive
18 Doug:	[It's (1)documented.= *(1) hands open wide as if to offer*
19 Cal:	=By, by uh, ((*person A*)).
20 Doug:	By, yeh, well actually it was ((*firm*)) [((*person A*)) was the ((*A*)) of ((*firm*)).
21 Cal:	[So we, ok, ok, so you're saying if its over a little bit we have to fight the battle ... for::: the library, I guess with the city, I [guess.
22 Doug:	[Yeh, now ... in terms of the public opinion ... I've experienced this in Sonoma, where you have this ... this opinion about ... reactions.

Charles directly asks Doug how he can document this kind argument, and Doug interrupts with a curt response that "It's documented" (18(1)) in the earlier studies. Cal seemingly agrees but then repeats (with audible uncertainty) that they will need to "fight the battle" with the city (turn 21). Doug simply moves on to another context of justification (turn 22), arguing by analogy to a project in Sonoma (another county in Northern California) that they will need to educate the public (not just the city architect) about different standards of safety.

Finding an Alternative Attachment to Existing Code

If the JCArchitects design scheme were to move forward with a Taraval library design that exceeds code thresholds, then the structural engineering firm would take on a substantial liability. Whereas Doug expands on his proposal to challenge existing code, describing how other communities have accepted these problems, Gary begins to describe an alternative proposal for how they might be able to "LIVE with" the excess test values (Episode 2, turns at talk not shown). By adding "seismic resisting elements" in the center of the library and strapping these to the ceilings, he proposes a design alternative that will distribute forces away from an over stressed wall during an earthquake. This scheme should allow the team to attach the Taraval library to an adequately stringent code (the UCBC general procedure; see Fig. 5.5[a]).

Gary's scheme to redistribute forces is another example of complex animation, and it shows seismic resisting elements as anchors within the existing building. Gary holds these anchors with one hand while his other hand traces out existing walls or regions that will be strapped to these anchors. These detailed, embodied renderings produce a structural design scheme, and they contrast sharply with the more selective, abstracted animations of attachment used by Charles to anticipate trouble with the city (turn 6 and Fig. 5.6[b]) or Doug to challenge "arbitrary" assumptions used when developing code guidelines (turns 12 to 16 and Fig. 5.6[c]).

However, Gary's alternative locations are also watched closely by the historical preservationists who later (not shown) evaluate the historical fabric in these locations to find a spot that will minimize damage. Also important for our analysis of interaction across disciplines, Charles writes furiously on a yellow pad (3(1)) as Gary animates his design alternative. In a public meeting at the library several days later, Charles reproduces much of what Gary says in this exchange, substituting less technical language but effectively replaying the same structural proposal (Stevens, 1999).

COMPARING CASES OF DISRUPTING INFRASTRUCTURE

As we argued in the beginning, changes in representational infrastructure have been analyzed as historical developments within scientific or technical fields, but work remains for understanding how these changes play out in interaction between participants from different disciplines. In our view, this is a central research problem for studies of distributed cognition.

We described and analyzed the BugHouse episode as a case of disrupting representational infrastructure by displacement, in the sense that entomologists meet with a statistician to develop a new technology for comparing chemical profiles. We described and analyzed the JCArchitects episodes as a case of disrupting representational infrastructure under dissent, in the sense that historical preservationists propose going against existing code guidelines that classify buildings as being safe for public use. In this comparative section, we address two questions across both cases: (a) What is common about the structure of interaction across these two cases of interdisciplinary consulting? and (b) What role do disruptions play in scientific or technical classification as an ongoing historical achievement?

Our approach to the first question (structures of interaction) examines several phenomena that are clearly illustrated in the case materials, including (a) common interactional structures of animation for assembling, juxtaposing, and evaluating representational states; (b) differences in discipline-specific forms of perception and action with what otherwise might appear to be the same represented world; and (c) the conditions under which these differences become remarkable and lead to conflict during consulting meetings. Our approach to the second question (historical achievement) asks how common structures of interaction might develop as conflicts are resolved in ongoing interaction. Several alternatives are considered, leading toward a view of development as dynamic processes of hybridization, coordination, and selective visibility. Within this view, "lightly equipped" humans (Latour, 1996, p. 56) become substantially thicker, and this leads to a rich set of new questions about the distributed (and distributable) character of cognition in scientific and technical work.

Common Structures of Interaction in Consultations Across Disciplines

At the level of interactional structure, consulting meetings appear quite similar. Media in both sites are densely heterogeneous, and problems of coordination arise both within and across disciplines. In our view, similar phenomena appear whenever people who are differently knowledgeable work together on projects in which represented worlds are only accessible through layers of representational technology (Goodwin, 1994; Hall, 1999; Hall & Stevens, 1995, 1996; Hutchins & Klausen, 1996; Latour, 1987, 1996).

Assembling, Juxtaposing, and Evaluating Representational States. Both the BugHouse and JCArchitects are dense with cultural artifacts, but representational states depicting termites, colonies, buildings, or remodeling schemes only appear when resources in the environment are selectively put into coordinated use (Greeno & Hall, 1997; Hall, 1996; Latour, 1987; Roth & McGinn, 1998). In both cases, we find interactional processes for animating representational states in working conversations. Specifically, Mark, Leah, and Doug assemble and explicitly contrast representational states that are important for ongoing design projects. At the BugHouse, Mark and Leah work carefully to create an environment for a contribution from the consulting statistician (Fig. 5.2). They assemble two instances that fall to either side of a classification they would like to make, and these are contrasted as different representational states within their current approach to classification (i.e., different vs. same termite colony). At JCArchitects, Doug's extended challenge to "arbitrary" code thresholds takes the entire team back in time, juxtaposing large margins set in earlier research with an excess test value that is "this close" for the Taraval library (Fig. 5.6[c]). He assembles a visual juxtaposition of two quantitative intervals, one a relatively small "excess" test value (i.e., their current state) and the other a comparatively large "backup" of acceptable values hidden by arbitrary assumptions. This contrast is then projected forward in time as the outcome of a successful classification (i.e., the library is judge safe despite excess test values).

Animation by Shifting Time, Space, and Agency. There are also common processes of animation in these examples, although the evidence is more complicated to follow. Looking within the details of how representational states are assembled, both Mark and Doug systematically shift who is taking action (agency), where objects are located or action takes place (space/place), and when these things happen (time). Shifts along these dimensions (agency, space, and time) provide basic resources for animating how representational states arise, what is going on within these states, and what their comparison might mean.

As Mark sets out to assemble a good comparing, he shifts among three distinct kinds of agency: (a) a collective voice describes work done by the BugHouse team (e.g., "*we* have three species"; turn 1b, italics added), (b) termites appear as protagonists who "interact" and "shift" in the BugHouse foraging study (e.g., "how might *they* interact"; turn 1b, italics added), and (c) representational devices are themselves described as participants in finding termites (e.g., "if *this station*, one day has species A"; turn 1c, italics added) and classifying them (e.g., "*things* that JUMP out at you ... say *these two peaks*"; turn 7c, italics added).

Along a spatial dimension, Mark shifts from the field to laboratory when describing how samples are put through chemical analysis (e.g., Turn 1c(5) and 1c(6)), then he also shifts from labeled CGs on the table to narrated or drawn accounts of distance in the field (e.g., the reversible mapping from laboratory to field depicted in Fig. 5.2[b]). Finally, along a temporal dimension, Mark shifts between present time in the consulting meeting (e.g., he sets a topic and question for the current meeting in turn 1) and future oriented or ongoing time when he describes types of analysis that the BugHouse group would like to be able to do (e.g., "how many different COLONIES of each species *might we have* in this plot"; turn 1b, italics added).

Similar types of shifting are evident in Doug's animation of "arbitrary margins" that undermine the adequacy of city code guidelines. Doug's use of narrative agency shifts from a nationally recognized engineering firm, to local municipalities making arbitrary political decisions, to the JCArchitects team struggling with excess test values, and finally to meetings in which city officials and members of the public take action. In space, Doug's challenge shifts from engineering studies in Southern California, to sites in Northern California where city officials adopted the results of these studies, to the JCArchitects consulting meeting, and finally to the city architect's office. In time, his challenge moves into the distant past (i.e., engineering studies done approximately 15 years earlier), to a more recent past when city officials adopted code guidelines, to their present design dilemma, and finally to future time review meetings.

Discipline-Specific Forms of Perception and Action. Although participants in both cases assemble representational states using common structures of interaction, the content or meaning of these states can be quite different across disciplines. This should not be surprising because participants in both cases were recruited from different disciplines to provide complementary perspectives on the same project. Again, we briefly revisit two examples.

In the BugHouse, entomologists and the consulting statistician had very different approaches to comparing CGs. Mark calculated distances from labels on these graphs, then he used visual and manual practices for picking out graphical peaks to find qualitative and quantitative differences (Fig. 5.2). Browning, in contrast, constructed a layer of statistical operations over the same CGs, treating these as composite objects and proposing a completely different technical meaning for distance

(Fig. 5.3). In a similar fashion at JCArchitects, the structural engineer (Gary) narrated and used his hands to demonstrate calculations from his structural analysis (Fig. 5.6[a]) as well as objects and their physical connections in a proposed force distribution scheme. In contrast, Charles and Doug questioned or challenged the results of Gary's structural analysis by assembling and comparing test values along a single quantitative scale, effectively animating the perspective of users or clients of structural analysis. These examples provide evidence that participants from different disciplines understand and use what might appear to be the same objects or concepts in quite different ways.

Disciplinary Differences and Conflict. When do differences across disciplines matter, and what role do they play in disrupting or changing representational infrastructure? The extended conflicts in these cases provide good instances for examining this question, particularly when contrasted with differences across disciplines that go unremarked and do not lead to disagreements about project activities. These conflicts are (BugHouse) Mark's resistance to a statistical meaning for distance as a form of displacement that would compete with or undermine existing entomological field practices (i.e., "When I'm talking about *distance*, I'm talking about *geography*"; italics added), and (JCArchitects) Doug's elaborate challenge to safety thresholds in existing code guidelines as a form of dissent that would have the team making very different proposals to city officials (i.e., "those numbers were driven down, has ... some *arbitrary* margins of safety"; turn 14, italics added).

These conflicts arise in a larger historical frame for each case. In the BugHouse, a representational infrastructure for classifying insects on the basis of their chemical fingerprints is still in the making (Latour, 1987), and this new infrastructure is at the center of an ongoing scientific controversy with more traditional methods of classification by morphological characters. The status of insects in general, not just termites and not just the particular samples that Mark and Leah select, hangs in the balance. Within this surrounding historical frame, the statistician is being asked to help stabilize the infrastructure of chemical taxonomy. His proposal to replace relatively nonstandard visual comparisons with a distance measure becomes the occasion for a sustained conflict. This conflict in which Mark even attempts to edit the statistician's language is one historical moment in an ongoing process of scientific development.

At JCArchitects, a representational infrastructure for classifying public buildings already exists, and the design meeting has been called to find a scheme that will attach a particular instance (the Taraval library) to this infrastructure (i.e., code guidelines for structural analysis). Just what counts as a safe building during a major earthquake is an ongoing controversy, but that controversy does not originate within the design firm we studied. When the structural engineers animate future-time trouble with the city over test values, the historical preservationists step in on their proposal, animating trouble in the past-time development of code guidelines and proposing that they convince the city (and the public) to accept minor code violations. This conflict, which we have analyzed as a form of dissent against the existing codes, is itself the disruption to representational infrastructure.

What of differences in discipline-specific perception and action that do not lead to conflicts? For example, Browning animates a complex layer of statistical operations over the surface of two CGs, but his exotic treatment of CGs as composite objects is unremarkable for the entomologists. What is remarkable and what leads to a sustained conflict is Browning's use of a technical term that may be incompatible with the way the BugHouse inscription device currently operates and the way other entomologists will read their work. If their purpose is to stabilize chemical taxonomy by introducing a new representational device, it will not do to have this new device unravel or destabilize other parts of their work. Likewise in the JCArchitects case. When Gary briefly animates the dimensional features of his structural calculations and their approximate status (i.e., Fig. 5.6[a]), none of the other specialists remark or (in the video image) note his activity. What is remarkable is Cal's future-time animation of trouble with the city, which the historical preservationists contest as a given outcome for the library. To meet existing code guidelines, they will need to strengthen the library using more invasive structural design schemes (Fig. 5.5[a]), and this is exactly what the preservationists would like to avoid.

Conflict arises then not because disciplinary specialists see the same project in different ways and fail to understand each other. Instead, conflict arises when discipline-specific perception and action from one discipline threatens to destabilize existing practices in another (i.e., statistics vs. entomology at the BugHouse) or to undermine the prevailing interests of another (i.e., structural engineering vs. historical preservation at JCArchitects).

Infrastructure as an Ongoing Historical Achievement

One view of interdisciplinary collaboration might be that people from different disciplines make distinct and complementary contributions, then these are somehow linked together as the project moves forward, and all participants need to reach the same understandings about proposed activity and its meaning. We find evidence at the level of ongoing interaction that this is not what happens in either case, or at least that what constitutes shared understanding is considerably more complex.

What Develops at the Level of Interactional Structure? In the BugHouse, Browning's (statistician) operations on CGs (i.e., threshold, superimpose, and collect differences) are compressed by Mark into a single quantitative scale, animated in gesture and talk as empirically derived intervals and a "standard" threshold along a horizontal axis. Although there are many reasons to think that Browning and Mark have reached a provisional agreement about a new representational device for comparing CGs, they do not as an outcome of this agreement appear to have the same understandings of or interests in the new device.

Similarly at JCArchitects, Gary's (structural engineering) brief sketch of test calculations for code thresholds on the Taraval library are questioned by Charles (principal architect) using a single quantitative scale, which Charles animates in talk and gesture as a range, limit, and excess. The projected outcome of Gary's tests is later taken up by Doug (historical preservation) who reproduces Charles' vertical scale, this time comparing safety thresholds set during engineering studies with what he calls "arbitrary" decisions to lower this threshold by municipal governments. Still within this vertical quantitative scale, Doug shows the excess h/t value for the Taraval library to be quite small when compared with arbitrary safety margins in existing code guidelines.

It could be that disciplinary specialists make representational devices that participants from other disciplines then use. Makers understand and animate the interior mechanisms making up a new device (e.g., statistical operations or structural calculations), whereas users simply understand and animate activity with outputs from this device (e.g., comparing intervals and thresholds along a single quantitative scale). If this were so, the fruit of interdisciplinary consulting would be a set of "black boxes" (Latour, 1987, p. 131) that only specialists could open (or design) but that anyone with suitable training could use. Rather than adopting this binary explanation of the development of representational infrastructure in interdisciplinary work (i.e., makers vs. users of black boxes), we find interactional evidence for what Star and Griesemer (1989) called an "n-way" (p. 389) perspective on representational technology. Under this view, classification systems and the devices that make them up are infrastructural resources at the boundary between disciplinary communities. These "boundary objects" provide an infrastructure that enables coordinated action between communities with different interests, needs, and accountabilities.

We argued earlier that (a) there are striking differences in how participants from different disciplines orient to and act on the same representational states, and (b) some of these differences appear to reflect the orientations of makers versus users of representational devices. However, we also find evidence in these cases that participants who might be called users can selectively unpack or open up the black boxes of makers. In the BugHouse, for example, the statistician (Browning) at several points mentions the possibility that his proposed distance measure would ignore peaks that the entomologists pick out as being biologically significant (e.g., turns 29b and 29d). Also, when reformulating his distance proposal in the face of Mark's challenges, Browning refers to and gesturally depicts compressing multiple "differences" into a single number (turns 27, 29a, and 37). We interpret Browning's selective use and mention of entomologists' orientations to "peaks" and "differences" as evidence that specialists can cross over disciplinary divides in the course of disrupting or changing representational infrastructure. Similar kinds of selective crossover occur in the JCArchitects episodes.

We have now considered three possible answers to the question of what develops in interaction as new devices are proposed, old devices are challenged, and conflicts across disciplines are resolved. One answer was that people from different disciplines needed to reach the same understanding, presumably including embodied forms of perception and action, for projects to move forward (i.e., a unary perspective). A second possibility was that sharp divisions between producers and users of representational technology would develop (i.e., a binary perspective), and we found some striking evidence that this can happen. The answer we find most compelling, however, is that participants develop hybrid and selective forms of perception and action as an outcome of working across disciplines (Engestrom, Engestrom, & Karkkainen, 1995; Galison, 1997). Under this view, even as new representational devices are being proposed or existing devices challenged, participants across disciplines are working out relations of selective visibility in terms of their own, local work practices and the kinds of organizational accountabilities they must meet (Engestrom, 1999b; Star, 1991; Suchman, 1995).

Back to the Future: Running Disruptions Forward in Time. Whereas conversations in these consulting meetings regularly (and systematically) bring past and future into the present, our analysis of disrupting infrastructure is suspended inside these complex, local interactions. We end the chapter by considering what actually changes, prospectively, in each case. Our studies continued well beyond the episodes analyzed in this chapter, and we followed these projects through to what participants considered points of completion.

In the BugHouse, work on chemical taxonomy continues as we write, now over 4 years since the consulting meeting. Following the statistician's advice at other points in the consulting meeting, the BugHouse team did adopt new data encoding and archiving procedures to facilitate sample comparison and statistical analysis. However, even by the end of the foraging study (i.e., after 2 years), Browning's proposal for a statistical distance had not been adopted or seriously investigated despite Mark's provisional elaboration and use at the end of the episode we analyzed. As we have shown in JCArchitects, Doug's dissent against existing code was dropped for a design scheme that relieved an over stressed wall and promised minimal damage to historical fabric within the Taraval library. Review meetings with the city architect were positive, but structural problems discovered with windows in the library pushed their design scheme into a substantially more invasive set of retrofitting procedures. Although the city originally hoped to keep the library open during construction, the more invasive scheme required that the library be closed, the stacks relocated, and all interior finishes removed to install steel braces within the library walls (i.e., a different line of attachment; see Fig. 5.5[a]).

Although bringing about lasting changes in representational infrastructure may be costly or difficult (Becker, 1995; Engestrom et al., 1991/1997; Star & Ruhleder, 1996), we think the kind of situations we have studied are fairly common—situations in which infrastructure breaks down (e.g., Hutchins' (1995) analysis of a navigational crisis), in which people challenge the existing infrastructure (e.g., Doug's dissent in the JCArchitects case), or when they agree to work on new representational devices (e.g., Browning's proposal in the BugHouse case). If we pay careful attention to how relevant historical material is densely present in people's interaction, disruptions to infrastructure are easy to find. Studying representational change requires that we examine how people make their actions accountable to each other's ongoing projects as the object of cognitive analysis. We think of this as research on distributing cognition.

ACKNOWLEDGMENTS

This work was supported by a grant from the National Science Foundation (ESI–94552771) to R. Hall, a National Academy of Education/Spencer Foundation dissertation fellowship to R. Stevens, and a University of California Mentored Fellowship award to T. Torralba. We thank participants in the Math@Work Project, Sharon Derry, Charles Goodwin, James Greeno, Leigh Star, and Karen Wieckert for helpful comments on this chapter.

REFERENCES

Anderson, J. (1990). *The adaptive character of thought.* Hillsdale, NJ: Lawrence Erlbaum Associates.

Becker, H. (1995). The power of inertia. *Qualitative Sociology 18,* 301–309.

Bowker, G., & Star, S. L. (1994). Knowledge and infrastructure in international information management. In L. Bud-Frierman (Ed.), *Information acumen: The understanding and use of knowledge in modern business* (pp. 187–213). London: Routledge.

Bowker, G., & Star, S. L. (1999). *Sorting things out: Classification and its consequences.* Cambridge, MA: MIT Press.

Bowker, G., Timmermans, S., & Star, S. L. (1995). Infrastructure and organizational transformation: Classifying nurses' work. In W. Orlikowski, G. Walsham, M. Jones, & J. DeGross (Eds.), Information technology and changes in organizational work (pp. 344–370). London: Chapman & Hall.

Bucciarelli, L. L. (1994). *Designing engineers.* Cambridge, MA: MIT Press.

Cole, M. (1996). *Cultural psychology: A once and future discipline.* Cambridge, MA: Belknap Press of Harvard University Press.

Cole, M., Engestrom, Y., & Vasquez, O. (1997). Introduction. In M. Cole, Y. Engestrom, & O. Vasquez (Eds.), *Mind, culture and activity: Seminal papers from the Laboratory of Comparative Human Cognition* (pp. 1–21). Cambridge, England: Cambridge University Press.

Engestrom, Y. (1999a). Communication, discourse and activity. *The Communication Review, 3,* 165–185.

Engestrom, Y. (1999b). Expansive visibilization at work: An activity-theoretical perspective. *Computer Supported Cooperative Work, 8,* 63–93.

Engestrom, Y., Brown, K., Christopher, L., & Gregory, J. (1997). Coordination, cooperation, and communication in the courts: Expansive transitions in legal work. In M. Cole, Y. Engestrom, & O. Vasquez (Eds.), *Mind, culture and activity: Seminal papers from the Laboratory of Comparative Human Cognition* (pp. 369–385). Cambridge, England: Cambridge University Press. (Original work published 1991)

Engestrom, Y., Engestrom, R., & Karkkainen, M. (1995). Polycontextuality and boundary crossing in expert cognition: Learning and problem solving in complex work activities. *Learning and Instruction, 5,* 319–336.

Galison, P. (1997). *Image and logic: A material culture of microphysics.* Chicago: The University of Chicago Press.

Goodwin, C. (1994). Professional vision. *American Anthropologist, 96,* 606–633.

Goodwin, C. (1997). The blackness of black: Color categories as situated practice. In L. Resnick, R. Saljo, C. Pontecorvo, & B. Burge (Eds.), *Discourse, tools, and reasoning: Essays on situated cognition* (pp. 111–140). Heidelberg: Springer.

Goodwin, M. H. (1990). *He-said-she-said: Talk as social organization among Black children.* Bloomington: Indiana University Press.

Greeno, J. G., & Hall, R. P. (1997, January). Practicing representation: Learning with and about representational forms. *Phi Delta Kappan,* 361–367.

Hall, R. (1996). Representation as shared activity: Situated cognition and Dewey's cartography of experience. *The Journal of the Learning Sciences, 5,* 209–238.

Hall, R. (1998). Following mathematical practices in design-oriented work. In C. Hoyles, C. Morgan, & G. Woodhouse (Eds.), *Rethinking the mathematics curriculum: Volume 10. Studies in mathematics education series* (pp. 29–47). London: Falmer.

Hall, R. (1999). The organization and development of discursive practices for "having a theory." *Discourse Processes, 27,* 187–218.

Hall, R., & Stevens, R. (1995). Making space: A comparison of mathematical work in school and professional design practices. In S. L. Star (Ed.), *The cultures of computing* (pp. 118–145). London: Basil Blackwell.

Hall, R., & Stevens, R. (1996). Teaching/learning events in the workplace: A comparative analysis of their organizational and interactional structure. In G. W. Cottrell (Ed.), *Proceedings of the eighteenth annual conference of the Cognitive Science Society* (pp. 160–165). Hillsdale, NJ: Lawrence Erlbaum Associates.

Hutchins, E. (1995). *Cognition in the wild.* Cambridge, MA: MIT Press.

Hutchins, E., & Klausen, T. (1996). Distributed cognition in an airline cockpit. In Y. Engestrom & D. Middleton (Eds.), *Cognition and communication at work* (pp. 15–34). Cambridge, England: Cambridge University Press.

Lakoff, G. (1987). *Women, fire, and dangerous things: What categories reveal about the mind.* Chicago: University of Chicago Press.

Latour, B. (1987). *Science in action: How to follow scientists and engineers through society.* Cambridge, MA: Harvard University Press.

Latour, B. (1996). Cogito ergo sumus! Or psychology swept inside out by the fresh air of the upper deck ... Review symposium: *Cognition in the Wild*. E. Hutchins. *Mind, Culture, and Activity, 3,* 54–63.

Lynch, M. (1991). Method: Measurement—ordinary and scientific measurement as ethnomethodological phenomena. In G. Button (Ed.), *Ethnomethodology and the human sciences* (pp. 77–108). Cambridge, England: Cambridge University Press.

Neisser, U. (1987). *Concepts and conceptual development: Ecological and intellectual factors in categorization*. Cambridge, England: Cambridge University Press.

Ochs, E., Jacoby, S., & Gonzales, P. (1994). Interpretive journeys: How physicists talk and travel through graphic space. *Configurations, 1,* 151–171.

Roberts, K. H. (1993). *New challenges to understanding organizations.* New York: Macmillan.

Rochlin, G. I., La Porte, T. R., & Roberts, K. H. (1987). The self-designing high-reliability organization: Aircraft carrier flight operations at sea. *Naval War College Review, 40,* 76–90.

Roth, W.-M., & McGinn, M. K. (1998). Inscriptions: Toward a theory of representing as social practice. *Review of Educational Research, 68,* 35–59.

Schegloff, E. A. (1992). On talk and its institutional occasions. In P. Drew & J. Heritage (Eds.), *Talk at work* (pp. 101–134). Cambridge, England: Cambridge University Press.

Smith, E. E., & Medin, D. L. (1981). *Categories and concepts.* Cambridge, MA: Harvard University Press.

Sperber, D. (1994). The modularity of thought and the epidemiology of representations. In L. A. Hirschfeld & S. A. Gelman (Eds.), *Mapping the mind: Domain specificity in cognition and culture* (pp. 39–67). Cambridge, England: Cambridge University Press.

Star, S. L. (1991). The sociology of the invisible: The primacy of work in the writings of Anselm Strauss. In D. Maines (Ed.), *Social organization and social processes: essays in honor of Anselm Strauss* (pp. 265–283). Hawthorne, NY: de Gruyter.

Star, S. L. (1996). Working together: Symbolic interactionism, activity theory, and information systems. In Y. Engestrom & D. Middleton (Eds.), *Cognition and communication at work* (pp. 296–318). Cambridge, England: Cambridge University Press.

Star, S. L., & Griesemer, J. R. (1989). Institutional ecology, "translations" and boundary objects: Amateurs and professionals in Berkeley's museum of vertebrate zoology, 1907–39. *Social Studies of Science, 19,* 387–420.

Star, S. L., & Ruhleder, K. (1996). Steps toward an ecology of infrastructure: Design and access for large information spaces. *Information Systems Research 7,* 111–134.

Stevens, R. (1999). *Disciplined perception.* Unpublished doctoral dissertation, University of California, Berkeley.

Stevens, R., & Hall, R. (1998). Disciplined perception: Learning to see in technoscience. In M. Lampert & M. Blunk (Eds.), *Talking mathematics in school: Studies of teaching and learning* (pp. 107–149). New York: Cambridge University Press.

Suchman, L. (1987). *Plans and situated action: The problem of human–machine communication.* Cambridge, England: Cambridge University Press.

Suchman, L. (1995, September). Making work visible. *Communications of the ACM 38*(9), 56–64.

APPENDIX A
TRANSCRIPT CONVENTIONS

Transcript conventions include the following:

- Numbered turns at talk start with named speakers.
- Numbers in parentheses index the (**n**)onset of activity within the turn.
- Matching, numbered ***(n)** activity descriptions* appear in italics below each turn.
- Within activity descriptions, R and L refer to right and left hands.
- Identification of points, beats, and descriptions of iconic/metaphoric depiction follow McNeill's (1992) descriptive categories for gesture.
- Italicized text in double parens provides ((*general description*)).
- Spoken EMPHASIS is shown by uppercase.
- Stre:::tched enunciation is shown with repeated colons.
- Words that were (difficult to transcribe) are in single parentheses.
- Matching brackets show the [onset and termination] of overlapping talk across turns.
- Matching =equal =signs mark turn boundaries with no audible silence.

6

Categories and Cognition: Material and Conceptual Aspects of Large-Scale Category Systems

Susan Leigh Star
Santa Clara University

The classic "fish scale model" developed by Donald Campbell, and reproduced in this volume, suggests that the boundaries between disciplines are not discrete fences, but rather broad areas of overlap. Knowledge is never complete within a discipline; however, an important robustness derives from the shared "scales" taken across all fields.

Since Campbell's article first appeared, a lively area of empirical research in the sociology of scientific knowledge has developed, especially over the last twenty years. One way to view some of this research (and it has included extensive work on discipline formation), is that it fleshes out the fish scale model. What is it exactly that overlaps? Ideas by themselves do not travel through the air, or even the zeitgeist. They require hosts—researchers, students, journals, experimental animals, and bureaucracy.

This way of thinking about ideas—as embodied, historical, and part of institutions—is another development that has been making much headway since the 1970s. Where ideas were previously modeled as "inside the head" or pure logic, the intellectual movement known variously as "situated cognition" or "cognition and practice" or "distributed cognition" reforms that location. Ideas are in the head—and in the hands, the tools, the networks of human and non-human actors that ecologically form the complex we call knowledge.

This chapter examines one facet of the multifaceted apparatus that mediate the fish scaling of ideas between disciplines—*categories and standards*. It draws on both the sociology of science, and the cognition and practice perspectives.

LEARNING FROM ANOMALIES: HOW CATEGORIES APPEAR FROM UNCERTAINTY

John Dewey observed that inquiry begins in doubt, and ends when that tension is relieved. Following the cognition and practice angle of vision, I wanted to see how people form categories and make knowledge in real world settings. My initial inquiries into the nature of scientific knowledge thus began as ethnographic journeys—examining the way that scientists work together, in the context of their allies and institutions. Because I come from a tradition in sociology that has tended to study people from all walks of life, I was predisposed to look at the ecology of the workplace—all of the things that are involved in the mediation of knowledge, from the janitor to the Nobel prize winner (Star, 1995a).

I teach students in my fieldwork classes to listen and look for two things: first, for the special language used in the location, metaphors, mots justes, turns of phrase, private codes used by one group and not another; second, for things that strike them as strange, weird, anomalous. What is causing them doubt? The strength of fieldwork is its anthropological strangeness, and nowhere is that more important than in the beginning stages of inquiry.

Over the past several years, in studies of various groups of scientists, technicians, doctors, nurses, and patients, I have often encountered that funny feeling of finding an anomaly, sometimes embedded in the distinct language of a workplace or health care venue. It's a little irritating feeling, kind of a pre-sneeze sensation—and it is also exciting. Learning to trust this message is the toughest lesson I have had to teach my students—as well as myself.

I'm going to highlight five anomalies that have tickled my nose, and use them to form the basis for this discussion of infrastructure and classification, and the program of research I have been developing over a number of years.

The first three anomalies come from a study I did some years ago of neurophysiology and brain surgery. After a field study in an electroencephalogram (EEG) lab, I wrote a historical book on a group of 19th century British researchers, administrators, and patients who were trying to locate functional areas in the brain (Star, 1989). In fact, they invented modern brain surgery. The period I studied spanned 40 years. At the beginning the mortality rate was 100%; by the end it had fallen to approximately 60%. I read hospital records, letters from patients, lab notebooks, administrative records, and published documents.

I noticed the first anomaly while I was looking at one of the physician/physiologist David Ferrier's notebooks in the Royal College of Physicians archives in London. The archives are royally housed in an imposing building overlooking Hyde Park, lushly carpeted in deep red, and bookcases filled with leather-bound gold trimmed volumes. After carefully divesting myself of anything toxic, like a pen or food that might damage the materials, I was seated at a mahogany table, and Ferrier's lab notebooks were brought out to me—literally—on a silver platter. Gingerly lifting them up (hoping I wasn't sweating or anything), I opened to one experiment where he was trying to measure the effect of a lesion he had produced earlier in the day, on the brain of an ape. The ape is less than cooperative—Ferrier's handwriting occasionally flies off the page, wobbles and trails off in what clearly is a chase around the room after the hapless ape. The pages, in sharp contrast to my chapel like surrounds, are stained with blood, tissue preservative, and other undocumented fluids. By contrast—and this is a finding often repeated in sociology of science—the report of the experiment is clean, deleting mention of the vicissitudes of this experimental setting (Star, 1983; Latour & Woolgar, 1979). This anomaly drew my attention to two things: the magnitude of invisible work that subtends any scientific experiment or representation, and the materiality that acts to mediate the conduct of science. I went on to develop models of invisible work for computer systems development (Star & Strauss, 1999), and to examine the kinds of materiality involved in museum representations.

The second anomaly came from the same study, this time from a set of clinical data on epileptic patients. The same researchers who were doing nasty things to monkeys were also looking at human patients—

those with brain tumors, epilepsy, syphilis, and other "nervous disorders." They were not well funded. Absent modern medical telemetry, they enlisted the families of epileptic patients to record information about seizures on what they called "fits sheets," or printed forms which had checklists of symptoms, timing, and other data. Family members, desperate as they were, tried desperately to comply in the data collection effort. The forms are moving documents revealing the relations of class and medicine in late 19th century England, penciled in, misspelled, and assiduously brought to the doctor's files. They also tell another story. All around the edges of the documents are scribbled messages to the doctor that don't fit the actual form: "Had too much hot soup yesterday," "exposed to night air," "rode alone in carriage." A whole folk medicine existed invisibly in the side comments entered alongside the filled-in forms. However, this wealth of information was discarded as unimportant—lost in the files—even though in a sense the patients were acting as research assistants to the clinicians. This anomaly drew my attention to the problem of collecting, disciplining and coordinating distributed knowledge. How does delegated work—called "hired hand research" by Julius Roth—affect data quality? How do forms shape and squeeze out what can be known and collected? The current web-based patient information exchange groups face conceptually similar problems of group memory, language differences, and what fits on the form of traditional medicine vs. what the patients really know in their lives. I went on to analyze this problem (with Geoffrey Bowker) in our model of the management of data collection in the international classification of diseases, and the tensions between traditional systems of medical knowledge and the forms distributed by the World Health Organization (Bowker & Star, 1994; Bowker & Star, 1999).

The third anomaly, and my final brain example, comes from another set of documents in another posh British archive. In the archives of the Royal Society, sticking again with the red velvet trope but not the silver platter, I found a curious set of referees' reports of a paper David Ferrier submitted to the Society for publication. To explain fully how weird this was, you have to realize first that monkey brains and human brains are very different in size and shape, and presumably, function. Ferrier had been trying to plot, at the millimeter level of scale, differences in function when he administered electricity to the surface of the monkey's brain. The article is about human brain function. Lacking a human subject (Penfield's famous surgical experiments with epileptics were almost a century later), Ferrier took the expedient step of simply taking the monkey map and transposing the circles marking functional areas to the human brain sketch. Anatomically, this is the functional equivalent of taking a map of the Paris subway and superimposing it on Cleveland, and using it to talk about travelling around Cleveland, reasoning that all large cities essentially have the same sort of transportation infrastructure.

Ferrier's paper was published, and was an enormous success. Why? I asked myself. The answer seemed to be that the map didn't *need* to be accurate in order to be useful. It could serve as the basis for conversation, for sharing data, for pointing to things without actually demarcating any real territory. It was a good communicative device across, for example, the worlds of clinical and of basic research. Its mediational qualities seemed to be that it "sat in the middle" between different groups, very ill-structured or sketchy in the common usage. But when a clinician or physiologist needed a real map, they would take the lineaments of Ferrier's diagram and adjust it to their own needs for surgery or the study of lesions. Later, in a related study of amateurs and professionals in a zoological museum, I came to call this class of arrangements *boundary objects* (Star & Griesemer, 1989). These are objects that are weakly structured in common usage, and more tightly tailored in the use in one particular line of work. They are ambiguous, but they are part of a durable cooperation across social worlds. They facilitate cooperation without consensus.

The next éclat came while working in another archive, the Bancroft Library at the University of California, Berkeley. This archive requires the same sort of hushed rite de passage as had the British ones. I had to leave my lunch, my pens, and my backpack in a specially provided locker before entering. This time, now some years later, I was allowed to take in a pencil and a personal computer. In California, one fills out little slips of paper and applies for boxes, which you yourself then cart over to the table you're using. No silver platters, but the lighting is much better. I was examining the letters, field notebooks and accounts of the development of the Museum of Vertebrate Zoology, founded in 1906. This was a fascinating venue in which to pick up the boundary objects idea—amateur naturalists, trappers, professional biologists, philanthropists and university administrators all left their imprint on the museum's development. I was able to flesh out the triangulation, mediation and translation issues much more thoroughly here. Here is the anomaly: One day, as I was reading a particularly dull bit of the accounts and receipts from an expedition to the Mojave to document gopher behavior, I lifted up one of the manila folders, opened it, and much to my astonishment, a dead bluebird (totally desiccated) fell out. A letter was also in the folder: "Dear Dr. Grinnell, I found this in my yard and I want to know what this is. I know you are the man who knows about these things. Can you help me?"

Grinnell, being a courteous man, probably answered the person's query, although there was no record of his reply in the archives. At the same time, the image of the bird stuck with me forcefully. This thing didn't fit his categories. In natural history, if you collect something without a proper label, or documentation of its habitat, it essentially is useless for the professional biologist (or as one respondent at the museum told me, "without a label, a specimen is just dead meat.") But Grinnell was also a "birder," active in amateur circles. Perhaps he knew the man who wrote the letter. He didn't throw out the bird; instead, he found a file folder and stuck it in there with a bunch of miscellaneous receipts. This fourth anomaly drew my attention to those things that don't fit categories or standards, which literally or figuratively get shoved into the nearest file folder or functional equivalent. Strictures and standards, and the exercise of brute force solutions to inter-category problems, continued to fascinate me for the next many years. This would come to include people as the objects of both scientific and political marginality or "otherness" (Star, 1991).

My final example is from a more recent study, a field study I conducted of a community of biologists who are sequencing the genome of a nematode. I worked as a partner with a computer scientist/systems developer to make sure that the system, an electronic data sharing/publication "virtual lab" matched the work needs of the biologists (reported in Star and Ruhleder, 1996). This work was done just at the advent of the World Wide Web, and the software was not web-based. (I might add that I was contacted by this computer scientist after he read the brain book and recognized all of the workplace challenges involved in building a system to communicate across social worlds!) The anomaly here occurred during the course of traveling to more than 40 laboratories and interviewing biologists about their use of the prototype system. A typical interaction: I would call up and say, "I'm Leigh Star, and I'm doing requirements analysis for the Worm Community system. Are you using the system? May I come and watch you work and interview you about it?" They would say, "yes, sure, we love the system, come on over." So I would go over—where over sometimes meant flying from England to Vancouver—and arrive in the lab, yellow legal pad and pen at the ready. I would begin to ask them to show me how they had installed the system, and where it fit in the flow of their work. On several occasions the interaction unfolded like this:

Me:	So, show me how you use WCS.
Lab:	Um, well, I know it's here somewhere. Let me just check. No, that's right, there's a postdoc who's using it. She's not in today. (yelling) Anybody here using WCS?

I would, very patiently I thought, point out that they had said they were using the system. Where was it? Then the phrase that made my ethnographer's nose twitch: "Oh, we are using it. We're just about to use it." Where was the conflation of future and present coming from? Were they just trying to spare my feelings? These were not otherwise mendacious people, and they were not afraid to criticize the system or give me feedback on it. As I delved deeper into the communicative processes occurring between developers and users, it became clear that what was occurring was a kind was a communicative tangle, including several Batesonian double binds. The messages that were coming at level one from the systems developers were not being heard on that level by the users, and vice versa. What was obvious to one was a mystery to another. What was trivial to one was a barrier to another. Yet clarifying this was never easy. The users liked the interface when they were sat in front of it. Yet they didn't know how to make a reliable working infrastructure out of it. They would ask the WCS team, who would reply in terms alien to them. I began to see this as a *problem of infrastructure*—and its relative nature.

The Materiality of Categories

Each of the anomalies described above served two purposes in the research I was developing. First, they helped me form some core categories of my own: boundary objects, invisible work, the importance of infrastructure, the problem of residual categories, and the materiality of knowledge. Second, they showed me how the scientists I was studying developed their categories in the context of very nitty-gritty work—the work of problem solving. As I explored each of these in a series of studies, I became interested in the problematics of studying categories.

How do categories arise? How do they form the borders of the communities that use them? Is it possible systematically to analyze something almost intangible, something buried between the material and the conceptual, as a classification system?

Many fields face this problem, including sociology, anthropology, history, and linguistics. All are concerned in a sense with invisible relationships—things that must be studied through indicators such as actions, productions, and inscriptions. One cannot directly *see* relations such as membership, learning, ignoring or categorizing. They are patterns, or indicators. One acquires membership in a given discipline, or any social group, by becoming comfortable with the language, categories, and things used by the group. Thus categories come from action, and in turn from membership. They do not last forever, but are rather continually re-negotiated, often when anomalies such as those described above arise.

However, the very notions of categories and classification have often been seen as distinct from work or from material culture. They have been seen as part of an abstract sense of "mind," and not as linked with the exigencies of work or politics. The tasks involved in framing categories in the context of politics and culture, and the ways in which those categories are ordered into systems, is often overlooked.

Categories are often built into our standardized environments. For example, the range of furnishings we select is institutionalized by the retail stores to which we have access, traditions of craft, and so forth. This institutionalization of categorical work across multiple communities of practice (Wenger, 1998; Lave & Wenger, 1991), over time, produces the structures of our lives, from furniture to scientific laboratories. They often are embedded in the built environment. In this sense, categories are historically situated artifacts. They are learned as part of community membership. This brings categories well beyond the individual mind, task, or the small scale. Looked at in this way, classifications qua technologies are powerful things that may link thousands of settings, and span complex disciplinary boundaries. The situated cognition/practice movement has helped to both ground and scale up traditional notions of categories and classification (Suchman, 1988; Hutchins, 1995; Keller & Keller, 1996; Lave, 1988; Hall & Stevens, 1995), making clear the role of materiality of categories in shaping action.

Revisioning Cognition

Since the 1980s, scholars from several disciplines (including anthropology, communication, ecological psychology, and the sociology of science) have been questioning many of the fundamental assumptions of cognitive science. They are questioning the notion of cognition as only individual, mental and non-social, and instead finding ways to describe thinking and knowledge situated, collectively produced and historically specific. Michael Cole, for example, has conceptualized tools and artifacts as:

> An artifact is an aspect of the material world that has been modified over the history of its incorporation into goal-directed human action. By virtue of the changes wrought in the process of their creation and use, artifacts are simultaneously ideal (conceptual) and material. They are ideal in that their material form has been shaped by their participation in the interactions of which they were previously a part and which they mediate in the present. (Cole, 1996, p. 117)

This double conceptual/material model is as well useful for analyzing the role of categories and classification in knowledge production. Categories are both ideal and material. They are also human products, and may be considered as tools or artifacts. Viewed this way, they become easier to analyze as part of practice. They represent persistent patterns of change and action, and are as well resources for organizing abstractions, thus ideal. They are inscribed, transported, and affixed to stuff, and are thus material. Cole's emphasis is on the conceptual and symbolic sides of things often taken as only materials, tools and other artifacts. Turning this emphasis around, one can equally well emphasize the material force of that which has been considered ideal, such as categories. Below, I consider related themes in Pragmatist philosophy that have informed sociology of science.

The Pragmatist Turn

One of the central tenets of Pragmatism, and one of the most radical, is the idea that consequences, not antecedents, are the most important aspects of arguments. This position was held by philosophers John Dewey and Arthur Bentley, and was picked up by later Pragmatist-influenced sociology such as W. I. Thomas, Dorothy Thomas and Everett Hughes. It is related, both historically and conceptually, to the cognitive reforms detailed above (see e.g., Dewey, 1916). Thomas and Thomas put forward the now-famous maxim, "things considered as real are real in their consequences." That is, what is important about any finding is whether or not people take it to be true, and then what they do with it (Thomas & Thomas, 1970/1917). Later sociologists turned this axiom into a way of viewing collective behavior, to develop "labeling theory"—the idea that definitions of a group should be viewed from within the group, not as labels applied from outside. Conversely, however, a label applied from outside may be internalized by the group. Thus, the definition of a situation—whether it is the label "deviant" or any other behavior—is what people will shape their behavior towards.

This is a profound form of social construction, and it goes beyond the notion that people construct their own realities. Where the definition of the situation may come from is not central. Its source may be human or nonhuman, structure or process, group or individual. This model underscores the idea that the materiality of anything (whether it be action, idea, definition, hammer, gun or school grade) is drawn from the consequences of its situation.

Both pragmatism and Cole's activity theory (or cultural-historical psychology) emphasize the ways in which things are perceived as real may mediate action (Star, 1996). If someone is taken to be demonically possessed, and the bureaucratic or technical apparatus for exorcism is developed, then the reality of possession appears as part of the consequences of the model. People are certified to perform exorcisms, rites developed, and so forth.

Below, I consider two cases of the development of categories as part of this "defining of the situation." To the extent that this reality becomes shared across settings, approaches, or schools of thought, it subtends interdisciplinary work.

THE EXAMPLE OF MEDICAL CLASSIFICATION

The International Classification of Diseases (ICD) is a classification scheme which began in the late 19th century. For the past one hundred years, it has been revised about once a decade through a series of conferences at which representatives from all over the globe meet and negotiate the categories to be used in the collection of mortality and morbidity data. It is still in use today; indeed, it is ubiquitous in medical bureaucracy and medical information systems. It forms the basis for the international death certificate. In one sense, it is nearly the largest and oldest of interdisciplinary tools, drawing together statisticians, epidemiologists, physicians, public health officials, insurance analysts, and many other groups. For the background research for understanding these international, interdisciplinary processes of classification, my colleague Geoffrey Bowker and I went to Geneva and studied the archives of the World Health Organization (WHO) and its predecessors such as the League of Nations and the Office Internationale d'Hygiène Publique. We were especially interested in the negotiation of categories as interdisciplinary tools. The United Nations and the WHO have kept some records of the process of revision; others are to be found in the file cabinets of individuals involved in the revision process.

At first we expected to find a progression of increasing consensus over time, with perhaps more reification of categories. What we found was far from that. There are arguments embedded in the innocuous, list-like ICD that have remained unsolved for a century. The international public health bureaucracy grew rapidly, encompassing more and more points of view and differing national medical cultures. Religion, epistemology, and technology are but a few of the factors that representatives disagreed about. However, most agreed that data, especially mortality data, had to be collected in order to track global changes and epidemics. The disagreements are usually buried in the paperwork involved, but at times the seams of the compromises show. For example, one of the extreme differences is between traditional Western medicine that denies or ignores the world of spirits, and traditional African cultures who believe that one may die as a result of activities in the spirit world. The two never reconcile in the ICD. The ICD designers went so far as to include a category for cause of death: "death from a non-existent disease!"

Religion and ontology are similarly at war in determining the moment when life begins (as with modern American abortion wars). If life is said to begin at the moment of conception, abortion is murder. As well, if a woman miscarries when she is three months pregnant, then it will be encoded as a case of infant mortality, and thus entered into one of the most important statistics collected about a country.

A means must be found to work around such higher-order issues if any data are to be collected, or some working definition must be crafted. In one example of such compromising, a statistical committee, assigned with determining the exact moment of the beginning of life, ends up setting a threshold for life: the number of attempted breaths and the exact weight of fetus or infant. How many breaths or attempted breaths will equal 'life'? It almost seems absurd. Yet if it remains unspecified and up to the discretion and religion of the certifying official, the quality of the data will remain moot.

These moral questions of course are not resolved by these formulae. The question of when life begins is hotly debated all over the planet. What does happen in forms like the death certificate, when aggregated, form a case of what Kirk and Kutchins (1992) call "the substitution of precision for validity" (see also Star, 1989). That is, when a seemingly neutral data collection mechanism is substituted for ethical consensus about the contents of the forms, the moral debate is partially erased. One may get ever more precise knowledge, without having resolved deeper questions, and indeed, by burying them.

There is no simple way around this from the point of view of practical information gathering. Making all knowledge retrievable, and thus re-debatable, is an appealing solution in a sense from a purely information science point of view. It would seem to be the ideal interdisciplinary tool, the key to unlocking the overlaps of disciplinary fish scales. However, from a practical organizational viewpoint the situation is more complex. As we know from studies of work of all sorts, people do not do the ideal job, but the doable job. When faced with too many alternatives and too much information, they do the best they can. As an indicator of this, studies of the validity of codes on death certificates repeatedly show that doctors have favorite categories; these are regionally biased; and autopsies (which are rarely done anyway) have a low rate of agreement with the code on the death certificate (Fagot-Largeault, 1989).

Even when there *is* relatively simple consensus about the cause of death, the act of assigning a classification can be in itself socially problematic. Thus, in some countries the death certificate has two faces: a public certificate which is handed to the funeral director in order that arrangements be made quickly and discreetly, and a statistical cause which is filed anonymously with the public health department. In this case, the doctor is not faced with telling the family of a socially unacceptable form of death: syphilis can become heart failure, or suicide can become a stroke. Thus, the process of moving to an anonymous statistical record may reveal hidden moral stances in the reporting of death. When the death certificate is public, then stigma and the desire to protect the feelings of the family may reign over scientific accuracy.

The ICD classification is in many ways an ideal mirror of how those doing interdisciplinary work struggle with uncertainty, ambiguity, standardization, and the practicalities of data quality. Deeper differences may be smoothed over by sharing common tools (boundary objects). The process is often in obscure places in the historical record, in the materiality of instruction manuals, or lists, or forms that do invisible work and provide infrastructure. It is thus important for the understanding of interdisciplinary work to recover some of these compromises by going into these archives, and reading tools like ICD closely through their changes. This may reveal some of the negotiation process. Below, I turn to the second case of widespread classification, one that became interdisciplinary in a more nefarious sense.

THE EXAMPLE OF RACE CLASSIFICATION AND RECLASSIFICATION UNDER APARTHEID IN SOUTH AFRICA

Under apartheid in South Africa, race classification provided the bureaucratic underpinnings for a vicious racism. Here too, the attempt to create a normalized, systemic bookkeeping system was embedded in a larger program of human destruction.

Background

From the early days of Dutch settlement of South Africa, the *de jure* separation and inequality of people co-existed with interracial relationships. When the Nationalists came to power in 1948, a much more detailed and restrictive policy, apartheid, was put into place. In 1950, two key pieces of legislation, the Population Registration Act and the Group Areas Act, were passed. These laws required that people be strictly classified by racial group, and that those classifications determine where they could live and work. Other areas controlled de jure by apartheid laws included: political rights, voting, freedom of movement and settlement, property rights, right to choose the nature of one's work, education, criminal law, social rights including the right to drink alcohol, use of public services including transport, social security, taxation, immigration (Cornell, 1960; United Nations, 1968). The ultimate idea was that the races shall remain utterly apart. These laws were in effect for more than four decades.

The racial classification which was so structured in the 1950s sought to divide people into four basic groups: Europeans, Asiatics, persons of mixed race or "Coloreds," and "Natives" or "pure-blooded individuals of the Bantu race" (Cornell, 1960). The Bantu classification was subdivided into eight main groups, with Xhosa and Zulu the most numerous. The Colored classification was also complexly subdivided, partially by ethnic criteria. The terribly fraught (and anthropologically inaccurate) word Bantu was chosen over preference to African (or Black African), partly to underscore Nationalist desires to be recognized as "really African."

For Black South Africans, the system was horrific. No Black was allowed to be in a White area for more than 72 hours without special permission, including government authorization for a work contract (such as that for a live-in servant). Failure to comply meant prison and probable death therein (Mathabane, 1986).

Scientifically, apartheid was in fact an interdisciplinary undertaking. Anthropologists, historians, criminologists, legal scholars and physicians all collaborated in making a eugenical science to justify these social practices. Tomes were written on why separate development was scientifically justified, and in fact natural.

For apartheid to function in this way, people had to be unambiguously categorizable by race. However, despite the legal requirement for certainty in race identification, this task was to prove just as elusive as finding consensus for the ICD. Many people did not conform to the typologies constructed under the law, especially those of mixed race, or who spoke a different language from the assigned group, or had some other historical deviation from the pure type. New laws and amendments were constantly being debated and passed. By 1985, the corpus of racial law in South Africa exceeded 3,000 pages (Lelyveld, 1985).

The scientific theories about race and the common sense of terms were conflated. The original official sorting by race after the 1950 Population Registration Act derived from the categories checked on the 1951 census returns. An identity number was given to each individual at that time (Horrell, 1958). The Director of Census was in charge of deciding everyone's racial classification using the census data and, where necessary, other records of vital statistics. When the record seemed wrong, the ultimate deciding factors on a person's race were "appearance and general repute." Horrell notes: "But this classification is by no means formal. Section *Five(3)* of the Population Registration Act provides that if *at any time* it appears to the Director that the classification of a person is incorrect, after giving notice to the person concerned, specifying in which respect the classification is incorrect, and affording him or her an opportunity of being heard, he may alter the classification in the register" (1958, p. 4). So in the case of apartheid, we have the scientistic belief in race difference on the everyday level, and an elaborate formal legal apparatus enforcing separation. At the same time, a much less formal, more prototypical approach uses an amalgam of appearance and acceptance, and the on-the-spot visual judgments of everyone from police and tram drivers to judges, to perform the sorting process on the street. Within the legal system, a person could appeal to their appearance or reputation if they wished to protest their classification.

The South African Colored Population and Reclassification

Approximately 1,500,000 South Africans fell into the Colored (multiracial) category at the time of the Population Registration and Group Areas Acts. Within this group, there were many cases where people chose to protest their classification. Most frequently, a person labeled Colored desired to be labeled White (or European). Local administration boards were set up to hear these borderline cases and reconsider classifications. These decisions could be appealed up to the level of the Supreme Court, a costly and time-consuming business. Often there were long bureaucratic delays in these reclassification hearings. Approximately 100,000 people applied for reclassification (Horrell, 1958). As of 1967, only one in ten were approved. The reclassification process was fraught in myriad ways and was internally inconsistent.

There was no clear onus of proof about the meaning of general acceptance as White. In ambiguous cases the Race Reclassification Board would decide after conducting hearings and administering a range of tests of race. Like child custody hearings in American courts, such painful (and often shameful) tests were not stable or guaranteed of permanence. The actual process of reclassifying was done behind closed doors, and kept highly secret by those working at the Population Registration Office. People could also be reclassified involuntarily. The Nationalist Government fostered a system of informers and surveillance that turned people in for passing as another race, or where they decided for any reason that a person had been misclassified. This resulted in several infamous cases of a person suddenly being declared another race, with all the attendant consequences for their life possibilities.

Intertwined with the system of classification were a set of crude technologies meant to sort out borderline cases. Combs and pencils were sometimes used to test how curly a person's hair was, with the idea that very curly hair indicated that the person was "Bantu." Horrell (1968) notes that barbers were sometimes called as witnesses to testify about the texture of the person's hair. One source mentioned expert testimony from the South African Trichological Institute (presumably an organization for the scientific study of hair). Affidavits were taken from employers, clergy, neighbors and others in order to establish general acceptance or repute. "The official may summon any living relative, including grandparents, and question them in a similar way." (Horrell, 1958, p. 32) Complexion, eyes, hair, features and bone structure were examined by board officials, and they could summon any relative and examine them in this way as well. Horrell (1958) notes: "It is reported that some were even asked 'Do you eat porridge? Do you sleep on the floor or in a bed?' Some Coloured people said that they had been told to turn sideways so that the officials could study their profiles" (p. 62). Folk theories about race abounded—differences in cheekbones, even the notion that blacks have softer earlobes than whites was takes seriously. A newspaper account notes that some Coloured people had reported that "the officials fingered the lobes of their ears—the theory is that Natives have soft lobes" (Sunday Times, 1955).

Reclassification Cases

"In one family, one twin was classified as Coloured and the other as African" (Horrell, 1958, p. 70).

As mentioned above, there were many famous cases of involuntary reclassification, reported in both the South African and international media. One such was the case of Ronnie van der Walt, a famous boxer in South Africa. One the eve of a big boxing match, he was suddenly reclassified from White to Colored by local officials. The local race classification board's decision "was based on an inspection of Ronnie, Rachel and their two children ... One man there, Ronnie recalls, walked around us peering at us from every angle like you do when you buy an animal. He said nothing, just looked ... Interior Minister P. M. K. Leroux insisted that the ruling on Ronnie would stand. 'He has never been a White person,' sniffed Le Roux. Then, with logic reminiscent of the Mad Hatter the minister added 'And I do not believe he will ever become one' " (*Newsweek*, 2/27/67: 42).

In another set of cases, the race borders were crossed in another direction. Two white children, Jane-Anne Pepler and Johanna de Bruin, developed Addison's disease (a severe malfunction of the adrenal glands, which causes the skin to turn brown). Jane-Anne had an operation to remove her diseased glands when she was fifteen years old. Suddenly, her skin and hair went from fair to dark brown. Her mother reported that: "only close friends and family who knew her before the operation know she is White" (*Newsweek*, 7/3/70: 31). Her mother said with unconscious irony, "Some of her school friends have ostracized her completely—just as though she were a real non-White" (*Newsweek*,

7/3/70: 30). Johanna contracted Addison's disease in infancy. "No White school would accept her when she reached school-going age. Her father told a reporter that he intended applying to the Education Department for a tutor to teach her at home until she had passed Standard V, after which she would be able to take correspondence courses" (Horrell, 1970, p. 26). Her mother "lives in constant fear that, because of the past difficulties, 'someone' will come and take her daughter away from her" (Wannenburgh, 1969, p. 15). In both cases, the children are stuck with the rigidity of the apartheid classification of race.

The case of Sandra Laing drew sustained international attention. Sandra was born to White parents. She had several features traditionally considered Black African. At the age of ten, she was abruptly expelled from her White school because she was considered colored. The United Nations reported that when she was expelled by school officials under the Population Register Act, it became illegal for Sandra to attend her Piet Retief boarding school, which was all-White (United Nations Office of Public Information, 1969, p. 4). The only way Sandra could continue living with her family was by being registered as a servant.

Although the family appealed the decision successfully up to the Supreme Court, Sandra's life was permanently torn apart. In 1983 the newspaper, *Rand Daily Mail* reported that Sandra had become completely alienated from her family and community. She "eventually lived with a Black man and, ironically, applied to be reclassified so she could live legally with her lover" (*Rand Daily Mail*, 7/23/83, p. 10).

CONCLUSION: HOW TO FIND A COMMON LANGUAGE FOR CATEGORIES AND PRACTICES

The two examples above demonstrate some of the ways that categories may stand between areas of knowledge and political practice serving as boundary objects, doing invisible work, and providing infrastructure. Although few examples are as dramatic as the reclassification of people under apartheid, the more general problem of the interweaving of beliefs, politics, ethics and culture directly into disciplinary infrastructure is important for the cognitive and social analysis of interdisciplinarity.

Classifications are important for several reasons that appear in all category systems. They are *ubiquitous* in modern society. Every organization, every building, journal, network and other infrastructures underlying disciplinary and interdisciplinary communication contains a myriad of categories and standards. The crafting of these standards and category systems are always imbued with local, practical politics and disparate viewpoints. At the same time, incomplete records are kept of the process of creating categories and often of how instances are classified into those categories. Thus, classifications remain historically indeterminate as traceable records. This in effect erases much of the history of interdisciplinary work. Finally, rather than being an undesirable side effect of inadequate science, ambiguity and "the other," or residual categories, play complex parts in managing organizational practice. Sometimes this is justified, sometimes not—but it is always important.

Future work on interdisciplinarity, in cognitive science and other fields, should take up the job of understanding the interplay between design and use in classification systems, and in standardization.

REFERENCES

Bowker, G. C., & Star, S. L. (1994). Knowledge and infrastructure in international information management: Problems of classification and coding. In L. Bud-Frierman (Ed.), *Information acumen: The understanding and use of knowledge in modern business* (pp. 187–216). London: Routledge.

Bowker, G. C., & Star, S. L. (1999). *Sorting things out: Classification and its consequences.* Cambridge, MA: MIT Press.

Cole, M. (1996). *Cultural psychology: A once and future discipline.* Cambridge, MA: Harvard University Press.

Cornell, M. (1960). The statutory background of apartheid: A chronological survey of South African legislation. *The World Today, 16,* 181–194.

Dewey, J. (1916). *Logic: A theory of inquiry*. New York: Open Court.

Engeström, Y. (1990). *Learning, working and imagining*. Helsinki: Orienta Konsultit Oy.

Fagot-Largeault, A. (1989). *Causes de la mort: Histoire naturelle et facteurs de risque*. Paris: Librairie Philosophique J. Vrin.

Hall, R., & Stevens, R. (1995). Making space: A comparison of mathematical work in school and professional design practices. In S. L. Star (Ed.), *The cultures of computing*. Oxford: Blackwell.

Horrell, M. (1958). *Race classification in South Africa: Its effects on human beings* (p. 19). A Fact Paper published by the S. A. Institute of Race Relations, No. 2.

Horrell, M. (1968). *A survey of race relations.* Johannesburg: South African Institute of Race Relations.

Hutchins, E. (1995). *Cognition in the wild*. Cambridge, MA: MIT Press.

Keller, C. M., & Keller, J. D. (1996). *Cognition and tool use: The blacksmith at work*. Cambridge: Cambridge University Press.

Kirk, S. A., & Kutchins, H. (1992). *The selling of the DSM: The rhetoric of science in psychiatry*. New York: Aldine de Gruyter.

Latour, B., & Woolgar, S. (1979). *Laboratory life: The construction of scientific facts*. Thousand Oaks, CA: Sage.

Lave, J. (1988). *Cognition in practice: Mind, mathematics, and culture in everyday life*. Cambridge: Cambridge University Press.

Lave, J., & Wenger, E. (1991). *Situated earning: Legitimate peripheral participation*. Cambridge: Cambridge University Press.

Lelyveld, J. (1985). *Move your shadow: South Africa, Black and White* (p. 82). New York: Times Books.
Mathabane, M. (1986). *Kaffir boy: The true story of a Black youth's coming of age in apartheid South Africa*. New York: Macmillan.
South Africa: A shade of difference. (1967, February). *Newsweek, 69*.
South Africa: White at heart. (1970, July 3). *Newsweek, 17*.
Race laws are condemned. (1966, December 14). *Rand Daily Mail, 17*.
Star, S. L. (1983). Simplification in scientific work: An example from neuroscience research. *Social Studies of Science, 13,* 205–228.
Star, S. L. (1991). Power, technologies and the phenomenology of standards: On being allergic to onions. In J. Law (Ed.), *A sociology of monsters? Power, technology and the modern world: Sociological review monograph, 38*, 27–57. Oxford: Basil Blackwell.
Star, S. L. (1992). Craft vs. commodity, mess vs. transcendence: How the right tool became the wrong one in the case of taxidermy and natural history. In A. Clarke & J. Fujimura (Eds.), *The right tools for the job: At work in twentieth century life sciences* (pp. 257–286). Princeton, NJ: Princeton University Press.
Star, S. L. (1989). *Regions of the mind: Brain research and the quest for scientific certainty.* Stanford, CA: Stanford University Press.
Star, S. L. (Ed.). (1995a). *Ecologies of knowledge: Work and politics in science and technology*. Albany, NY: SUNY Press.
Star, S. L. (Ed.). (1995b). *The cultures of computing. Sociological review monograph series*. Oxford: Basil Blackwell.
Star, S. L. (1995c). The politics of formal representations: Wizards, gurus, and organizational complexity. In S. L. Star (Ed.), *Ecologies of knowledge: Work and politics in science and technology* (pp. 88–118). Albany, NY: SUNY Press.
Star, S. L., & Griesemer, J. R. (1989). Institutional ecology, translations and boundary objects: Amateurs and professionals in Berkeley's Museum of Vertebrate Zoology, 1907–39. *Social Studies of Science, 19,* 387–420.
Star, S. L. (1996). From Hestia to home page: Feminism and the concept of home in cyberspace. In N. Lykke & R. Braidotti (Eds.), *Between monsters, goddesses and cyborgs: Feminist confrontations with science, medicine and cyberspace* (pp. 30–46). London: ZED-Books.
Star, S. L., & Ruhleder, K. (1996). Steps toward an ecology of infrastructure: Design and access for large information spaces. *Information Systems Research, 7*(1), 111–134.
Star, S. L., & Strauss, A. (1999). Layers of silence, arenas of voice: The ecology of visible and invisible work *Computer-Supported Cooperative Work: The Journal of Collaborative Computing, 8,* 9–30.
Suchman, L. (1988). Representing practice in cognitive science. *Human Studies, 11,* 305–325.
Sunday Times. (Johannesburg). (1995, August 21). Concern at methods of classifying coloureds, p. 4.
Thomas, W. I., & Thomas, D. S. (1970). Situations defined as real are real in their consequences. In G. P. Stone & H. A. Farberman (Eds.), *Social psychology through symbolic interaction* (pp. 54–155). Waltham, MA: Xerox Publishers. (Originally published 1917)
United Nations Commission on Human Rights. (1968, December 17). *Study of apartheid and racial discrimination in Southern Africa: Report of the special rapporteur of the Commission on Human Rights* (United Nations Documents E/CN.4/979 and Add. 1 and 6).

Wannenburgh, A. J. (1969, March 9). Parents fight to have daughter accepted as White. *Sunday Times,* (Johannesburg), p. 15.

Wenger, E. (1998). *Communities of practice: Learning, meaning and identity.* Cambridge, England: Cambridge University Press.

7

Schema (Mis)Alignment in Interdisciplinary Teamwork

Lori Adams DuRussel
Sharon J. Derry
University of Wisconsin–Madison

Everything is vague to a degree you do not realize till you have tried to make it precise.
—Bertrand Russell

Interdisciplinary teams often form to bring multiple disciplinary perspectives to bear on a single, domain-spanning problem. However, because individuals from different disciplines bring different disciplinary-based prior knowledge to their problem conceptualizations and solution processes, it may be very difficult for such groups to reach a point at which members see the team and its work in a similar way. This can be problematic, for both research and common sense indicate that successful team-based problem solving requires a substantial degree of congruence among team members' conceptualizations of the team and its work. In this chapter, we present a qualitative interpretive case study of one interdisciplinary team that did not successfully complete its task, seeking evidence for and against the hypothesis that their difficulties were associated with disciplinary-based misalignments in their conceptualizations of the team and its work. We also discuss whether certain observed features of the team's process and organization might have hindered its ability to become aware of misalignments and overcome them through communication.

Some prior research on interdisciplinary teamwork has focused on participants' mental models, the dynamic internal representations that comprise each person's evolving understanding of a particular situation such as a team meeting. The cognitively based concept of a mental model (e.g., Gentner & Stevens, 1983; Johnson-Laird, 1983) as well as other similar concepts from organizational literature such as situation awareness (Salas, Prince, Baker, & Shrestha, 1995) are here treated as particular instantiations of general, reusable mental representational systems that we call *schemas* (e.g., Bartlett, 1932; Brewer & Nakamura, 1984; Derry, 1996; Rentsch & Hall, 1994). An individual's schematic prior knowledge of past similar circumstances helps shape his or her mental model of a particular situation. The particularized mental model serves as a scheme within which a person understands a particular situation and learns within it (Alba & Hasher, 1983). Mental models assimilate and account for situational stimuli and are dynamic and changing over the course of a situation such as a team meeting. As situational learning occurs, mental models also influence and change the content and structure of the more permanent schemas they instantiate, thus possibly reshaping an individual's expectations for future situations. As a person communicates and accumulates new experiences within a particular situation, his or her more permanent, overarching schemas evolve to reflect any newly acquired general understandings of a more enduring nature.

We propose that the ability of team members to work together successfully over time is dependent to an extent on the compatibility, or alignment, of situational mental models as well as the evolving schemas they bring to bear on their work. Schema alignment can be conceptualized as team members' sharing similar enduring representations of a team's work; mental model alignment in particular situations is more likely to occur when similar schemas are brought to bear (Rentsch & Hall, 1994). If team members have schemas and mental models that are reasonably aligned, they will likely be more successful in coordinating team work (Cannon-Bowers, Salas, & Converse, 1993; Orasanu & Salas, 1993). Alignment enables team members to approach their problem from complementary perspectives (see Bolman as cited in Salas et al., 1995; Stout, Cannon-Bowers, & Salas, 1994). The existence of common features among team members' perspectives has been shown to influence team decision making (Cannon-Bowers et al., 1993; Orasanu & Salas, 1993; Salas et al., 1995), with "complementary" mental models being associated with more successful performance (Endsley as cited in Salas et al., 1995; Hartel, Smith, & Prince as cited in Salas et al., 1995).

Interdisciplinary teams, which often are designed with maximum participant diversity, face special challenges in facilitating the negotiation of adequate similarity among participants' mental models and developing work-related schemas. The process of communicating within a team itself may increase schema and mental model alignment among team members as participants share common experiences, learn from one another, and are exposed to similar stimuli in the team environment. However, this is not guaranteed. Without explicit discussions of team goals—which are not common in most teams (Hackman, 1987)—team members may not be aware when they hold and must overcome different schema-based models related to their work (Bettenhausen & Murnighan as cited in Hinsz, Tindale, & Vollrath, 1997; Kim as cited in Hinsz et al., 1997). Interdisciplinary teams would seem particularly vulnerable because different disciplines have different goals, different normative practices for communication and work, and different technical languages that may in fact employ identical terms that have different meanings in different fields.

It is not clear what kinds of schemas team members must hold in common to be successful. Previously cited work by Cannon-Bowers et al. (1993) implicated misalignments in task-related and team-interaction models. In defining the nature of task and team schemas (and models) for this study, we were influenced by situated cognition and activity theory frameworks (Engeström & Middleton, 1996; Lave, 1988; Nardi, 1996), which conceptualize teamwork as an activity system with the following key interacting components. In addition to the team as a whole (a community) and individuals who make up the team, there are the goals or objectives of work; work distribution, which mediates between the team and the team's objectives and specifies who will do what part of the work; tools including conceptual and other analytical tools such as computer programs that mediate between team members' work and team objectives, shaping the outcome; and both implicit and explicit norms and rules including the various social and organizational structures implicitly governing communication patterns. Thus, in examining team members' schemas about their group and individuals' tasks and the team's interactions, we looked for alignments and misalignments in how each member conceptualized the team's goals, the distribution of the team's work, and the conceptual and analytical tools at their disposal. We also report evidence of implicit social rules and structures governing the team's interactions that seemed to influence the formation and evolution (or lack of evolution) of these schemas, as well as members' awareness of whether they were aligned or not.

DESCRIPTION OF THE STUDY

Subject of Study

The Group and Their Task. In this case study, we examined a small interdisciplinary working group within a large "think tank"-type institute. The institute's goal was to examine a variety of issues in the domain of science, technology, engineering, and mathematics (STEM) education. The institute was composed of several teams, each of which conducted research on a different area of STEM education. The team under study, "Team A," was involved in two related tasks pertaining to analysis of 1st-year college STEM education: (a) to identify the major "pathways" through STEM curricula for undergraduate majors and (b) to examine equity issues in 1st-year college STEM education. The first task—identifying student pathways—provided the context for this study. The goal of the task was to use student transcript data (which included information about courses taken and dates completed, grades, and final major) to identify major patterns of course taking in STEM majors. The team assumed that identifying these major pathways would enable the team to identify patterns of interest to STEM educators such as how minority pathways diverged from pathways frequently taken by members of majority groups or how women who dropped out diverged from those who completed their majors.

Team A's participants represented a variety of disciplinary backgrounds including engineering physics, math education, mathematics, social work, sociology, and statistics. The team leader was a male professor in mathematics education. Other members included a male professor of social work and a female assistant professor of engineering, a male academic staff member, and several graduate students, with regular membership ranging from five to six members during the period of observation. The person with primary responsibility for accessing and analyzing the student transcript database was a male academic staff data analyst with a PhD in mathematics and experience working with university databases including transcript databases. Participation on the project was part-time for all team members, with reimbursement ranging from half-time (for the data analyst) to zero-time (volunteer).

Development of a Conflict. At the start of the observation period (about 1 year), team members appeared to be at the stage of defining their goals and desired analyses for the pathways task. The first observed meeting early in the project involved three team members (not including the analyst) reviewing initial data reports provided by the analyst and outlining goals and possible approaches for the analysis. At subsequent meetings, the analyst brought to the group analyses and reports of student transcript data for discussion.

However, these reports were not well received by other team members. They seemed to have difficulty understanding the reports, and they expressed disagreement with the approach taken by the analyst. Approximately 2½ months into the observation period, the analyst left the team. After the initial analyst's departure, responsibility for the transcript pathway analysis was passed to other analysts (graduate students). However, no additional progress was made in data analysis, and the transcript analysis was never completed. The group as a whole was disbanded approximately 1 year after the start of our observations.

Goal of Our Study

The focus of the work reported in this chapter was to examine and compare team members' individual schemas about the team and its work during the period of the perceived conflict during the tenure of the initial data analyst. In this chapter, we summarize our findings based on a qualitative interpretive examination of team meeting discourse, interviews with individual team members, and e-mail and interview feedback from some team members who read and responded to drafts of this chapter in development. We specifically attempted to assess and describe the extent to which individual conceptualizations regarding key components of teamwork as those components are specified in others' past research and within an activity theory framework were either aligned or misaligned during this period. From this, we attempted to draw informed hypotheses for future work about how potential misalignment during an early development period might impact a team's outcome and to speculate and generalize about how similar misalignments might in future be avoided, or recognized and overcome, through team management.

Data Collection

Data about team interaction were collected through audio recordings of team meetings and observer notes (collected by either one of us or a fellow researcher). After an initial 3-month period of observation, interviews were conducted with team members to help uncover their views about the team and the observed conflict. Additional interviews were conducted both as a part of the ongoing study of the team (which continued for approximately 1 year) and in conjunction with the evolution of this analysis (as drafts of this analysis were distributed and feedback was obtained). All recorded data from team meetings and interviews were transcribed in preparation for analysis.

Approach to Analysis

Our analysis of team members' schemas is based on a qualitative examination of issues and topics raised in team meetings and interviews. Because this analysis focused on the pathways study and the conflict that arose between the initial data analyst and other team members during the early part of the team's work, we focused on transcripts related to that issue, which were from early (the first eight) team meetings. Team conversations on the topic of the equity paper (the team's second main task) were not included in the analysis unless they specifically mentioned the pathways work or appropriated pathway-related terminology and thus were considered to have potential links to the team's pathways study.

Transcripts of team meetings and interviews were coded by conversational turn to identify talk on specific topics of interest. Because the pathway metaphor was explicitly adopted as a name for the team and the analysis itself, we looked for discourse regarding participants' conceptualizations of a pathway and how it was related to the transcript analysis task. Based on our previously described activity theory framing and others' past research, we looked for talk about goals and scope of the team's pathways task, approaches to analysis, distribution of work among team members, and group social norms for working. Participants' talk on each of these topics produced a reduced transcript, which was further analyzed to characterize schemas that team members appeared to be bringing to the group work.

In characterizing participants' conceptual viewpoints, we connect our work to earlier efforts by Cannon-Bowers et al. (1993), who analyzed what they referred to as task-related and team interaction models. Their analysis of task-related models is equivalent to our analysis of how team members characterized the pathways task, including an understanding of task goals, the appropriate or expected ways of approaching these goals, and the tools needed to accomplish them. As implied by activity theory, members' schemas about the task should be highly interactive with schemas related to what Cannon-Bowers et al. termed *team interaction models*. For example, a team member's use of particular conceptual and analytical tools in defining a task becomes important if that team member is either seen by others as responsible for accomplishing that task (task distribution) and/or if team members with different points of view are involved in judging the adequacy of task performance (group norms). Schemas about work distribution and group norms thus were expected to interact with task-related schemas in interesting ways, and our analysis sought evidence for and described such interaction.

We constructed our conception of participants' evolving schemas based largely on comments they made during meetings and interviews and also (as we discuss later) by attending to observable but sometimes unvoiced social norms within which participants operated. Consider, for example, the exchange in Protocol 1, which was coded as dealing with task distribution (it also deals with analytical approach and group norms):

Protocol 1: Meeting Observation

KU (analyst):	[We should] look at some of it and as you see it, say, "Oh, wait a minute. Let's do a graph of this." ... It's a question of how the group wants to go through those figures....
DL:	Right, that's what I was getting at.
HC:	I think that part of what we should be relying on [the analyst] for is to go through the manuals for us and present some of the more interesting stuff. We can be, as a group, involved in the data analysis, [but] that's not efficient and that's using [the analyst] as just doing the data prep, which is a lot less than what he can be doing.

Based on Protocol 1, participant KU's schema of task distribution assigns the job of analyzing individual transcripts to all team members, whereas

HC's schema assigns analysis of individual transcripts to the analyst alone. Foreshadowing our results, we found that the most significant differences were between this analyst (KU) and the other team members. Excluding the analyst, most team members expressed views that reflected relatively compatible schemas related to task and team. However, the analyst's comments and actions indicated that his schemas deviated substantially from the group norm. As a result, in this chapter we often contrast the analyst's schemas with those held by most other participants. This characterization of other members as a singular unit with identical viewpoints downplays some of the differences between their individual schemas that we felt were unimportant for this analysis. However, we do not claim that all other team members were identical in their representations of the task and team interactions, although they were similar in some key respects related to our hypotheses.

MISALIGNMENTS IN TASK DISTRIBUTION SCHEMAS

We defined a *schema of task distribution* as a participant's collective ideas and expectations for the distribution of responsibilities within the team. In certain respects, team members' schemas regarding task distribution (and in particular, their view and justification of the analyst's role in the team) were compatible: The analyst had an advanced degree in mathematics and experience in working with university data sets. He was hired to do detailed work with the transcript analysis. Team members acknowledged the analyst as the expert on the transcript database, and it was agreed that he, not individual team members, would take primary responsibility for much of the work on the analysis.

Notwithstanding elements of alignment among participants' schemas involving the analyst's role, interactions at initial meetings and comments from retrospective interviews revealed significant differences with the analysts' viewpoint. Table 7.1 illustrates some major elements of alignment and misalignment in participants' task distribution schemas. The data columns indicate the extent to which the participants viewed each task as fully distributed (among all team members), distributed among all members except the analyst, or predominantly the responsibility of the analyst. Highlighted, italicized rows designate the areas of widest misalignment between the analyst and other team members.

As shown, although schemas were aligned in some important ways, the analyst's task distribution schema differed from other team mem-

TABLE 7.1
Schemas of Task Distribution

Task Type	*Task*	*Task Distribution According to* *Analyst's Schema*	*Others' Schema*
Non-data-related	*Conceptual discussions*	*Distributed among nonanalysts*	*Fully distributed, including analyst*
	Literature review	*Distributed among nonanalysts*	*Fully distributed including analyst*
Data-related	Data access	Mainly analyst; some distribution	Analyst only
	View and summarize data on individual transcripts	*Fully distributed*	*Analyst only*
	Analysis of summary data	Fully distributed	Fully distributed
	Review of analyses	Fully distributed	Fully distributed
	Investigation of new analytical techniques	*Not applicable*	*Analyst only*

bers' schemas particularly in terms of how they represented the analyst's involvement in important conceptual and interpretation stages of the analysis.

Analyst's Role in Nondata Issues

Team members differed from the analyst in their views of how the analyst—an acknowledged expert in database analysis and a PhD-level mathematician—would participate in various tasks within the team. For example, in the following excerpt from an interview below, the analyst described his role in purely technical terms, indicating that he left the job when the group wanted him to expand his duties to include reviewing literature:

Protocol 2: Interview

KU (analyst): I was hired to uh do database stuff ... and analysis and statistics ... just as somebody who ... knows the [university] database, and being familiar with those prob-

> lems [specific to the database].... [Then] there was a big shift [to sorting through literature] .... At that point ... I said, "Well, look, I'm not the person you want for that job. I'm a mathematician. I don't know anything about science and education research, ... I don't even know the name of one journal."

The analyst thus defined his role in relatively narrow terms, which did not include responsibility for non-data-related tasks such as literature searches. Such protocol data helped support our view of the analyst's self-characterization as a technician.

Other team members' schemas included a more expanded view of the capabilities and potential of the data analyst. These schemas were shaped both by their knowledge of the analyst's background as a PhD-level mathematician and by their general expectations for team member participation in multiple tasks. Team members expressed that the analyst's work did not match the quality or scope that they had expected from a professional mathematician (see Protocols 3, 4, and 8). They expected that given his math background, the analyst would be able to provide information and expertise that extended to tasks beyond crunching data. For example, although the analyst did not expect to have to do nontechnical work (such as a literature review), at least one other team member explicitly questioned his lack of participation in broader conceptual discussions.

In addition, other team members expected the analyst to provide expert knowledge to the group about various mathematically valid approaches to analysis. This view was expressed by one team member in the interview segment shown in Protocol 3:

Protocol 3: Interview

DL:	From our professional experience [we knew] that pattern classification is a whole discipline in itself. [It includes things like] neural networks, and fuzzy logic. There is a whole bank of mathematical techniques.... Our first impression was that he was gonna kind of figure this thing out.... There is a body of mathematical literature, and he was a mathematician and so we were looking for ... something along, along those lines.

It was not clear whether group members clearly communicated their expectations to the analyst about his role in providing or developing tools for the analysis. The analyst never mentioned new tool development or identification as one of his responsibilities. This topic was never explicitly discussed at any recorded meetings, and in interviews, no team member mentioned that specific requests on this topic had been made. Based on our data (which did not cover all possible team communications), it was likely that this mismatch of expectations was not explicitly identified or discussed; clearly it was never resolved.

Division of Labor in Data-Related Tasks

Although all team members recognized that the analyst was the designated data expert, it was less clear how other group members were to participate in data-related tasks. Team members' models of the task distribution were misaligned in their portrayal of which tasks would be shared by all team members and which were the responsibility of the analyst alone.

Initially, the data analyst was the only team member with access to or detailed knowledge of the transcript database. However, in an interview, he indicated that he expected other team members also to work directly with the data. The analyst's model of the task distribution thus claimed primary responsibility for data access, but it also included the possibility of distribution of data access. In addition, he brought unprocessed data to team meetings for others to review. This indicated that his model included the view that all team members would be involved to some extent in discussions of detailed, unsummarized student transcript data.

Other participants' models of the team limited the participation of team members in the handling of detailed data. Only one other team member indicated any expectation of working with the database software, but she did not indicate that she would be responsible for that task. Other participants did not indicate that having more people familiar with the software would have strengthened the group's ability to complete the task. In their models of the team, the analyst alone was responsible for accessing student transcript data.

This misalignment between the analyst's model and other team members' models contributed to the conflict because it resulted in mismatched expectations for how data analysis would proceed. Whereas the analyst produced preliminary reports with the expectation that other team members would aid in the detailed interpretation, other team members—expecting that the analyst would be producing complete analyses independently—judged the quality and quantity of the work as inadequate. For example, at the first meeting at which data were presented (Meeting 2), the analyst presented one report in the form of individual transcript records with the expectation that other team members would help him identify interesting patterns in the data. Other team members seemed to want the analyst to process the information more fully before presenting it to the group. An expansion of Protocol 1, Protocol 4 illustrates the analyst's expectation that the group would aid the interpretive process. It shows how other group members preferred that the analyst process the data more thoroughly before presenting it:

Protocol 4: Meeting

KU (analyst):	We need to decide what we want to go into [pause] Yeah, I think we want to get a feel for the data before we say, let's [pause] I could spend hours generating graphs....
KU:	... I think looking at the transcripts is good because you get ideas of "hmmm, this pattern seems to be coming up, I wonder how many ...." ... It's a question of how the group wants to go through those figures....
DL:	Right, that's what I was getting at.
HC:	I think that part of what we should be relying on [the analyst] for is to go through the manuals for us and present some of the more interesting stuff. We can be, as a group, involved in the data analysis [but] that's not efficient and that's using [the analyst] as just doing the data prep, which is a lot less than what he can be doing.

In Protocol 4, the analyst (KU) noted that generating graphs and other aspects of data preparation could be time consuming, but he did not question that his role included those tasks. However, he seemed to expect more guidance in identifying the elements that should be selected for additional investigation. For example, he expected group members to jointly participate in pattern identification and thus determine which patterns to graph. The analyst's model of the group as including multiple participants in detailed analysis was supported in a subsequent interview:

Protocol 5: Interview

KU (analyst):	I confronted people directly and I said, ... "Let's look at transcripts, let's actually look at transcripts." You know that was an idea I had .... Like for example I said, "There is only 40 grad, seniors, female seniors who started in engineering that are graduating this semester. Why don't we just ... look at the path that they took .... [But] nobody wanted to do it ....
LD (interviewer):	They never indicated why, or?
KU:	Nope.... I thought everybody on this team should look at course histories. Uh, and it the it pretty well resounded of "No, we are not."

As Protocol 5 indicates, the analyst's view of task distribution involved everyone looking at detailed data. The group's refusal to look at data was unexpected and frustrating for the analyst.

In contrast, other team members' schemas gave the analyst more responsibility for doing independent work. For example, in Protocol 4, HC explains how an efficient team organization would involve the analyst processing information more fully before presenting data to the group. Based on this model of task distribution, the analyst's presentation of minimally summarized data constituted a lack of progress and attention to his task.

Influences on Schemas of Task Distribution

Participants' views of the scope of the data analyst's role were shaped both by their knowledge of the analyst's background and by the team organizational structure. As mentioned previously (and as shown in Protocol 3), other participants believed that as a professional mathematician, the analyst would be able to use his math expertise—not just technical experience—in multiple tasks. In addition, team members' schemas of the analyst's role were influenced by the fact that he was paid, whereas other members were only minimally reimbursed. Given his official assignment within the team, they expected that he would do more work than members who were not paid. The analyst, in contrast, seemed to base his expectations for task distribution mainly on his status as a technical expert, thus narrowing his scope of responsibility.

The meetings involved some interactions that may have been attempts to resolve these misalignments. For example, in Meetings 1, 2, and 3, team members requested that the analyst bring more data summaries to the group. This reflected their expectations that the analyst process data more fully before presenting it to the group. The analyst complied with this request to an extent. To Meetings 2 and 3, he brought both the requested summary statistics and individual transcripts for the group to look at, and by Meeting 7, he no longer brought any individual transcripts for the group's perusal. This may have reflected some progress toward alignment in views regarding task distribution and the analyst's role. However, interview and meeting data showed that the summaries produced by the analyst did not meet group expectations. We discuss this issue further with presentations of Protocols 7, 8, and 12. The mismatched expectations about the analyst's scope of work were never sufficiently resolved in the group. One team member reported that team norms of politeness precluded public criticism of the analyst, so the issue was never explicitly mentioned in team meetings. Although it is possible that the issue was discussed outside of team meetings and never mentioned in an interview, none of our data indicate that there was any communication on this issue. Consequently, the analyst continued to think of himself in a support role as a technician rather than as researcher.

These unresolved misalignments between the analyst and other team members contributed to the analyst leaving the group. Both the analyst and other team members expressed frustration at what each saw was a lack of response on the part of the other with respect to the amount and type of work done. Had there been awareness of possible misalignments regarding task distribution, and had this possibility been explicitly discussed, it is likely that team members' task-distribution schemas would have become more aligned, and the transcript task might have continued more successfully. Possible outcomes might have included the analyst's more carefully analyzing the goals and ideas of other team members. Or both the group and the analyst might have come to accept the analyst as having authority for choosing opening analytical approaches. Or the group might have agreed on the necessity for everyone's devoting time to collaborating in the processing of individual transcript information and conceptualizing next steps. Or the group might have begun to view the analyst more as technical support, accepting more responsibility for directing him. Many resolutions for the task-distribution problem were possible. However, this aspect of the conflict never surfaced in team awareness or discussions, and there were still significant differences in task-distribution schemas when the analyst left the group.

MISALIGNMENTS IN SCHEMAS OF THE PATHWAYS ANALYSIS TASK

In addition to misalignments in their task distribution schemas, participants in Team A operated with misaligned schemas of the task itself, that is, the team's research goal and how it would be accomplished. The shaded, italicized rows of Table 7.2 illustrate the participants' differing views of what the task actually was.

As Table 7.2 shows, we concluded that participants' task-related models were misaligned on two elements. Although members agreed that the broad task was to identify pathways, we found that they did not have a consistent view of what a pathway was and thus did not have a clear view of what approach would be used and what the units of analysis would be. Also, participants' models of the task differed in their description of how software would be used in the task. The role of software is important; both situated cognition (e.g., Lave, 1988) and activity (e.g., Nardi, 1996) theory predict that tools have a significant shaping effect on how a task is conceptualized. Differences in these views contributed to conflict between the analyst and other team members as well as to lack of progress on the transcript pathways analysis task.

TABLE 7.2
Some Features of Participants' Schemas of Team Task

	Analyst's Model	*Team Leader's Model*	*Others' Model*
Overall goal	Identify major "pathways"	Identify major "pathways"	Identify major "pathways"
Type of pathways to be identified	Aggregate (groups)	Aggregate	Aggregate
Appropriate unit of analysis	*Ambiguous*	*Individual paths*	*Individual paths*
Role of software tool in shaping task/method	*Use existing software*	*Find/use graphical software*	*Find or develop new analytical software*

Ill-Defined Definitions of Pathways

In reasoning with analogy, a base idea (i.e., pathway) is applied to a target concept (i.e., student trajectory through school) to clarify the target. Participants' schemas of the task included a representation of what pathways, the base idea, were and how they should be identified in transcript data. The quality of the analogy between a pathway and education thus was a factor influencing the clarity and content of participants' task schemas. The task-related schemas were unclear in part because of ambiguous definitions of what a pathway was and how this analogy could be used to describe progress in school.

Ambiguous Base Analogy. Team A's participants shared a view of a generic "commonsense" pathway that included a person moving in a direction while passing through a series of steps. Protocol 6 (taken from interviews) illustrates typical definitions:

Protocol 6: Interviews

BC:	I've always thought of it as the route that students take from starting college to when they get their major.... Sort of where they start and where they end. And the metaphor really is the path they took from start to finish.
DL:	[A pathway is a student going] course to course, decision point to decision point of their major. [The pathway] is seen as this totally rigid unchangeable thing.... I enter here, I exit here.

Although participants thus shared a common model of generic paths, they varied in their characterizations of specific types of pathways. For example, some characteristics—such as the degree of flexibility or rigidity implied by a path—were mentioned in multiple meetings and by multiple participants. However, others—including determinism, directionality, and continuity—were shared infrequently and by only selected members of the group.

The variance in descriptions of pathways indicated that participants' characterizations of pathways did not comprise a definition of a single phenomenon (i.e., "what a pathway is") but rather views of various types of possible pathways (i.e., "what types of pathways exist"). As such, it was difficult to attach particular characteristics to a singular view of what a pathway was, and the term was periodically ambiguous.

Team members' different interpretations and applications of the pathway analogy seemed linked to differing degrees of experience with applying the pathway concept to educational issues. For instance, the team leader had extensive experience in reflecting on the pathway analogy and its relation to educational domains. He thus represented an "expert" in the pathway analogy, and his representation of a pathway was conceptually rich and detailed. In addition, in his professional domain (educational research), the pathways concept was common and was viewed as important and interesting. Other participants' work was not closely linked to the idea of a pathway, and their descriptions of pathways were correspondingly less articulate. Several members indicated that other metaphors (those used in their professional domains) provided more insight into student experiences. These different initial representations in turn resulted in differing incentives to apply the pathway analogy. As a result, although the pathway analogy was theoretically discipline neutral, people in different disciplines interpreted it differently. Because the analogy served as a base for conceptualizing the team task, ambiguities in definitions of pathways greatly "fuzzified" the team's task.

Multiple Analogical Targets. Participants' target application of the pathway analogy also was ambiguous. The term *pathway* was used to describe two levels of student experience: individual pathways and nonindividual or aggregate pathways. In our analysis, a comment was identified as referring to an *individual pathway* if it indicated that a pathway was unique to an individual or determined by an individual's choices. An *aggregate pathway* was identified in comments implying that the path was not based on experiences of a single individual. This included references to group pathways (common paths taken by a group of students) and institutional pathways (expected pathways defined by a major department or university). For example, references to "STEM major paths" or to paths in which the "owner" was a group or institution were counted as aggregate pathway references. Although both types of analogical targets were mentioned throughout the team meetings, team members typically did not specify which target (individual or aggregate) was implied when the term *pathway* was used. This made references to pathways potentially ambiguous.

This ambiguity in the target's level of specificity contributed to difficulty in applying the analogy. In particular, the team's research questions (requiring an aggregate view of pathways) and the intended analytical approach (requiring an "individually based" conceptualization) implied different levels of analysis. The team's primary research questions involved analyzing aggregate pathways. For example, participants wanted to compare paths through various majors and to examine the impact that a major path's flexibility or rigidity had on retention of students from different backgrounds. In general, the group's focus on equity drew attention to the pathways of groups of people (such as minorities). Thus, the research questions invoked a sense of searching for aggregate pathways both in the sense of identifying minority paths through school and in the sense of looking at deviations from expected major paths. As Table 7.2 indicated, participants seemed to agree on this interpretation.

However, four out of seven team members described pathways as individually based, and at the first meeting, participants indicated that an individually based approach should be taken in the analysis. This exchange is shown in Protocol 7:

Protocol 7: Meeting

DL:	What I was thinking is what [the analyst's] doing is looking at percentages. He's not tracking the person throughout.
HC:	Right, it's not really path-oriented.
SY (leader):	It's clusters.
DL:	It's clusters, and it's, you know, one thing, people approximate retention measures that way.... And if we could get him to think about whether he can truly plot data sets per person and really measure a longitudinal path-based retention, and also along these lines we can see where the dropouts occur along the path.
SY: (leader)	So what will we do? Tell him to go work with essentially engineering majors or with physics majors and say, okay, look at the pool that clicks off these two or three things and begin to track them individually? What will we ask him to do?

DL:	... Let's see how he framed it: "The 60% who get B or above in calculus, they go on and receive their engineering major." That's how he framed the query.
HC:	Yeah, but that's just looking at things as, that's not looking at a path, that's just a static thing and he's just looking at the percentage ... without looking at the individual paths.

At this meeting, group members criticized the initial analyst's reports for providing summary statistics (aggregates) instead of tracking individuals and thus not being "path oriented." Because these team members' task representations at this time were based on a view of individual pathways, the analyst's work did not meet their expectations.

However, interviews with both the initial data analyst and subsequent ones indicated that it was not clear how to obtain an aggregate level conclusion (to answer the research questions) while maintaining an individually based analysis. As the first analyst indicated, "There's so many paths that students take through the university [and] almost everybody has a unique path ... so you HAVE to group things." Thus, in the analyst's task model, aggregation was necessary to make sense out of individual pathways. In a later interview, a subsequent student analyst also expressed that "the object is to look at groups ... and how groups go through the pathways differently." In neither case was there clarification of the distinction between individual and aggregate pathways or of how the analysis might group students or use analytical techniques to bridge the gap between the two levels.

The misalignment between aggregate-level research goals and individual-level analytical approaches reflected internal ambiguity in team members' task schemas. Most members' task schemas defined the desired approach at one level (i.e., that the analysis should be individually based). This was then used as a criterion for judging the analyst's work. However, team members did not reconcile the discrepancy between their desire for an individually based approach and their goal of answering questions about aggregate pathways. When the analyst produced reports at one level (e.g., a report couched in summary statistics), it was easy for other team members to criticize it based on one element (their desire for an individual analysis) without seeing its alignment with another element (the aggregate nature of the research questions). The issue of how to use individual transcripts to obtain conclusions about aggregate paths was never clarified by any group member; there was no evidence that members' schemas of the task involved connecting the gap between research goals requiring aggregate data and analysis approaches requiring individual paths as basic units of analysis

Misalignments Related to Software's Role in the Task

Team members' task schemas also differed in their view of how much influence available software should have on the data analysis. Table 7.2 summarized these differences. The analyst initially had access to a tool that enabled queries of the transcript database. His comments and actions indicated that his task schema accepted the existing software as a constraint on the type of work that could be done. Other members' schemas of the task, however, included the possibility that different analytical software might provide an easier way to identify pathways in the transcript data.

For example, in an interview, the team leader described the ideal software as a program that would have converted the numerical data into graphical pathways. This potentially would have enabled the data analyst to present results in a format that matched stereotypical graphical views of what a pathway is, thus facilitating the analysis for the entire group. The team leader perceived the lack of such software as an impediment.

Other team members (those with experience in engineering and statistics), however, pointed out in interviews that analytical software (not necessarily graphical in nature) was already available for a variety of mathematical and statistical analyses. As a person with an advanced degree in mathematics, the data analyst was expected by those team members to have familiarity with and access to a variety of mathematical tools that could be applied to this type of problem. For example, Protocol 3 illustrated one participant's expectations that the analyst would be able to find alternative tools. In Protocol 8, the same participant expressed the view that even if specific applications did not exist, the analyst should have been able to create what he needed:

Protocol 8: Interview

DL:	A couple of us had a lot of experience, just with software mathematical modeling [or] whatever. And so ...

> the fact that there weren't computer codes available was a nonissue. Because anybody who does any computer modeling takes what he or she has and makes it into what he or she wants....
> And myself and the graduate students who had our hands deep in computer and mathematical modeling ... [we know how to] write our own programs ... that that's what we do. And so, we were seeing this as a nonissue.

Participants' schemas thus differed with regard to the role of software in the pathways task. The analyst's schema was influenced by the abilities of the current database access software. It is unclear whether he was unfamiliar with other potential applications or whether he had discounted their use for some reason. In contrast, the team leader expected that additional tools—particularly graphical ones—would have facilitated the analysis. A third participant—DL—had a schema that allowed for the analyst creating his own tools. These three members, representing three categories of membership (analyst, leader, participant), thus had task schemas that were misaligned.

The misalignments in participants' task schemas seemed related to their differing professional backgrounds. The analyst, who came to the group with experience using the existing database software, selected analyses that could be done easily given that software. Participants such as DL, with backgrounds in computer modeling or engineering (which often involved tool development), had task schemas that discounted the limitations of existing software and included the possibility of the analyst developing or identifying other analytical tools. The team leader, an educational researcher whose main experiences with software were as an end user rather than developer, initially stressed the need for graphical applications.[1] In each case, the participant's prior experience with software—as technician, developer, or user—was reflected in his or her schema of the appropriate role of software in the pathways task.

These initial misalignments likely were exacerbated by a lack of communication among team members. In our records, the analyst did not let other members know which types of tools (if any) had been consid-

[1]In a subsequent response to an initial draft of the analysis, the team leader clarified that analytic software—be it graphical or not—was his highest priority.

ered and rejected. In addition, other group members did not seem to clarify that the analyst was expected to find and utilize other appropriate tools from the mathematics community.

These misalignments may not have caused problems for the team were it not for the fact that they remained uncommunicatative and shaped participants' expectations for the analyst's actions. The analyst limited his actions to those expected within his viewpoint, which was constrained by the existing tools. Other group members perceived the analyst's emphasis on existing software capabilities as inappropriate and indicative of a lack of desire or ability to develop other analytical tools and thus perform what was part of his task according to their schemas.

Misaligned Elements in Startup Approach

Table 7.3 highlights additional elements of participants' task schemas pertaining to the desired initial analytical approach which were revealed in comments made in Meetings 1 and 2. As unshaded rows indicate, participants' views were aligned in philosophy. The shaded italicized row shows they were misaligned in terms of implementation method.

As Table 7.3 indicates, both the analyst and other team members indicated that an inductive approach was appropriate and that the initial goal was to identify individual paths through a selected major. However, the analyst's model defined the task as a manual effort involving his going through the database to identify patterns. Two other participants, however, indicated that they believed the analyst should develop a

TABLE 7.3
Elements of Startup Approach

	Analyst's Schema	*Others' Schema*
Overall approach to be taken in task	Inductive	Inductive
Formulation of original task	Look at small sample of individual data	Look at small sample of individual data
Method of analysis	*Go through records manually using existing software tools*	*Discover or develop automated analytical technique*

method of automatically identifying patterns. As a result, although participants' models were aligned in one important respect—overall desired approach—they were misaligned in their expectations for how the analyst would accomplish the task.

This particular misalignment contributed to participants' rejecting some of the analyst's products that (to the researchers) appeared to match the group's requests. For example, at the end of the first meeting, team members decided to have the analyst track the paths taken by a small group of students in a particular major. At the second meeting, the analyst presented several reports to the group, including a description of patterns that he had identified in a small subset of data. His interpretation of the data and the group's subsequent response are shown in Protocol 9:

Protocol 9: Meeting

KU (analyst):	Two major patterns formed ... [in] the ones that I looked at. Those two major patterns are: ... the majority of the 30 females ... were A students, and they left after one or two semesters.... And the other pattern I was seeing was that people struggled for two, three, four semesters and were getting low grades in key courses and so they probably said, ... "Okay, this is enough, I should probably do something else." So those were the two major patterns. [*In the next segment, the group focused on reading the data tables and understanding who was included in the data set. The conversation is picked up again after the group seems to understand the data presented.*]
DL:	Now, [pause] what are we going to do with this? That's my question.
KU:	Well, I just brought it to [pause].... You can see, for example, that there seems to be two major patterns. Females who left with high grades and females who left after struggling.
DL:	Right, but I need to compare this with the males before I make any kind of—
BC:	Here he is just presenting it as an illustration.
KU:	Yeah ... I'm just presenting it as an illustration. But there's value in that.... Maybe one thing we should say, ... My guess would be that the females who were strug-

	gling got some help and they still ended up struggling and decided to leave. But maybe, some effort should be put to getting some of our strong students to stay.
DL:	Yeah but ... Before we make conclusions like that, I want to know how to really read this thing without ... I really like this but I don't want a conjecture on looking at a few samples here.

Subsequent to Protocol 9, other team members basically ignored the analyst's interpretations of patterns in the data. Instead, they questioned the validity of the analysis by questioning his sampling procedures and classification procedures.

Protocol 9 reflected how misalignment of one element can outweigh alignments in other elements of participants' schemas of the task. In this case, the analyst presented individual transcript data in which he identified a trend or pattern. This represented an inductive process (an aligned element of team members' schemas) that focused on individual data points (also an aligned element). However, the analyst's report was the result of *manually* identifying patterns in a nonstatistically significant data sample. As a result, despite alignments on some elements, misalignments in other schema elements led participants to reject the analyst's product.

TEAM ORGANIZATION AND MISALIGNED WORK-RELATED SCHEMAS

Status and Perceptions of the Pathways Analogy

An analysis of participants' evaluations of the pathway analogy indicated that high-status participants (who were not involved in data analysis details) held either positive or marginally positive views of the analogy, whereas lower status members were more skeptical regarding its value as a framework for analysis. Positive comments—predominantly mentioned by the team leader—concerned the overall value of the analogy as a tool for highlighting global issues and questions. In the leader's view, "Pathways as an analytical tool ... will get us to start thinking about the appropriate [research] questions.... Pathways as a metaphor is appropriate."

In contrast, most of the negative comments about the metaphor expressed dissatisfaction with the analogy's power for detailed transcript analysis. The original analyst never specifically discussed the pathways concept in his interview, and subsequent analysts stated that they did not find the analogy particularly useful. As one indicated, "I didn't think about pathways much at all. It just didn't do a lot for me." Even a higher status participant recognized that the analogy might need to be better defined before it would aid in the analysis.

Participants' roles and status within the group (and their resulting personal goals for participation) thus exacerbated ambivalence toward the pathway analogy. For the team leader and high-status participants involved in conceptualizing and planning the analytical approach, the pathway analogy was seen as a potentially useful tool and was influential in defining the task and setting directions for the research project. However, for team members involved in detailed aspects of the analysis (e.g., working with the transcript database), the analogy was not specific enough to help them in their task, so they did not incorporate it into their schemas for the task. However, their lack of status seemed to prevent them from expressing their doubts in meetings.

Access to Tools

Their differing levels of exposure to the database software and other analytical tools also influenced team members' task schemas. A growing body of cognitive research (e.g., Latour & Woolgar, 1986), based in part on Vygotskian theory (Vygotsky, 1978), indicates that when a tool is used in conjunction with a task, the nature of the tool influences a person's conceptualizations of and performance on that task. Technological tools such as software have been found to have significant effects on group work and learning processes (Herbsleb et al., 1995; Olson, Card, et al., 1993; Olson, Olson, Storrøsten, & Carter, 1993). Thus, distribution of tools within a group may lead to different evolution of task-related conceptualizations. In Team A in which only an analyst accessed the database, software influenced only the analyst's approach, which in turn caused his schema of the task to deviate from those of team members who did not have experience with the database tools. In addition, the analyst's involvement on a related project in which the software was used informed the development of his viewpoint in a way that other team members did not share or appreciate.

Impact of Software Use on Task Schemas. A variety of conversational data indicated that software played different roles in different participants' task schemas. The analyst's discussions in both meetings and interviews included multiple references to what the software could or could not do. Other participants rarely mentioned the software unless they were asking a question about its capabilities. For example, in Meeting 2, the analyst spontaneously referred to the capabilities of the database software six times, whereas other members mentioned the software only in response to the analyst's comments. In addition, in an interview, the analyst's description of his work contained many references to the data and database software including comments about its limitations, the format it presented data in, and the ease with which it could be used. In contrast, other team members mentioned software only occasionally in their initial interviews (approximately one time each), and they did not elaborate on the topic.

At several points, the analyst's models of the analysis seemed to have been strongly influenced by the capabilities of the software. For example, one segment of conversation in the second meeting focused on data sampling procedures. One team member (DL) was concerned that the analyst (KU) was using too small a sample and asked several questions about the feasibility of selecting a larger sample. The analyst's initial response (shown in Protocol 10) comprised an explanation of how the software enabled a smaller sample rather than a statistical justification for the sample that was chosen:

Protocol 10: Meeting

DL:	And so what's the size of the record again? This is it? Or is there more than that?
KU (analyst):	Well it's as big as you want. I chose some stuff that would fit on one page easily enough and be relevant to look at.
DL:	How big would it be if ...
KU:	I mean if you wanted to pull it off from.... Well ... That's the nice thing about Info Access. You don't have to pull it all off at once. You don't have to get it on your PC. You can work with it on there.

Later comments in the meeting indicated that DL's concern about the size of the sample was motivated by statistical reasons. The analyst's response, in contrast, was linked to the capabilities of the tools. At several other points, he also justified choices and selected actions based on what the tools would allow rather than on methodological concerns.

The team leader, as shown in Protocol 11, interpreted the analyst's emphasis on the software tool negatively:

Protocol 11: Interview

SY (leader):	I felt at one point that [the analyst] was more interested in what Microsoft Access could do than he was with what the problem was.
Interviewer:	He was influenced by the tool?
SY: (leader)	I thought that the tool was driving his work. So that he would say to us that "Well I can do this with Access and that with Access." So that it was [what] the tool can produce. And I don't know if he was asking me "Make your question fit with what the tool says" or was saying to me "Well this is all you are going to get. Because this is all the tool can produce." I would look at this as being like many doctoral students where the method drives the question rather than the question drives the method. And I have the sense that the tools [were] driving the way he has been doing things.

The team leader's comments revealed that in his schema of group work, the question, not tools, should drive the analysis. It also showed a frustration with the influence that the software seemed to have on the analyst's work.

The fact that use of technology was limited to the analyst thus set the stage for the development of incompatible task schemas. As the software tool shaped the analyst's schema, other team members' task schemas were influenced more by their general visions of the question and possibly suitable analytical methodologies. Their schemas were not shaped through exposure to database tools, and the resulting differences gave team members different expectations for how work would be done.

Impact of a Concurrent Experience

Although the database software was one tool that shaped participants' schemas of the task, less tangible tools, such as methodological knowledge based on experience in other projects, also influenced participants' task schemas. In particular, the analyst's concurrent experience on a similar project, including the tools and approaches used in that context, influenced the task schema that he brought to Team A's work. Unfortunately, the team leader interpreted the resulting similarity between his work for Team A and his other work negatively. One of his main criticisms was that the analyst's reports were extremely similar to work that the analyst was doing on another project related to university databases. The leader felt that the analyst was not producing enough original work for Team A .

It is possible that the analyst was not devoting as much time to the project as expected and did not adequately respond to team members' requests. However, although we have no direct evidence of the amount of time spent by the analyst on work related to Team A, several of the analyst's comments indicated what appeared to be a deep knowledge of both the database software and of various data-specific issues that would be known only to someone who had spent considerable time exploring the database, generating various reports, and manipulating data. This supported the analyst's representation of having spent considerable time on project-related work.

The similarity of the analyst's work within Team A and his work on a related project can be explained by examining influences on the analyst's task-related schemas. Prior to his joining Team A, the analyst had worked on a similar project that involved analyzing student transcripts in a university database. His experiences on that project shaped his conceptualization of the types of analyses that were possible given the data. This task schema then was brought to bear on Team A's project. As the analyst continued to have experiences both with Team A and with the related project, his concept of the task (including his understanding of feasible approaches and acceptable analyses) was shaped by exposure to both teams. Also, given the lack of specificity of Team A's research questions and ill-defined approach to analysis, Team A did not provide the analyst with much information that would significantly alter his initial schemas for this type of analytical task. It is unsurprising, therefore, that his work from another project was transferred and informed his work for Team A.

This illustrated how methodological tools, in this case, the methodologies introduced to the analyst through his exposure to the other project, can influence schema development. The analyst's task schema was shaped in part by his experiences within Team A and in part by his experiences on the related project. As a result, he took a similar approach in both situations. However, the team leader interpreted this similarity as a lack of dedication to Team A's work and as an inappropriate borrowing of the tools and approaches used in the other project.

Inconsistent Attendance

Another factor contributing to misalignments among participants' schemas of the team and task was inconsistent attendance at initial meetings. Table 7.4 shows attendance at the first 8 meetings, which occurred during the first 3 months of observation of the team. An "X" in the table designates that the person (represented by initials at the top of the column) attended the meeting. As indicated, the analyst and the team leader were together only at one of the eight meetings that occurred before the analyst left the team. We believe that the lack of concurrent attendance contributed to the analyst and team leader constructing radically different work-related schemas.

In a retrospective interview, the team leader indicated that one factor contributing to the team conflict was that the analyst was producing too many reports at an aggregate level and thus was not responding to the needs and requests of the team. In an e-mail reaction to an initial analysis in this study that we circulated to team members (excerpted in Protocol 12), he indicated that the group had decided not to pursue such summary statistics.

TABLE 7.4
Participation at Team Meetings

Meeting	*SY (Team Leader)*	*KU (Analyst)*	*BC*	*DL*	*HC*	*HL*
1	X			X	X	
2		X	X	X	X	X
3		X	X	X	X	X
4		X	X	X	X	X
5		X		X	X	X
	X					
6	X			X		X
7	X	X		X		X
8	X		X	X	X	X

Protocol 12: Interview

SY (leader): Early in the life of this effort ... the group specifically discussed and rejected the utility of the sorts of aggregations that [the analyst] was doing as relevant to our efforts for mapping out pathways. We had communicated that fact to [the analyst].

Although the leader's retrospective model of the situation involved team members requesting nonaggregate data, transcripts of early group meetings (including Meetings 1, 2, and 3) in fact contained comments in which team members specifically asked for summary, or aggregated, data. The team leader was present at one of those meetings (Meeting 1), but he was not present at the others. This suggests that the leader's perception of the group's decision did not match the message the group actually gave to the analyst. Was the team leader basing his view of the situation on those meetings that he had attended, but the analyst was generating reports based on (potentially different) messages that he was receiving at other meetings? Because of the lack of concurrent attendance on the part of the analyst and the team leader, it is likely that their schemas of the task were evolving based on different experiences and thus were not brought into alignment based on common experiences.

CONCLUSION

We believe that team members' unfavorable judgments of the analyst's work and the resulting conflict were due to a combination of factors. First, team members' schemas of task distribution contributed to their differing judgments of the analyst's work. The analyst clearly expected the group to help him make sense of complex data. When the analyst presented work in progress (which he saw as acceptable given his view of the role of the team), it did not meet the expectations of other group members (who had expected him to work independently toward a more finished product).

Second, misalignments in participants' schemas of the task itself contributed to the conflict. Ambiguities in participants' definitions of what pathways were and how they should be investigated resulted both in inconsistencies within individuals' schemas of the task and in differences across team members. In addition, because participants' schemas of software's role in the task were not well aligned, the analyst used existing software in a way that did not match other participants' expectations, and his work was criticized by other team members. This problem was compounded by the fact that the analyst's task models were shaped in part by his experiences from a related team.

Even important philosophical alignments—as in participants' agreement that an inductive approach was necessary—did not mitigate the conflict in points of view. The negative impact of misalignments related to implementation appeared to overcome apparent agreements.

In this study, we identified several factors contributing to schema misalignments. First, as expected, individuals' differing backgrounds (including their disciplinary history and experience with tools such as software) influenced the schemas they brought to the task. In addition, the distribution of tasks among team members caused each participant's schemas to be shaped by different goals, tools, and task experiences. This may have resulted in participants' schemas evolving in different directions. Although these misalignments could have been resolved productively through group communication processes, this did not occur. The lack of articulation of the task (including the ambiguous nature of the pathway analogy) and of specific research approaches hampered team members' attempts to work together successfully. Inconsistent attendance also may have prevented significant alignment from occurring. And while the team was aware of conflict, they did not appear to be aware of the *nature* of their misaligned understandings and *did not discuss or explore* them.

Failure to communicate, or even to be aware of the need for communication, was particularly striking. Theorists writing about interdisciplinary learning have long recognized that achievement of excellence requires openness to controversy and debate surrounding rival ideas (e.g., Flower, 2000; Graff, 2003; Wells & Claxton, 2002). Thus rival schemas are essential for knowledge growth, but so are norms and communication practices that lead to deeper analysis and understanding and integration of rival points of view. However, the team we observed did not discuss their problems or their processes openly.

This analysis has several implications for the development and management of interdisciplinary teams. Given the diverse nature of interdisciplinary teams, some degree of schema misalignment is to be expected and desired because it theoretically enhances the wealth of perspectives that can be brought to bear on a problem. However, it is crucial for interdisciplinary teams to bring team members' viewpoints into some degree of alignment so participants can work together successfully. It is important that *members' task and teamwork schemas be articulated and discussed such that participants become aware of different understandings and misalignments in understandings*. In addition, *task distribution should be clarified* so that all team members are aware of who is responsible for what parts of the project. Without adequate intrateam communication on these issues, the team may face similar problems to those observed in Team A in which products were being produced and judged based on different criteria. One key to ensuring adequate communication is as simple as *arranging for consistent attendance at team meetings*, thus ensuring that all team members are influenced by the same interactions and increasing the chances that participants' team and task schemas will evolve in compatible ways.

Additional research on the role of communication and management in schema negotiation in interdisciplinary teamwork is needed. Team A provided a unique example of a team in conflict in which schema misalignment and failure to recognize and resolve it likely contributed to low productivity. Studies of successful teams would allow for comparative analyses showing whether and how initially disparate schemas are brought into alignment through communication processes. This would provide both theoretically and pragmatically useful information because it would illustrate the cognitive processes involved in interdisciplinary knowledge construction and highlight techniques to facilitate the productive management and design of interdisciplinary groups.

ACKNOWLEDGMENT

Preparation of this chapter was funded in part through the National Institute for Science Education, a partnership of the University of Wisconsin–Madison and the National Center for Improving Science Education, Washington, DC, with funding from the National Science Foundation (Cooperative Agreement No. RED-9452971). However, the ideas expressed herein are not endorsed by and may not be representative of positions endorsed by the sponsoring agencies.

REFERENCES

Alba, J. W., & Hasher, L. (1983). Is memory schematic? *Psychological Bulletin, 43,* 202–231.

Bartlett, F. C. (1932). *Remembering*. Cambridge, England: Cambridge University Press.

Brewer, W. F., & Nakamura, G. V. (1984). The nature and functions of schemas. In R. S. Wyer, Jr., & T. K. Srull (Eds.), *Handbook of social cognition* (pp. 119–160). Hillsdale, NJ: Lawrence Erlbaum Associates.

Cannon-Bowers, J. A., Salas, E., & Converse, S. A. (1993). Shared mental models in expert team decision making. In N. J. Castellan, Jr. (Ed.), *Current issues in individual and group decision making* (pp. 221–246). Hillsdale, NJ: Lawrence Erlbaum Associates.
Derry, S. J. (1996). Cognitive schema theory in the constructivist debate. *Educational Psychologist, 31,* 163–171.
Engeström, Y., & Middleton, D. (Eds.). (1996). *Cognition and communication at work.* New York: Cambridge University Press.
Flower, L. (2000). The rival hypothesis stance and the practice of inquiry. In L. Flower, E. Long, & L. Higgins (Eds.), *Learning to rival A literate practice for intercultural inquiry* (pp. 27–48). Mahwah, NJ: Lawrence Erlbaum Associates.
Gentner, D., & Stevens, A. (Eds.). (1983). *Mental models.* Hillsdale, NJ: Lawrence Erlbaum Associates.
Graff, G. (2003). *Clueless in academe: How schooling obscures the life of the mind.* New Haven, CT: Yale University Press.
Hackman, J. R. (1987). The design of work teams. In J. W. Lorsch (Ed.), *Handbook of organizational behavior* (pp. 315–342). Englewood Cliffs, NJ: Prentice Hall.
Herbsleb, J. D., Klein, H., Olson, G. M., Brunner, H., Olson, J. S., & Harding, J. (1995). Object-oriented analysis and design in software project teams. *Human–Computer Interaction, 10,* 249–292.
Hinsz, V. B., Tindale, R. S., & Vollrath, D. A. (1997). The emerging conceptualization of groups as information processors. *Psychological Bulletin, 121,* 43–64.
Johnson-Laird, P. N. (1983). *Mental models.* Cambridge, MA: Harvard University Press.
Latour, B., & Woolgar, S. (1986). *Laboratory life: The construction of scientific facts.* Princeton, NJ: Princeton University Press.
Lave, J. (1988). *Cognition in practice.* Cambridge, England: Cambridge University Press.
Nardi, B. A. (1996). Studying context: A comparison of activity theory, situated action models, and distributed cognition. In B. Nardi (Ed.), *Context and consciousness: Activity theory and human–computer interaction* (pp. 69–102). Cambridge, MA: MIT Press.
Olson, J. S., Card, S. K., Landauer, T. K., Olson, G. M., Malone, T., & Leggett, J. (1993). Computer-supported co-operative work: Research issues for the 90s. *Behavior and Information Technology, 12,* 115–129.
Olson, J. S., Olson, G. M., Storrøsten, M., & Carter, M. (1993). Groupwork close up: A comparison of the group design process with and without a simple group editor. *ACM Transactions on Information Systems, 11,* 321–348.
Orasanu, J., & Salas, E. (1993). Team decision making in complex environments. In G. A. Klein, J. Orasanu, R. Calderwood, & C. E. Zsambok (Eds.), *Decision making in action: Models and methods* (pp. 327–345). Norwood, NJ: Ablex.
Rentsch, J. R., & Hall, R. J. (1994). Members of great teams think alike: A model of team effectiveness and schema similarity among team members. In M. M. Beyerlein & D. A. Johnson (Eds.), *Advances in interdisciplinary studies of work teams: Theories of self-managing work teams* (Vol. 1, pp. 223–262). Greenwich, CT: JAI.
Russell, B. (1985). *The philosophy of logical atomism.* Peru, IL: Open Court. (Original work published in 1918)
Salas, E., Prince, C., Baker, D. P., & Shrestha, L. (1995). Situation awareness in team performance: Implications for measurement and training. *Human Factors, 37,* 123–136.
Stout, R., Cannon-Bowers, J. A., & Salas, E. (1994). The role of shared mental models in developing shared situational awareness. In R. D. Gilson, D. J. Garland, &

J. M. Koonce (Eds.), *Situational awareness in complex systems* (pp. 297–304). Daytona Beach, FL: Embry-Riddle Aeronautical University Press.

Vygotsky, L. S. (1978). *Mind in society: The development of higher psychological processes*. Cambridge, MA: Harvard University Press.

Wells, G. & Claxton, G. (Eds.). (2002). *Learning for life in the 21st century: Sociocultural perspectives on the future of education*. Malden, MA: Blackwell Publishing.

III

Studies on Cognitive Science

8

Cognitive Science: Interdisciplinary and Intradisciplinary Collaboration

John T. Bruer
James S. McDonnell Foundation

Since 1956, when the cognitive revolution in psychology began, the science of the mind has grown into a mature discipline (Gardner, 1985). In 1977, the first volume of the journal *Cognitive Science* appeared and 2 years later the Cognitive Science Society was incorporated in Massachusetts and the Society convened its first meeting in La Jolla, California. The emergence of a journal and a society indicated that 20 years after the revolution, cognitive science had become established as an academic discipline. Over the past 20 years, cognitive science has continued to develop. One mark of a mature science, as Glaser (1988) has said many times, is that a mature science can apply its theories and methods to problems—sometimes even practical ones—outside its field. Furthermore, attempts at such extradisciplinary problem solving can sometimes feed back into the ongoing refinement of basic theories. Mature sciences also share a second characteristic. Although a mature science occupies and gives order to a well consolidated intellectual territory, there is still lively action at the research front where issues, theories, and methods develop as the discipline attempts to move into new territory.

Both of these characteristics of mature science point to the importance of collaboration. Attempting to apply one's science to solve problems in other fields creates opportunities for interdisciplinary collaboration. At the research front where the action might tend to be more heterodox than orthodox, there are not only opportunities but there is also the need for intradisciplinary discussion, debate, and collaboration. Thus, as a mature cognitive science moves into its second 50 years, interdisciplinary and intradisciplinary collaboration is an appropriate theme for research in cognitive science.

For the past 15 years, the James S. McDonnell Foundation has supported programs that encourage cognitive scientists to engage in interdisciplinary collaborations with educators (Cognitive Studies for Educational Practice), systems neuroscientists (McDonnell–Pew Program in Cognitive Neuroscience), and more recently, and on a smaller scale, with geneticists (Panel and Pilot Studies: Underlying Cognitive, Neural and Genetic Bases of Social Behavior) and rehabilitation scientists (Program in Cognitive Rehabilitation). In supporting these interdisciplinary collaborations, the Foundation has also seen the importance of collaboration within the discipline as the science continues to mature. I rely on examples from the Foundation's experience to address the interdisciplinary collaboration theme. Because the Foundation's experience is limited, so are the examples. There are, of course, many other fruitful areas for important interdisciplinary collaboration that lie outside the Foundation's interests of which I am not aware, but I would hope they share some features with the examples I discuss.

Most of these examples will illustrate one traditional strength of the cognitive perspective in psychology. Traditionally, cognitive scientists have upheld a methodological commitment to carefully analyze tasks and to develop computational models based on those analyses to show how mental constructs guide human behavior. The fine-grained understandings of behavior that cognitive science provides are fundamental to current and future work in numerous fields including education, cognitive neuroscience, behavioral genetics, and cognitive rehabilitation.

COGNITIVE SCIENCE AND EDUCATION

A fundamental insight that emerged early on in the cognitive revolution is that humans are not passive communication channels but active information processors. Over the past 40 years, cognitive research has provided a new framework for thinking about teaching and learning. Based on this research, cognitive scientists now appreciate that when we remember and learn, we tend to do so by associating new information with knowledge that we already possess. This understanding of how our minds work provides an empirical basis for claims that we actively construct our understanding of new information and a scientific basis for constructivist theories of learning. We know from cognitive research that effective instruction must build on students' prior knowledge. Furthermore, cognitive research has told us that when we are actively engaged in constructing our understandings, we can optimize our learning if we monitor and control our cognitive processes. Hence, we have come to appreciate the importance of metacognition in effective learning and teaching (Resnick & Hall, 1998).

In addition, however, because of its commitment to fine-grained analyses of behavior and building models of these behaviors, cognitive research has also revealed in some detail how knowledge, skills, and strategies must be orchestrated within a subject domain to result in expert-like performance. These fine-grained understandings of expertise and learning trajectories within subject domains give educators tools they can use to diagnose and treat learning problems that occur in classrooms everyday.

Numerical cognition and learning first formal arithmetic provides a simple but compelling example of what can emerge when cognitive scientists collaborate with educators to solve instructional problems. One widely acknowledged instructional problem is that minority students typically underperform their White majority classmates in mathematical achievement (Dossey, Mullis, Lindquist, & Chamber, 1988). What is at the root of this problem? How might educators overcome it?

In 1986, the National Science Foundation sponsored a survey on mathematical achievement in Montgomery County, Maryland. Montgomery County is a relatively affluent suburban-rural area that invests considerable resources in education. The survey (NSF, 1988) reported that children of all ethnic backgrounds started formal schooling with relatively small differences in mathematical understanding as measured by standardized tests. However, racial and ethnic differences in mathematics achievement began to appear as early as the first or second grade. The small initial disparities in mathematical understanding present at school entry increased as children moved through their elementary school years. By the time children reached sixth grade, nearly 50% of

Black students were performing below grade level in mathematics as were nearly 40% of Hispanic students. In contrast, only 10% to 15% of Asian and white sixth graders scored below grade level in mathematics.

A *Science* article that reported the results of the survey speculated about what might account for the increasing disparities across ethnic groups in math performance (National Science Foundation, 1988). As with most of our commonsense thinking about educational disparities, the possible explanations emphasized social, economic, and attitudinal factors. Positive indicators of high mathematical achievement during the elementary years included parental attitudes and encouragement, teacher encouragement, and students' liking for the subject matter. Negative factors that seemed to affect a minority student performance included family economic status, fragmentation of families, and teachers' beliefs about minority students' abilities.

Although these positive and negative indicators were no doubt correlated with students' mathematics achievement, they tell us very little that would help teachers identify students learning problems or that would tell them how to attack the problem in the classroom. If, as cognitive research suggests, what one can learn is based on one's prior knowledge, what is it that children might or might not know about numbers when they enter school? Are there ethnic or class differences in the informal number knowledge children acquire before they enter school? What do standard elementary school mathematical curricula assume about children's number knowledge at school entry? Answers to these questions derived from cognitive research might help teachers identify students likely to have learning problems at school entry and provide teachers with instructional tools to help those students.

Cognitive research on mental arithmetic has a history going back to the early 1960s. By the mid-1970s, there was already an expanding literature on how young children developed numerical and mathematical understanding (Gelman & Gallistel, 1978; Groen & Resnick,1977). My example picks up the story 1987, the year the McDonnell Foundation began its program, Cognitive Studies for Educational Practice.

In that program's first grant competition, the Foundation received an intradisciplinary collaborative proposal from Case and Siegler (Griffin, Case, & Siegler, 1994). The initial insight that triggered their collaboration was an observation Case had made about Siegler's (1981) study of children's performance on the balance scale task. Siegler had found that 6-year-old children used a rudimentary quantitative rule to solve balance scale problems. The 6-year-olds' rule involved counting the weights on each side of the fulcrum, comparing the number of weights on each side, and judging that the side with the greater number of weights would go down. To use this rule, the 6-year-olds had to use and coordinate both their skill at explicit counting and their skill at comparing numbers for size. In contrast, the 4-year-olds appeared to use qualitative rule to solve balance scale problems. The 4-year-olds estimated which side of the fulcrum had more weight and judged that the side with more weight would go down. The 4-year-olds' rule did not rely on explicit counting but on a judgment of more versus less. Yet, Case knew that 4-year-olds can count small arrays of objects and can also make magnitude comparisons (as is implicit in the rule they used). Thus, it appeared that 4-year-olds could not explicitly count the number of weights on each side of the fulcrum and compare the resulting numbers for size. Four-year-olds had not yet fully coordinated their abilities to count and compare. From these observations, Case and Siegler concluded that 6-year-olds but not 4-year-olds had fully acquired the central conceptual structure required for quantitative reasoning (Griffin, Case, & Siegler, 1994).

What is this central conceptual structure? Earlier cognitive and developmental research has suggested that one could think of this central conceptual structure as represented by a mental number line (Case, 1996). This simple conceptual structure captures concepts that are fundamental to numerical reasoning. Numerical reasoning requires knowledge of written number symbols (Arabic numerals) and number words. In counting, objects are tagged in a one-to-one mapping with a number word or symbol. Numbers name cardinal set values, and if the counting procedure is done correctly, the last number name or symbol used gives the cardinality of the set. The mental number line shows that number names (either words or numerals) occur in a fixed order. It shows that each succeeding number name refers to a cardinality one unit larger than the cardinality names by the previous number name in the sequence. Furthermore, by envisioning the number line in the mind's eye, one can "see" which of two numbers is larger or smaller by noting their relative positions along the line. Although the mental number line is a rather simple structure, it provides a rather fine-grained model of the knowledge and skills that underlie counting and basic arithmetical cognition.

Case and Siegler reasoned that what differentiated 4-year-olds from 6-year-olds was their grasp of this central conceptual structure (Griffin et al., 1994). They expected that children who lacked all or parts of this conceptual structure would most likely be poor at numerical comparison and at solving simple arithmetic problems.

Case, in collaboration with Griffin (1996), developed a Number Knowledge test to assess children's grasp of the mental number line. Originally, they believed that acquiring this conceptual structure was a developmental phenomenon that occurred in all children between the ages of 4 and 6 years. However, as Griffin and Case (1996) attempted to norm their test on various student populations in the United States and Canada, they were surprised to find that children's socioeconomic status (SES) influenced their Number Knowledge test performance. They found that at school entry, children from lower SES homes appeared to be delayed one developmental level in their numerical understanding. That is, 6-year-olds from low-SES homes performed like 4-year-olds from higher SES homes on the Number Knowledge test. The most pronounced differences between the high and low-SES 6-year-olds appeared on numerical comparison questions. Among the higher SES children, 96% could determine which of two Arabic digits was larger or smaller, but only 18% of the lower SES children could do so. The ability to make correct numerical comparisons also correlated with children's ability to solve simple number problems. Thus, Griffin and Case (1996) found that acquiring the mental number line structure was not simply a developmental phenomenon but that it was influenced by children's home environment. More important, their results suggested that some lower SES children entered school not having acquired the central conceptual structure that most standard, first formal arithmetic curricula assumed students to have.

Now, think back to the Montgomery County survey. That survey suggested that low SES was associated with decreasing mathematical achievement during the elementary school years. However, the association between SES and standardized test performance did not say much about why small initial differences turn into larger ones or what do to about it. A more detailed understanding of numerical cognition and its component knowledge and skills yields insights that allow us to identify a specific learning problem: Some low-SES children arrive at school without the central conceptual structure that first formal arithmetic assumes. The Number Knowledge test can identify these children. Furthermore, once cognitive scientists have identified children and the gaps in their numerical understanding, they can also use their understanding of numerical cognition to guide curriculum development.

In collaboration with classroom teachers in Worcester, Massachusetts public schools, Griffin and Case (1996) developed such a curriculum, the Number Worlds Curriculum. This interdisciplinary collaboration required both that the researchers provide professional development to the teachers and that the researchers learn from the teachers about what does and does not work in a public school classroom. The Number Worlds Curriculum explicitly teaches all the skills and understandings captured in the mental number line representation. The group learning sessions, using multiple physical representations of number, are focused around number games. Group learning within a game context encourages both the development of number skills as well as active, constructive discussions about numbers, counting, and comparison.

In a series of studies, Griffin and Case (1996) evaluated the effect of their curriculum on students from low-SES homes who attended neighborhood schools, students who had been identified as at risk for school failure based on their Number Knowledge test performance. After 1 year in the curriculum, these at-risk students were indistinguishable from or superior to middle-class students who attended a well staffed magnet school on solving word problems, writing number stories, performing successive mental arithmetic operations, and solving formal equations. On tests of numerical computation such as those used in international comparisons, the Number Worlds students compared favorably with Japanese students. At-risk children who received the standard mathematics curriculum continued to lag one developmental level behind the Number Worlds students on the Number Knowledge test and also lagged behind those students in mastery of elementary arithmetic.

The genesis and success of the Number Worlds Curriculum illustrates how fine-grained cognitive models of component knowledge and skills that underlie expert performance within a subject domain allow educators to identify specific gaps in children's conceptual and procedural knowledge that inhibit learning. Once the gaps are identified, these fine-grained understandings can also inform the development of curricula and teaching tools that can fill those gaps. One can point to similar examples from other school domains—reading, writing, science education, social studies—that illustrate the contributions cognitive research can make to solving instructional problems in the classroom (Bruer, 1993; McGilly, 1994).

With a little effort, one could no doubt generate a lengthy list of other examples. However, one should not be deceived by the length of the list. The job is far from over. Learning is complex. There are many learning domains and subdomains, even for school subjects, which continue to present learning obstacles to students and instructional challenges to teachers. For most of these domains and subdomains, we do not yet have a fine-grained understanding of their underlying knowledge and skill structures. In all these domains and subdomains, we know neither how to diagnose nor remedy specific learning problems, but we do have the means to find out. As Siegler and Klahr (1982) wrote over 20 years ago, cognitive science "provides an empirically based technology for determining people's existing knowledge, for specifying the form of likely future knowledge states, and for choosing types of problems that lead from present to future knowledge" (p. 134). Education provides abundant opportunities for both interdisciplinary and intradisciplinary collaborations that can apply and extend this technology.

COGNITIVE NEUROSCIENCE

Historically, neuroscientists, as biological scientists, were interested in brain anatomy and physiology. They had little interest in the mind. Psychologists, as behavioral scientists, were interested in understanding human behavior, sometimes by employing theories and hypotheses about the mind. They had little interest in the brain. Early in the cognitive revolution, cognitive scientists opted to eschew physiological and neuropsychological models in favor of more abstract, symbolic, and functionalist models of cognition. Initially, this decision facilitated the development of cognitive theories and cognitive science. As Newell and Simon (1972) wrote in their book *Human Problem Solving*, "Perhaps the nonphysiological nature of the theory is not as disadvantageous as one might first believe, for the collection of mechanisms that are at present somewhat understood in neuropsychology is not at all adequate to the tasks dealt with in this book" (p. 5). During the 1970s and early 1980s, among some cognitive psychologists, ignoring the neural in attempts to understand the mental changed from a useful working hypothesis into a theoretical necessity and disciplinary article of faith. Thus, cognitive science continued to evolve independently from neuroscience based on the view that how the brain implemented cognition was largely irrelevant to a science of the mind.

By the mid-1980s, however, this strict division between mind and brain scientists began to dissolve. Neuroscience, particularly at the systems level, and neuropsychology had made considerable progress in understanding brain mechanisms that might be relevant to human cognition. Also, brain imaging and brain recording technologies emerged and developed as useful tools with which to study patterns of brain activity in normal, cognizing human participants.

These developments gave rise to a new interdisciplinary research front, cognitive neuroscience. Initially, this collaboration involved neuroscientists, cognitive psychologists, and computer scientists. Cognitive neuroscientists attempt to map mental functions onto neural structures and circuits. Much of their work revolves around brain imaging and brain recording technologies such as functional magnetic resonance imaging (fMRI), positron emission tomography (PET), magnetoencephelography, EEG, and evoked response potentials (ERP). Brain imaging technologies such as fMRI and PET have spatial resolutions down to the level of cortical columns, that is, to the level of millimeters; however, their temporal resolutions range from seconds to hours. Brain recording techniques such as ERP have temporal resolutions down to the level of milliseconds but have relatively poor spatial resolutions. What cognitive neuroscientists must do then is to find candidate relations between brain structures and mental functions that optimize the information they can derive from the imaging and recording techniques currently available. Posner and Raichle (1994) explained the cognitive neuroscientific challenge in this way: "The challenge for the future is to understand at a deeper level the actual mental operations assigned to the various areas of [brain] activation. Before this goal can be achieved, the experimental strategies used in PET studies must be refined so that more detailed components of the process can be isolated" (p. 97). Meeting the cognitive neuroscientific challenge thus depends on analyzing tasks and developing ever more fine-grained models of mental functions. In the continuing climate from the decade of the brain and amid popular (and scientific) fascination with brain images, cognitive scientists might feel neglected and unappreciated. However, as any cognitive neuroscientist will readily admit, the quality of a brain imaging study is only as good as its underlying task analyses and cognitive models. Studies of numerical cognition again provide a simple but highly illustrative example.

In 1985, Roland and Friberg reported results from one of the first brain imaging studies that measured changes in regional cerebral blood flow while patients performed a mathematical thinking task. In Roland and Friberg's study, they used a task taken from standard neuropsychiatric evaluations: Count backward from 50 by 3s. Roland and Friberg found that this task activated brain regions in the left and right angular gyri and in prefrontal cortex, and they noted that these activations looked "different" from the activations they recorded when their participants were "thinking about nothing" or "thinking about every third word of a well-known Danish jingle" (p. 1222).

This pioneering study (Roland & Friberg, 1985) did tell us something about brain areas involved in numerical reasoning. However, because the experimental task was rather crude, the resulting imaging data were not very precise. A cognitive psychologist would quickly point out that counting backward from 50 by 3s is a highly complex mental task. It involves several obvious subtasks such as retrieving arithmetic facts, borrowing and carrying, and holding information in verbal working memory. An experimental task that was more sensitive to such underlying cognitive complexity might yield more precise imaging data. With the right task, the right task analysis, and the right experimental design, it might be possible to discover which subcomponents of a numerical task are correlated with specific brain activations.

From my previous discussion of learning early arithmetic, the reader saw that numerical comparison is an important basic skill for learning first formal arithmetic. Cognitive psychologists have further analyzed numerical comparison into its subcomponents and developed cognitive models based on these analyses. Dehaene (1996) proposed a three-stage, serial model of numerical comparison: Comparing two numbers for size requires that one (a) identify the input stimuli, (b) compare the numerical magnitudes named by those stimuli, and (c) prepare and generate a response.

Following the logic of the additive factors, Dehaene (1996) reasoned that if his serial model were correct, then experimental manipulations that affected only one of the processing stages—identification, comparison, or response—should influence participants' reaction times only for that one stage. Dehaene manipulated the identification stage by using both written English number words and Arabic numerals as input stimuli for his comparison task. Dehaene influenced the comparison stage by requiring his participants to make both close (4 vs. 5) and far (9 vs. 5) numerical comparisons. This manipulation depends on the well known distance effect: Comparing two numbers for size is more difficult if they are closer together in magnitude than if they are farther apart. To manipulate the response phase, Dehaene's experiment required that his participants, all of whom were right-handed, respond with the right hand on half of the trials and with a left hand on half of the trials.

Dehaene's (1996) behavioral analysis of the reaction time data showed that each of the three experimental manipulations influenced participants' reaction times only for the intended processing stage, suggesting that his simple, serial model was at least an adequate first approximation for the numerical comparison task.

While his participants made numerical comparisons under the various experimental conditions, Dehaene (1996) made ERP recordings of their brain activity. These data allowed Dehaene to associate brain activations with each of the three serial processing components, thereby tracing the neural time course and the underlying neural circuitry for numerical comparison. Dehaene found that number word stimuli were identified by a neural subsystem in the brain's left posterior hemisphere, whereas Arabic numerals were identified by a neural subsystem that involved posterior areas of both the left and right hemispheres. Dehaene also found that the numerical comparison itself seemed to be computed in a brain region in the posterior right and left hemispheres and that this same "comparison" area was activated both when the input stimuli were number words and when they were Arabic digits. This suggested to Dehaene that when humans make numerical comparisons, we translate the input stimuli, either number words or Arabic numerals, into an abstract, analog magnitude representation, and we use that representation to make the comparison.

Using this same experimental design but replacing number words with number dot patterns, Temple and Posner (1998) studied numerical comparison in 5-year-old children. Temple and Posner found the same brain activation patterns in children that Dehaene (1996) had found in adults. They also found that as with number words and Arabic digits in adults, children's comparisons of dot-pattern inputs also seemed to be computed in a bilateral, inferior, parietal brain area. Furthermore, although children took longer to make comparisons, the adult brain activation pattern was present in children by 5 years of age.

Remember the mental number line conceptual structure. That conceptual model represents how number words and Arabic digits function as symbols for set sizes. Studies like Dehaene's (1996) and Temple and Posner's (1998) show that these distinct symbolic representations of number are associated with distinct brain areas. These studies also support the claim that there is yet a third representation of number and quantity, an analogue magnitude representation that humans use when we make numerical comparisons. This suggests further that the ability to integrate these three distinct representations is fundamental to arithmetic reasoning. Cognitive neuroscientific studies like these will allow cognitive scientists to ask and answer some interesting questions about development and learning. How are children's brains reorganized when they acquire the ability to make numerical comparisons between age 4 and 6 years? Does this reorganization somehow involve connecting the symbolic representations of number with the magnitude representation? How does the Number Worlds Curriculum affect and change children's neural circuitry?

Here, too, one could generate another lengthy list of additional examples—reading, attention, visual imagery, memory—in which cognitive analyses and models have been fundamental to advances in cognitive neuroscience. Here, too, despite the length of that list, far more remains unknown than is known about how neural structures support cognitive functions. Linking systems neuroscience with cognitive psychology—meeting the challenge of cognitive neuroscience—provides another research frontier that is rich in interdisciplinary research opportunities.

DELINEATING COGNITIVE PROFILES

The Montgomery County survey tried to understand patterns in children's mathematics performance by associating test scores with social and economic factors. Early in the development of brain imaging studies, Roland and Friberg (1985) tried to understand the neural basis of numerical cognition using a crude neuropsychological test. In both cases, although cognitive scientists learned something, we did not learn very much. We did not learn very much because the tests and measures brought to the task—standardized test scores, family incomes, a traditional neuropsychological test—were too crude. The tests and measure obscured rather than illuminated subtle, interesting, and important differences. Cognitive research provided more precise and sensitive tools that illuminated the differences and thus advanced cognitive science's understanding of learning first formal arithmetic and of the neural basis of numerical cognition.

There are other areas in which cognitive science can also provide more precise, sensitive tools and measures to advance research and address practical problems. One such area is the study of how genes affect brain development and cognition. A second such area is the study of the cognitive sequelae of aggressive therapies to treat brain cancer. In both these areas, geneticists, neuroscientists, and physicians tend to rely primarily on psychometric measures to assess their participants and patients. These rather crude measures can mask profound differences in the cognitive abilities of participants and patients. Measures and assessments based on cognitive models can contribute to research in these fields by allowing one to delineate more precisely the cognitive profiles of patients and participants.

Recent interdisciplinary research on Williams Syndrome illustrates what cognitive science can bring to genetic studies. Genetic deficiencies, such as Williams syndrome that affect cognitive performance provide a means by which humans can better understand how mind, brain, genes, and environment contribute to our behavior. Williams syndrome affects 1 in 20,000 births. Williams people are short in stature, have elfin-like facial features, and suffer from cardiac problems. Since this syndrome was first discovered 35 years ago, geneticists have established that it is caused by gene deletions on Chromosome 7. If researchers want to know specifically how these genes affect brain development and cognition, they will need reasonably accurate descriptions of the cognitive deficits associated with the syndrome. If we want to understand the genotype, we first need a precise description of the phenotype.

Traditionally, Williams people were described as severely mentally retarded. Using psychometric measures Williams people are indistinguishable from Downs people, having IQs of around 60. However, as recent cognitive studies have shown, characterizing Williams people on the basis of IQ alone is similar in its level of refinement to understanding mathematical cognition by using tasks such as counting backward from 50 by 3s.

Bellugi and her colleagues (Bellugi, Klima, & Wang, 1996; Lenhoff, Wang, Greenberg, & Bellugi, 1997), using assessments based on cognitive research, have attempted to delineate the cognitive profile that is associated with Williams syndrome; they have compared this cognitive profile to the profile for Downs syndrome patients. Bellugi et al. (1996) and Lenhoff et al. (1997) have found that if one looks more carefully, the superficial similarities in IQ and general cognitive ability between Williams and Downs people conceals profound differences in their cognitive profiles, differences in which skills are spared and which are lost because of their respective genetic deficiencies. Williams people are best known for their highly preserved linguistic abilities. Despite their very low IQs, they use correct and complex grammatical constructions, have large and unusual vocabularies, and typically use emotionally and affectively rich language. In contrast, Downs people show poor grammatical understanding, have limited vocabularies, and manifest diminished use of emotion and affect in speech. When engaged in hierarchical processing, Williams people show attention to local detail at the loss of global structure—they can see the tree but not the forest—whereas Downs people show attention to global structure at the loss of local detail; they can see the forest but not the trees. Williams people are very good at face recognition, whereas Downs people are highly impaired at this task. Other researchers are attempting to elaborate and refine this cognitive profile. Other research has shown that Williams people, in contrast with other low-IQ groups, appear to retain some ability to understand other minds (Tager-Flusberg, Boshart, & Baron-Cohen, 1998) and that although Williams people can learn by acquiring new facts, they do not seem to learn things that require conceptual change and reorganization as happens in normal children when they develop folk biological theories (Johnson & Carey, 1998). The cognitive dissection of Williams syndrome is a growing area of study because all the cognitive deficits and abilities that distinguish Williams people from other mentally retarded patient groups and from normal participants must be related in some way to the documented gene deletions (Wang, 1999).

What researchers have already learned from the cognitive profile of Williams syndrome is that there is no simple explanation of how the gene deletions on Chromosome 7 affect brain development. A highly simplistic and naive explanation might have been that the deletions spare the left hemisphere, where language is typically located, but disrupt right-hemisphere processing of visuospatial information. However, comparison of the cognitive profiles of Williams and Downs people provides a counterexample to this simple explanation. Williams people have heightened emotional expressiveness and are excellent at face recognition, both of which are thought to depend on right-hemisphere processes. Thus, whatever the effects of the gene deletions are on brain development, it is considerably more complicated than neuropsychologists initially surmised.

More refined cognitive profiles for individuals suffering from Williams syndrome as well as from other genetic deficiencies will be needed to advance this highly interdisciplinary research agenda. At present, one can look at research on Williams syndrome as a model system that might lead to insights in how to pursue interdisciplinary research at the intersection of genetics, neuroscience, and cognitive science. The Foundation-funded project "Panel and Pilot Studies: Underlying Cognitive, Neural, and Genetic Bases of Social Behavior," headed by Bellugi and Korenberg, a geneticist, is currently engaged in this activity.

Pediatric oncology is another area in which medical practitioners and researchers need more precise cognitive profiles for their patients and in which cognitive science could make a significant contribution. Central nervous system tumors are the third leading cause of death in children under 16 years of age. Better diagnosis and treatment have improved the survival rates of these children, but generally, pediatric neural oncologists pay little attention to other outcomes, particularly patients' cognitive outcomes, and the subsequent quality of life for the young survivors. Assessments of therapeutic outcomes rely heavily on psychometric measures that, as with Williams versus Downs people, may mask significant differences in posttherapeutic cognitive outcomes. Mueller, a pediatric neural oncologist who is currently involved in collaborations with cognitive scientists on these issues, noted that the psychometric approach, which dominates all tumor studies, is highly atheoretical and provides no insights into either the patients' specific cognitive deficits or their underlying neural mechanisms. Furthermore, the reliance on psychometric measures loses clinical information that might be valuable for rehabilitation programs. According to M. Mueller (personal communication), "we need to address this problem within a context of a theory of cognitive development."

Between 40% and 100% of children who survive brain cancer, depending on the tumor type and therapy regimen, have lasting cognitive deficits and long-term academic problems. Interestingly, Mueller (personal communication) found that these patients often exhibit learning and developmental delays similar to those that Griffin and Case (1996) found among low-SES children in the domain of numerical cognition. Full head radiation is one of the leading causes of these deficits. New techniques of confocal radiation therapy show promise of limiting damage to healthy brain tissue, but such therapeutic techniques are expensive and remain experimental. Families and physicians as well as insurers will want to know if the benefit of these new therapeutic techniques offsets their cost. The potential benefit here is not longer survival but improved quality of life and spared mental function. Answering cost–benefit questions like these will require careful assessments of patients' cognitive profiles. Pediatric neurology and neural oncology are not well prepared to address these issues. Neurologists and oncologists would no doubt welcome collaborations with cognitive scientists to address these questions. The answers will have significant implications for clinical practice, quality of life, and health care policy.

COGNITIVE REHABILITATION

Stroke and traumatic head injury also exact a substantial burden on patients, families, and the health care system. From a cognitive perspective, there are interesting parallels between the rehabilitation of head-injury patients and educational practice. Like educational interventions, rehabilitation regimens often are not well characterized in the literature and are replicated with low fidelity in the clinic. As in education, rehabilitation programs tend to be atheoretical. The knowledge we have about underlying cognitive and neural processes rarely contributes to rehabilitation program design or evaluation. Also, like education, there is a considerable gap between researchers and practitioners. Researchers tend to see atheoretical practitioners, and practitioners tend to see research that has only limited clinical relevance.

What if one were to look at rehabilitation following head injury as a learning problem? Might rehabilitation, like education, provide opportunities for interdisciplinary collaborations that could contribute to improved clinical practice as well as extend and refine the theories of cognition? An example of what might happen if one answers "yes" to these questions is a collaboration currently underway at Michigan State University between Carr, a cognitive psychologist, and Hinckley, a rehabilitation scientist.

Aphasia—loss or partial loss of speech production or comprehension skills following stroke or traumatic brain injury—affects one million people in the United States. Each year, in the United States alone, there are 85,000 new cases. Aphasia is much more prevalent than Parkinson's disease and results in considerable personal pain and financial cost.

Carr and Hinckley (1997) observed that there are two general approaches to aphasia therapy. The first they called the part-task, or the neuropsychological, approach. On this approach, therapy consists of analyzing tasks that the patient should master into their subcomponents and then helping the patient to practice and master these subcomponents separately. It is a "divide and conquer" approach. It assumes that the subcomponents are easier to learn in isolation and that subcomponents so learned readily transfer to other tasks.

Carr and Hinckley (1997) called the second approach to aphasia therapy the whole-task, or the contextual or functional, approach. Here the idea is to use entire real-world tasks in patient therapy and to teach those tasks in authentic contexts and situations in which the patient will have to perform the tasks. This approach assumes that it is preferable to provide the patient with real-world skills in real-world contexts. Making the therapy as much as possible like everyday situations will assure that the learning transfers from the therapeutic context to the real world the patient inhabits. Both approaches to aphasia rehabilitation have some commonsense appeal and are pursued religiously but largely atheoretically within different segments of the rehabilitation community.

Yet, as Carr and Hinckley (1997) pointed out, both approaches have affinities to cognitive theories and hypotheses. The part-task approach assumes, as do all the other examples I have discussed so far, that it is always legitimate to analyze tasks into subcomponents and then to treat these subcomponents as independent entities, that is, as tasks in their own right. It further assumes that practicing subcomponents reduces cognitive load, that this is best way to automatize the subcomponents, and that part-task learning will facilitate transfer to related tasks in other contexts.

However, cognitive science is a mature discipline, and as Carr and Hinckley (1997) pointed out, there are other cognitive theories currently at the research front that question this analytic, componential approach to rehabilitation and to learning. Not surprisingly, the whole-task approach to aphasia therapy has affinities with these cognitive theories. Applications of dynamical systems theory to cognition maintain that a task's component processes always and necessarily interact and that they do so in different ways in different situations (Van Gelder, 1998). If so, one should not assume that tasks can be usefully analyzed into context independent subcomponents. Instance-theoretic views of automaticity hold that tasks become efficiently automatized only when entire task performances are available in memory for retrieval. On this view, too, subcomponents of a task are different tasks and practicing them in isolation results in memory traces that are insufficient to automatize the whole task (Logan, 1988). Finally, episodic theories of memory retrieval claim that new learning is closely bound to the context in which it is acquired. If so, part-task learning would be self-defeating because the part task is learned out of context. It is not being learned in the context of the other required part tasks with which it must be combined. Part-task learning of necessity creates artificial contexts that do not facilitate transfer to real-world, authentic contexts (Jacoby & Brooks, 1984; McClelland, McNaughton, & O'Reilly, 1995).

The part-task versus whole-task dichotomy is also reminiscent of contrasting positions that are currently being debated within the cognition and education literature. There are similarities between the part-task approach and what has been called the cognitive perspective in the cognition and education literature (Anderson, Reder, & Simon, 1996, 1997). As applied to learning, the cognitive perspective takes tasks as the unit of analysis and holds that tasks are factorable or decomposable, that one can learn subtasks in isolation, and that such learning transfers across tasks. There are similarities between the whole-task approach and what has been called the situated perspective in education (Greeno, 1997). On this view, the (social) situation, not the task, is the unit of analysis, and one should not assume that situations are factorable. Furthermore, situated theorists claim that humans learn most effectively using whole, authentic problems and that learning is context specific. Thus, views about aphasia research contain within them implicit appeals to some quite fundamental theoretical differences within cognitive science and raise some important intradisciplinary issues.

To the extent that such theoretical differences are amenable to experimental adjudication, Carr and Hinckley (1997) saw aphasia research as providing an appropriate experimental context. Carr and Hinckley proposed to develop and assess two aphasia rehabilitation protocols, one grounded in the part-task perspective and the other in the whole-task perspective. Carr and Hinckley will develop two training regimens for a variety of tasks, such as ordering items from a mail order catalog, that might be used in rehabilitation and assess patient outcomes under the different regimens. Carr and Hinckley will also assess their patient-participants on severity of injury, age at injury, premorbid cognitive function, and the stage of recovery at which therapy begins. Carr and Hinckley may find that the part-task or whole-task approach is superior for all patients, or they may find that certain patient subgroups fare better under one of the therapeutic approaches than they do under the other. Either result will provide important insights into how to improve aphasia rehabilitation programs. It is also the case that either result will provide interesting feedback from the clinical context into cognitive theory, creating important opportunities for intradisciplinary collaboration, discussion, and debate and possibly for significant theory revision.

An additional advantage of looking at the rehabilitation clinic as a learning laboratory is that it provides a research context in which both cognitive scientists and rehabilitation specialists have considerable control over therapeutic interventions, research-based innovation, and theory-inspired assessment. In this regard, the clinic provides a real-world setting for research that is quite different from the classroom. Interdisciplinary and intradisciplinary collaborations in the clinic might not only contribute to improved clinical practice. They might also provide unique opportunities to address some fundamental issues within cognitive science about the nature of memory, automaticity, transfer, and learning.

This is exactly what one would hope for a mature discipline like cognitive science.

CONCLUSION: AVOIDING NAÏVE REDUCTIONISM

Several years ago, I attended a neuroscience retreat where the attendees were briefed about the implications of the Human Genome project. In the discussion that followed, one neuroscientist suggested that given the rapid progress in genetics, it would soon be possible to identify the gene for musical intelligence. This is an example, coming from an otherwise sophisticated scientist, of attempting to map a highly complex but unanalyzed human and culturally defined skill onto a simple structure, a gene. It is an example of naïve reductionism. Cognitive science provides a theoretical perspective and a research methodology that allows us to avoid naïve reductionism. The previous examples illustrate the importance of looking beneath gross, unanalyzed behaviors if we want to understand how we learn, think, remember, and behave. Looking beneath gross, unanalyzed performance on standardized math tests to find the details of numerical cognition allows us to improve learning and teaching. If cognitive scientists want to meet the challenge of cognitive neuroscience, we will also have to look beneath gross, unanalyzed behavior and develop high-resolution cognitive models. We have to look beneath the obvious and at a higher level of precision to develop cognitive profiles to advance behavioral-cognitive genetics and evaluate therapies. We can improve rehabilitation protocols if we are explicit about the cognitive theories and assumptions on which those protocols are based and explore the implications of those theories and assumptions.

These examples also illustrate that cognitive science inhabits a unique position within the scientific hierarchy or network. The methods and assumptions of cognitive science allow cognitive scientists to reach up the hierarchy or out into the network to understand and study significant social and cultural problems such as those surrounding education. Those same methods also allow us to move down the hierarchy or deep into the network to understand and study brain function and human genetics. Cognitive science thus provides a fundamental scientific link, if not the linchpin, between the behavioral-social sciences and the biological sciences. For this reason, cognitive science is well situated to make fundamental contributions to research in a wide variety of skills. Thus, there are many opportunities for interdisciplinary collaboration, opportunities that will no doubt continue to heat up the action at the research front.

REFERENCES

Anderson, J. R., Reder, L. M., & Simon, H. A. (1996). Situated learning and education. *Educational Researcher, 25*(4), 5–11.

Anderson, J. R., Reder, L. M., & Simon, H. A. (1997). Rejoinder: Situative versus cognitive perspectives: Form versus substance. *Educational Researcher, 26*(1), 18–21.

Bellugi, U., Klima, E. S., & Wang, P. W. (1996). Cognitive and neural development: Clues from genetically based syndromes. In D. Magnusson (Ed.), *The lifespan development of individuals: Behavioral, neurobiological, and psychosocial perspectives (Proceedings of the Nobel Symposium)* (pp. 223–243). New York: Cambridge University Press.

Bruer, J. T. (1993). *Schools for thought: A science of learning in the classroom.* Cambridge, MA: MIT Press.

Carr, T., & Hinckley, J. (1997). *Cognitive abilities, language processing, and aphasia rehabilitation.* Unpublished manuscript.

Case, R. (1996). Introduction: Reconceptualizing the nature of children's conceptual structures and their development in middle childhood. *Monographs of the Society for Research in Child Development, 51*(61, Serial No. 246).

Dehaene, S. (1996). The organization of brain activations in number comparison: Event related potentials and the additive-factors method. *Journal of Cognitive Neuroscience, 8,* 47–68.

Dossey, J. A., Mullis, I. V. S., Lindquist, M. M., & Chamber, D. L. (1988). *The mathematics report card: Are we measuring up?* Princeton, NJ: Educational Testing Service.

Gardner, H. (1985). *The mind's new science: A history of the cognitive revolution.* New York: Basic Books.

Gelman, R., & Gallistel, C. R. (1978). *The child's understanding of number.* Cambridge, MA: Harvard University Press.

Glaser, R. (1988). Cognitive science and education. *International Social Science Journal, 40,* 21–44.

Greeno, J. (1997). Response: On claims that answer the wrong questions. *Educational Researcher, 26*(1), 5–17.

Griffin, S. A., & Case, R. (1996). Evaluating the breadth and depth of training effects when central conceptual structures are taught. *Monographs of the Society for Research in Child Development, 61*(1–2, Serial No. 246).

Griffin, S. A., Case, R., & Siegler, R. S. (1994). Rightstart: Providing the central conceptual prerequisites for first formal learning of arithmetic to students at risk for school failure. In K. McGilly (Ed.), *Classroom lessons: Integrating cognitive theory and classroom practice* (pp. 25–50). Cambridge, MA: MIT Press.

Groen, G., & Resnick, L. B. (1977). Can preschool children invent addition algorithms? *Journal of Educational Psychology, 69,* 645–652.

Jacoby, L. L., & Brooks, L. R. (1984). Nonanalytic cognition: Memory, perception, and concept learning. In G. H. Bower (Ed.), *The psychology of learning and motivation: Advances in research and theory* (Vol. 18, pp. 1–47). San Diego, CA: Academic.

Johnson, S. C., & Carey, S. (1998). Knowledge enrichment and conceptual change in folkbiology: Evidence from Williams syndrome. *Cognitive Psychology, 37,* 156–200.

Lenhoff, H. M., Wang, P. W., Greenberg, F., & Bellugi, U. (1997). Williams syndrome and the brain. *Scientific American, 277*(6), 68–73.

Logan, G. D. (1988). Toward an instance theory of automatization. *Psychological Review, 95*, 492–527.

McClelland, J. L., McNaughton, B. L., & O'Reilly, R. C. (1995). Why there are complementary learning systems in the hippocampus and neo-cortex: Insights from the successes and failures of connectionist models of learning and memory. *Psychological Review, 102*, 419–457.

McGilly, K. (Ed.). (1994). *Classroom lessons: Integrating cognitive theory and classroom practice*. Cambridge, MA: MIT Press.

National Science Foundation. (1988). National Science Foundation Survey. *Science, 241,* 408.

Newell, A., & Simon, H. A. (1972). *Human problem solving*. Englewood Cliffs, NJ: Prentice Hall.

Posner, M. I., & Raichle, M. E. (1994). *Images of mind*. New York: Scientific American Library.

Resnick, L. R., & Hall, M. W. (1998, Fall). Learning organizations for sustainable education reform. *Daedalus, 127,* 89–118.

Roland, P. E., & Friberg, L. (1985). Localization of cortex areas activated by thinking. *Journal of Neurophysiology, 53,* 1219–1243.

Siegler, R. S. (1981). Developmental sequences between and within concepts. *Monographs of the Society for Research in Child Development, 46.*

Siegler, R. S., & Klahr, D. (1982). When do children learn? The relationship between existing knowledge and the acquisition of new knowledge. In R. Glaser (Ed.), *Advances in instructional psychology* (Vol. 2). Hillsdale, NJ: Lawrence Erlbaum Associates.

Tager-Flusberg, H., Boshart, J., & Baron-Cohen, S. (1998). Reading the windows to the soul: Evidence of domain-specific sparing in Williams syndrome. *Journal of Cognitive Neuroscience, 10*(5), 631–639.

Temple, E., & Posner, M. I. (1998). Brain mechanisms of quantity are similar in 5-year-old children and adults. *Proceedings of the National Academy of Sciences, 95*(13), 7836–7841.

Van Gelder, T. (1998). The dynamical hypothesis in cognitive science. *Behavioral and Brain Sciences, 21,* 615–628.

Wang, P. P. (1999). Cognitive dissection of Williams syndrome. *American Journal of Medical Genetics, 88,* 103–104.

9

Making Interdisciplinary Collaboration Work

Susan L. Epstein
Hunter College and The Graduate School of The City University of New York

Cognitive science is, by definition, an interdisciplinary endeavor, one where the investigators have their formal training in different academic disciplines. This chapter is an informal study of how 21 people enjoy and succeed at it with researchers from other disciplines. It is based upon interviews and discussions with computer scientists and neuroscientists, psychologists and linguists, even a philosopher or two. Each of them is, by the standards of publication and research funding, a successful and expert interdisciplinary collaborator.

This report is anecdotal; it summarizes hours of conversation and pages of electronic mail, interspersed with some personal experience. Wherever possible, quotations appear to let the respondents speak in their own experienced and authoritative voices. At their request, these are unattributed. A full list of the participants appears at the end of this chapter. Many of them generously volunteered details of failures, as well as successes, from their own experience at interdisciplinary collaboration. In this chapter, *collaboration* is assumed to be joint work among two or more researchers of similar status (Thagard, 1997). Collaboration is not just reading and writing about the other person's work. As one respondent said, "That's working in parallel." Nor is it collaboration when one person asks all the questions, and the other provides all the answers. The same person noted, "That's getting help." Collaboration often entails a degree of role play, in which the participants alternate as learner and instructor.

Although this chapter is focused upon the particular features of interdisciplinary collaboration in which the investigators have their formal training in different disciplines, there are, of course, ingredients essential to any good collaboration. An obvious, but non-trivial one is willingness to work with other people. As one scientists ruefully observed, "Some people just don't like to collaborate." Arrogance or vanity on the part of any participant makes collaboration very difficult. Healthy mutual respect and good will are also important. A successful collaborator pointed out, "You [need to] consider each other equals." Because disagreements, over authorship and funding, as well as more intellectual matters, are inevitable, "you need confidence in yourself and respect for the others." One scientist found that taking a long-range view is helpful: "You need to believe that there will be results beyond a single paper." And, of course, fundamental human relationship skills are essential. Another researcher suggested, "Tell [your readers] to be nice ... don't compete ... have a healthy respect for each other." Stability in one's work and personal life is also important; several tales of failure emphasized how difficult collaboration became when people changed affiliations or became parents, particularly for the first time. Successful collaborators are volunteers; collaboration imposed upon its participants by an institution or funding agency is unlikely to succeed.

Certainly, there is a remarkable amount of interdisciplinary work out there. Examples in this study included

- A psycholinguist and a computer scientist working in a computer science laboratory on the rapid detection of eye movement in natural settings.
- A team of pediatric neurologists, developmental neurobiologists, psychologists, and psycholinguists studying early childhood language development.

- Philosophical logicians and computer scientists formulating an ontology for geography complete with a Web site for ontology engineering in a philosophy department.
- A computer scientist and some psycholinguists shaping theoretical models to correlate with results on humans.

Researchers cite three primary motivations for interdisciplinary collaboration. First, the topic may demand it. As one scholar observed, "My area is inherently interdisciplinary ... we need a confluence since no one discipline has a sufficiently rich language or a sufficiently rich perspective, let alone an answer [to these questions]." Another noted, "My area by definition requires broad-based knowledge for success anyway ... by necessity you must reach out." The second common motivation for interdisciplinary collaboration is that it can provide a new perspective. One scientist asked, "Why get caught up in a single approach of your discipline?" Another said, "It's stimulating and thought provoking ... it gives me access to other techniques." The third motivation is that interdisciplinary collaboration provides additional knowledge. One person wrote, "I always was very goal-oriented and felt that if I wanted to try to study a particular topic that spanned more than one discipline, why should I reinvent the wheel when I can work with people who specialize in that area I want to include in my study?" Another said that he turned to people in other disciplines because "they had tools and skills I needed." Still another recalled, "I had been doing a lot of experimental work which focused on various phenomena in isolation, and was beginning to feel frustrated by the lack of a coherent underlying account of the various fragments." Pragmatism is indeed a driving factor. As one researcher put it, quite matter-of-factly, "they had skills I wanted to learn in order to answer some questions I was interested in."

The remainder of this chapter focuses on the keys to a good interdisciplinary collaboration, recounts some stories, and offers advice for those about to embark upon it.

KEY ELEMENTS

The key elements to a successful interdisciplinary collaboration are attitude, communication, time, proximity, institutional climate, funding, roles, appropriate topics, and publication. They are considered one at a time.

Attitude

The successful interdisciplinary collaborator is, of course, a good collaborator in the ways just described. The researchers should like each other. As one observed, "[It helps when you get along well because] you have so much else to surmount." They should also have complementary talents and expertise. One long term collaborator, in reflection, said, "We agreed a lot to begin with ... [but] both believed the other had valuable things to communicate." Another observed, "You have to feel like it adds to both your work." As a result, successful interdisciplinary collaborators appear to impose limits on the exchange of knowledge. One researcher asked, "It's too hard to learn that too—why should a psychologist learn to code a simulation?" Another insisted, "becoming an expert in each other's empirical techniques requires too many resources." It does help, however, if you share a theoretical framework and can use different empirical techniques. A scientist noted that "Wrapping your head around each other's theoretical perspective is essential." Nonetheless, some knowledge of other domains is useful. One respondent found interdisciplinary collaboration natural because "I'm enough of a dilettante." A good interdisciplinary collaborator is, first and foremost, receptive. One particularly prolific collaborator said, "Everybody who has touched my work has made it better." In addition, however, one must have a good reason to work outside one's disciplinary rules. As a thoughtful respondent observed, "My thinking style is basically analogical ... I like to look for convergent evidence.... It's easier for me to look for themes than to concentrate on detail."

Clearly, a collaborator must be open to different disciplines. A senior researcher suggested that "Security in your own work helps you open up to new ideas." Another said, "[It helps when] researchers take an interest in big questions." Several people indicated that their early experience with interdisciplinary work, as a graduate student or a postdoctoral researcher, had particular influence. One person had to form his own interdisciplinary thesis committee; another was a psychologist with a civil engineer on her committee. The collaborator may also need to be resourceful about locating colleagues. Several stories hinged on recognizing a need for an individual to do a particular task and then "find[ing] someone you know to do it ... [you have to choose a person] you like and trust ... someone who works hard and has a similar working style." (That last requirement, however, seems to be the one most often compromised.)

Once a team is formed, it is essential to know or learn about each collaborator's field. Recounting one collaboration, a respondent indicated that start-up was relatively easy because "I already knew a fair amount of linguistics anyway." Efforts to learn are recognized, and appreciated. A scientist observed of another, with great admiration, "[she] would go miles to learn." Many an interdisciplinary collaboration succeeds or fails based on the ability to bow to each other's expertise. Respondents said how important it was to view interdisciplinary collaboration as a partnership, not as the exploitation of someone's useful technical skills. One insisted that, if it is to last, "You have to feel you are both doing something new."

People often engage in several interdisciplinary collaborations "because you have lots of ideas." Some are enduring; others produce a single paper after a year or two and cease. Three of my respondents had extensive interdisciplinary collaborations with their spouses as well as with others. Most agreed that "marriage" was an apt metaphor for an interdisciplinary endeavor. One person suggested that, like marriage, an interdisciplinary collaboration "start[s] off tentative, careful" and that you gradually "build up shared agreements and ways to get out of shared disagreements." A sense of humor was frequently cited as essential. Collaborations were described with warmth: "It was fun"; "we were really having a good time ... humor diffuses things"; and "we kidded around a lot ... a good mix of having fun and working hard ... there's really got to be joy in it."

Communication

Even with the right attitude, however, interdisciplinary collaboration requires careful attention to language. Despite several years of work together, one participant noted, "[we] don't always understand each other the first time." A psychologist speaking with biologists and computer scientists about neural networks confronted what he termed "field-specific dialects. Thus although we're discussing a problem that may be relevant to cognitive science, our 'classic' training in specific disciplines makes it difficult to 'connect' appropriately." His computer scientist colleague reported a similar difficulty when she spoke with another psychologist. Fundamental terminology should be established early and reviewed regularly. This may mean that the minority chooses to speak the majority tongue. A computer scientist, for example, working with physiologists said, "I [learned and then] used their language to communicate." Alternatively, a common vocabulary can be deliberately constructed. A member of a group of collaborators reported, "We developed a third language [neither computer science nor neuroscience] that we could all use. [We] chose some known terms and then deliberately coined others that captured the ideas we were trying to address."

Discipline-specific jargon is easily acquired; it is the undetected, specialized use of the same words that creates real problems. For the first 2 or 3 years of our work together, every time we sat down to talk one of my own collaborators extracted from his shirt pocket a 3 × 5 card with a list of words and definitions supposedly in both our vocabularies, but ones that I defined in ways quite different from the way he would. "There," he would say, slapping the card on the table and scanning its contents, "that's Susan speak. Now we can talk." I attribute a great deal of our success together to that deliberate effort on his part. In contrast, one respondent spoke about what was, for him, a rare failure: "I thought we were talking the same language ... we had great conversations in the cafeteria ... it took me 6 months before I realized that [we were speaking different languages]." Another reported in frustration, "We had a real honeymoon for 6 months until we discovered that we had been using the most basic terms, such as 'symbol' with entirely different meanings." Other words that arose in similar stories were "representation," "schema," and "grammar."

Early vigilance is well repaid. Long-standing collaborators report no consciousness of any language problem whatsoever. They have formulated a common context. As one member of a pair said, "we have such a bandwidth together.... We share most of our models with each other.... [There is] no need to rebuild [a shared] structure." My own recounted collaboration remains alive and well, but I have not seen that 3 × 5 card in some time.

Time

Given the right attitude and accurate communication, time becomes an issue. Particularly in the first few years, interdisciplinary collaborations "take more time [than those within discipline] ... which I don't mind because I enjoy them more." There is learning to be done and a vocabulary to be established. One researcher said it was important to have students and collaborators present to each other regularly, another time-consuming activity. Even after the initial period, interdisciplinary work seems to be slower. One book authored by a group took, by one estimate, "three times as long to write [as it would have had I done it myself, and yet it was] a very positive experience." One respondent told of an interesting collaboration that he refused because of its fast-approaching deadline.

Academic obligations often make time allocation a real issue. One scientist, while happily cataloguing his many fruitful collaborations, repeatedly pointed out how much more freedom he had than his academic collaborators because his work situation does not depend on teaching responsibilities. In another collaboration there were "great connections" still influencing [the collaborators'] work, but the group never published because a key member was too busy. That cost them their funding and ended the collaboration. Even one overextended collaborator can create serious problems.

Proximity

Once a well-motivated, clearly communicating team makes time to work together, proximity is important. One researcher defined *face time* as time collaborators spend in the same room for the sole purpose of their mutual goals. During face time, he said, "You have their attention." Another observed, "You know, there really is no substitute for 2 hours of talk." Telephone conversations, in contrast, permit distractions. One respondent asked, "Did you ever hear someone typing at a keyboard while you spoke to them?" During face time, such behavior would be considered rude. Even during face time, care must be taken to control many potential office distractions: the telephone, alerts about new e-mail messages, and students and colleagues who stop by to visit. Homes present special distractions of their own. One group of collaborators met in the evening always at the home of the only member in the group without children. In another collaboration between spouses, meetings were deliberately scheduled outside their home, and often included a meal.

Face time provides more than undivided attention. It is motivational, because it puts social pressure on each person and leads to insights unavailable otherwise. Work that entails diagrams, vision, or spatial reasoning is particularly reliant on face time because drawing is intrinsic to communication.

Face time is particularly essential at the beginning of a collaboration. It enables you to learn the intonation and social aspects implicit in your communication early. Later, when your communication becomes primarily electronic, you can correctly interpret the nuances you learned during that initial contact. Many long-term interdisciplinary collaborations start out with a regular weekly meeting (an hour or an evening) dedicated to the purpose of mutual education, with the hope that appropriate topics will eventually surface. Although one respondent insisted that a 20-year, very fruitful collaboration had always been long distance, under repeated questioning, she eventually recalled quite a different start. Although they had been acquainted for some time, only later, when she was on sabbatical at her collaborator's institution, did they have 6 months of face time in which to brainstorm. Since then, their work has entailed "regular annual meetings, which lasted varying periods of time."

Face time not only begins an interdisciplinary collaboration, it keeps it alive. It should come as no surprise, then, that such collaboration is easier within a small geographical area. Visiting is expensive, time consuming, and stressful, typically involving hotel stays, extended child care, and air travel. One scholar recalled better days for a collaboration in which she had been involved for several years:

> [Two of us] had offices next to each other.... We talked about the research almost every day, we were [there to] answer questions, bounce ideas off of.... Once I left ... there was a definite decrease in the amount of work that [got] done on this project and also the clarity of the research question. Having [the third collaborator 500 miles away] wasn't really a problem, we used e-mail to tell him how things were going, but it was definitely better to have at least two of the 3 collaborators in the same building. I'm not saying it's impossible to work across institutions, but this leads to the other main factor: money.

Funding

Interdisciplinary collaboration tends to be expensive. Indeed, most respondents identified funding as their greatest problem in interdisciplinary research. One person noted, "No outside funding slows you down." Another acknowledged, "It is important to be able to pay your collaborator." Interdisciplinary collaboration often requires the services of programmers who are in great demand and command high salaries. One psychologist wrote, "the stimulus I use ... tends to be more complicated ... and I ... do not know how to program, so I need money to hire someone [or] the stimulus doesn't get developed." That required proximity is expensive too. Indeed, an experienced collaborator characterized a "good visit" as one that lasted 2 weeks to 3 months and required a comfortable place to stay, a kitchen, a rental car, and copy and telephone privileges.

Many collaborators said that it was far easier to secure funding for their own work within discipline than for interdisciplinary proposals. There were a few noteworthy exceptions. Linguists and computer scientists readily find support, presumably because natural language processing is a well-established subfield of artificial intelligence (AI). One scientist, much of whose work was "industrial related," had several collaborators who had moved to industry from which they could, and did, readily offer support. A larger team credited its success in part to its early development under the aegis of the MacArthur Foundation, a farsighted supporter of long-term, interdisciplinary research in the cognitive and behavioral sciences. Indeed, once funding is secured, interdisciplinary collaboration begins in earnest. Several people recounted stories in which preliminary investigations blossomed into 10 to 20 papers after funding.

Regrettably, the review process for interdisciplinary work appears to be particularly problematic. One respondent noted that, "In [our country] now they claim to want to fund interdisciplinary work, but a proposal goes to a panel in a single discipline, so it is necessary to subordinate the work of all to the work of one." If a review panel is constructed from experts in the individual disciplines, their disciplinary expertise may not be relevant, for example, the "wrong" branch of psychology. One researcher suggested that expertise in one of the fields involved in the proposal is not sufficient, that each reviewer should also have had experience as an interdisciplinary collaborator. Another told of a review panel that took the proposal's description of a successful prototype experiment as evidence that the group had already learned how to collaborate, and therefore did not require funding.

Institutional Climate

Interdisciplinary collaboration is particularly influenced by its environment. Institutions can deliberately become "homes for people who work across the traditional boundaries, and also people who, while rooted in a discipline, need support to spend time on stuff that isn't recognized by their discipline." This approach is often termed *horizontal*, because it encourages people to organize themselves around a problem rather than around a discipline or subdiscipline. In one deliberately interdisciplinary environment, newcomers must offer a 3-month tutorial on their fields, and there is a weekly internal afternoon symposium and an annual 4-day retreat to discuss mutual goals.

Because, as we have observed, interdisciplinary work takes more time, researchers involved in it must be relieved of other duties. Not surprisingly, many interdisciplinary collaborators teach at universities that reduce the course loads of active scholars. Supportive departments readily sponsor a collaborator as a research scientist, encourage visits, and invite talks from "neighboring" disciplines. Supportive communities are described as having a "multimethodological tradition." Regrettably, the relative scarcity of cognitive science programs and departments in universities suggests that a narrow perspective remains all too common.

Because physical proximity fosters interdisciplinary work, researchers in companionable disciplines should also be near each other. One respondent observed an environmental irony that "at small colleges you're more likely to share a building with philosophers and linguists and develop across discipline collaborations, but being at a small college there is less institutional support and more teaching and community service responsibilities." In a large institution, physical proximity can be dictated to deliberately foster interdisciplinary work. For example, at AT&T's famed Murray Hill laboratory, scientists of the same discipline were deliberately scattered in offices throughout the complex. A typical hallway housed a mathematician next to a chemist next to a physicist. Geographically, these were "people accustomed to playing in each others' backyards." The single, attractive dining room served excellent, inexpensive meals; all its tables were round and readily seated 8 to 10 people. The laboratory itself was relatively isolated, with few outside mealtime alternatives. Not surprisingly, lunch time conversation was wide ranging, fertile ground for future collaboration.

Another institution that deliberately nurtures interdisciplinary research is the Center for the Study of Language and Information, an interinstitutional laboratory housed at Stanford University. One scholar noted that, "One crucial thing ... is the building itself. It is almost impossible to move around in it without meeting people having a discussion about something. The 'meeting places' are at corners of the corridors, so that informal, and even sometimes formal, meetings have people wandering past all the time." Respondents cited examples of other places where interdisciplinary collaboration prospers: the University of California at San Diego, the Beckman Institute for Advanced Science and Technology at the University of Illinois, the Santa Fe Institute, and the Institute for the Interdisciplinary Study of Human & Machine Cognition at the University of West Florida.

Institutions that successfully foster interdisciplinary research address "the importance of nonintellectual factors ... [that can] provide means to flourish ... [and encourage] social bonding." One research facility takes care to begin everyone's contract on the same calendar day so that no newcomer has more seniority than any other. The same institution mandates attendance at 4 o'clock tea and coffee every afternoon, with cake provided by the author of any newly accepted or newly published paper. Thus, the researchers celebrate each other's achievements and are encouraged to collaborate on new ones.

The Canadian Institute for Advanced Research (CIAR) ran a program from 1985 to 1995 to encourage interdisciplinary research among about 15 top university researchers in a variety of disciplines. To give these scholars the requisite additional time, CIAR released them from all obligations except one graduate-level course per year. To provide face time, CIAR required them all to attend an annual 3-day meeting. At this meeting, and at the 3-day workshops for subgroups, "increasing levels of sophistication [in each others' disciplines] were encouraged" because participants were expected to present their work to each other. Many of the scholars found this a supportive experience.

Roles and Learning

Most collaborators report that their work is facilitated by allocating responsibilities. This "useful division of labor" maintains progress (Thagard, 1997). The respondents believed that collaborations required a leader to define the common problem and the language in which to discuss it, to set priorities, and even to target publications. In small teams, they reported, the leadership assignment may alternate, and it is important to recognize your role in a particular situation. One respondent, however, insisted that consensus could govern instead, substituting patience and listening skills for a single leader. The impressive history of that group is ample support for this alternative model.

A second important role identified by many respondents was that of the facilitator. This person frequently translates the language of each collaborator for that of the others, and must be trusted by everyone. As a rule, the facilitator is not the leader, and is often a graduate student. One respondent describes this as "a young, enthusiastic, intelligent person

... learning from a linguist and a psychologist and a computer scientist, and ask[ing] questions to each from the other's perspective, noticing links (and mismatches between usages and assumptions) that the teachers themselves would not otherwise discover." As another scholar indicated, "Either one of you develops mastery in the other's field, or one of you has a very smart student that does." For some very busy, senior researchers, such a student is essential for peer collaboration "because [otherwise] I'm too busy." In other peer collaborations, one discipline appears to intermediate between others, and its representative serves as the facilitator.

Even with a facilitator and a leader, learning is essential and "it's harder than you think [it is going to be]." As one researcher reported, "despite a great deal in common, we found that there were ... large areas where the differences in background meant that a great deal of mutual education was necessary."

Appropriate Topics and Publication

Given the right individuals, good communication, a supportive environment, and plenty of time and money, interdisciplinary collaboration still requires a topic. "Some topics just beg for collaboration." A good interdisciplinary topic is couched in several disciplines, none of which is particularly advanced toward a solution. It is also a problem of mutual interest. One successful collaborator recounted several different start-up attempts that failed. In each case, he had met regularly to "brainstorm with the explicit intention of collaborating, but ... [we] couldn't find an issue that grabbed both of us equally." Such complementarity is important. One successful collaboration began because, one scientist said, "Each of us felt we had half the answer." A good topic often demands new methods, costly facilities, or subjects available to only one of the collaborators. As one scholar indicated, "collaboration will let me solve problems I could not solve on my own ... because I lack tools that I see as relevant." Several people suggested that applied research was likely to be interdisciplinary.

As work is accomplished, one individual reported that "The major stumbling blocks ... have to do with who takes responsibility for writing the work up, where is it submitted, and who gets identified with the work." Typically, one person drafts the paper according to its targeted venue and is therefore the first author. (One team, however, chose always to publish alphabetically, to avoid "silly wrangles.") The drafter typically assigns sections to the others, or leaves "holes" for a collaborator to fill in with data, elaboration, or scholarly support from the other's discipline. The first author also becomes identified with the work in the discipline in which the publication appears. Very strong results double the places available to publish. For smaller interdisciplinary results, however, finding a venue is likely to be much more difficult.

CASE HISTORIES

This section chronicles some scenarios in interdisciplinary collaboration. John-Steiner (1998) described four patterns of collaboration: distributed, complementary, family, and integrative. The respondents provided many examples of each type; their assignment to categories is my own.

A Distributed Collaboration

Distributed collaborations are "characterized by informal, voluntary roles, similar interests, and spontaneous and responsive working methods" (John-Steiner, 1998). Participants in a distributed collaboration "exchange information and explore thoughts and opinions" (John-Steiner, 1998). One philosopher respondent told of a model he had promulgated in a paper at an annual meeting of the Cognitive Science Society. A psychologist disagreed with the model, and they ultimately collaborated in a series of computer-based simulations and experiments on people. Their results refuted the original model and appeared in a paper at a subsequent meeting of the same society. With the resolution of the issue, the collaboration ended, leaving both participants more knowledgeable.

A Complementary Collaboration

Complementary collaborations form the bulk of the work described here; the researchers partition the work according to their skills, knowledge, and temperaments. Individual strengths are combined to develop and disseminate a body of work. These strengths include mathematical skill, well-equipped laboratories, organizational talent, writing ability, and a network of colleagues. Participants in a complementary collaboration bring pieces of the same puzzle to the table. One respondent, a linguist by training, wrote

> I had been doing a lot of experimental work which focused on various phenomena is isolation, and was beginning to feel frustrated by the lack of a coherent underlying account of the various fragments.... So I began to hang out at [psychology] lab meetings [in my own university], and then began what turned into a long-term collaboration with [a psychologist]. Whereas I had a solid background in the domain of speech perception, but little experience developing behaviorally-oriented computational models, [he] knew more about modeling but little about speech perception.

The result was a set of ground-breaking papers.

A second example of a complementary collaboration occurred among a group of four scholars: a cognitive psychologist, a social psychologist, a computer scientist, and a philosopher. Initially they met weekly for a year, educating each other and establishing good communication. At the end of that year, they decided to write a book together. After the group outlined the project, one person was assigned to draft each chapter, but revision was thorough. One of the authors wonders, "Who knows if there were even sentences of the first chapter that wound up in the book?" One of the chapters had at least 20 versions. Nonetheless, he believes that the resulting book was only possible as a group project; "no one [of us] could have done it alone."

Several years later, two of those four authors wrote a second book together. This time it was "much easier ... [because] we had a common theoretical framework to start." They met only once in 18 months of writing, at a dinner to work on one problematic chapter. "Neither of us knew much about [it] so we each wrote a draft and then merged them—I think it's the most fun chapter in the book."

A Family Collaboration

In family collaborations, people interchange roles more often and share a common expertise. One remarkable example of a family collaboration is a developmental research center begun with long-term funding from the MacArthur Foundation and perpetuated through grants from the National Institute of Health. Fourteen researchers from psychology, pediatric neurology, and developmental neurobiology have collaborated within it for more than 10 years. The team works collectively, always seeking consensus in their weekly meetings, even on budgetary matters. They organize their efforts horizontally, along questions rather than disciplines. Their communication, after so many years, is effortless, and their publication record exemplary.

An Integrative Collaboration

John-Steiner (1998) believed that the epitome of collaboration is integrative, an intense, long-term relationship in which people share a vision, draft together, and ultimately construct "a new mode of thought." Collaborators in such relationships no longer distinguish substantial segments of ideas or results as the property of a single individual. The production of a recent book is an example of integrative collaboration in which a team of six authors forged a new perspective together. The group consisted of two psychologists, a developmental neuroscientist, a psycholinguist, a computational linguist, and a cognitive scientist. Each member had worked with at least one of the others previously, and had participated in a training program that involved both postdoctoral students and senior scientists. At the close of the training program, they decided to write a book. As one of the authors recounts, "Initially, we envisioned an edited book with each author contributing one or more chapters. But in our early planning sessions we discovered that there were some deep issues that needed discussion first. So we began what turned into a 3-year informal proseminar." The collaborators were spread around the world, but they met once or twice a year. "We secluded ourselves and met intensively for 6 to 8 hours a day, for 5 to 10 days.... The process was hard because, despite a great deal in common, we found that there were significant points of disagreement." Although they began as, and remained, good friends, frustrations arose. One participant recalls some theoretical disagreements so severe that people wanted to pull out, but the "solid commitment to the importance of what we were doing" held them together.

Eventually, the six had a joint outline that described the contents of each chapter. Then, in contrast to the book collaborations described previously, every author wrote a working paper for every chapter. One author noted, "I had a two-foot pile of stuff." After each person had read all the drafts, teams of two or three drafted each chapter. Everyone worked on at least five chapters, sharing a computer or writing fragments. One chapter in particular went through four full versions over 18 months. Finally, one person took the responsibility for unifying the text and style. His final verdict, "The outcome [after four years] ... was deeply satisfying. All of us felt changed by the process .... [It] was clearly something that none of us could have written alone."

SOME CONCLUDING ADVICE

Crossing disciplines is fraught with peril. In some sense, a discipline forms one's outlook. As one respondent put it, "A disciplinary affiliation is more than just a body of knowledge and expertise: it's also a set of cultural attitudes." Because a discipline defines what is expected of and valued in a professional, it may reject what is foreign on those terms. We are often narrowed in our education and narrowed once again as we become experts in our field. By definition, interdisciplinary collaboration contends that there is more than one way to address a problem. Therefore, one must be willing to step outside one's culture and training, to consider other approaches once again.

A good example of this challenge to one's perspective is the development of a course on computational models of learning, team taught in one school by a psychologist, a biologist, and two computer scientists. At a meeting to formulate the syllabus, the psychologist related, "We had a hard time agreeing on the appropriate way to structure the course. The biologist wanted to spend the first third of the course on the nature of brain and neuron structure, the CS people wanted to do basic computation, [and] I wanted to establish the relevance to modeling human behavior." The integration of ideas from different disciplines is nontrivial; it requires open and creative minds.

Make certain that the intellectual community is receptive to your intended work. Not only must the individuals be ready to collaborate but the disciplines must be amenable as well. Several respondents told stories of disciplinary arrogance, particularly of professionals in other areas who viewed computer scientists merely as support personnel, "just another techy." Even award-winning research in the others' area did not improve the computer scientists' status in the eyes of their collaborators. One respondent noted, "I started looking at [a computer application for a particular field] before the people I tried to work with even had accepted that computers were useful at all! Similar lack of readiness can also be found in more formal work—if one group has no formalism yet for describing what they do … and another group comes in saying they wish to formalize their domain, [that interdisciplinary collaboration] won't work."

Anticipate communication problems, and discuss your language carefully. At the beginning, there may be a brief, exciting honeymoon until you hit the methodological and terminological differences. Sometimes, as in the case of a neuropsychologist and an AI researcher trying to match their models for the same phenomenon, you may confront deep assumptions too late.

Remember that learning is required. One respondent noted that "You can't just read the other guy's papers," but you can ask for a few classic papers to read in the other's field. If you know that proximity is out of the question, start well and schedule face time, either on site or at conferences of mutual interest. Work on getting your travel schedules to overlap.

Given their demands, it is easier to break off an interdisciplinary collaboration than one within discipline. One also runs the risk of no results. A researcher who was not an academic pointed out that he could take more risks than his academic colleagues because he could "afford a project failure." Another respondent suggested that "senior folks ... have the luxury of playing around and taking risks." Senior faculty who do so thereby encourage their own students to be broad-minded, that is, serve as role models for them. Alternatively, another respondent suggested that graduate and postdoctoral students were ideal for interdisciplinary work because they "had the time and energy to broaden" themselves. There remains, however, the problem of tenure. One respondent described a successful facilitator who had "had his fun [and then had to] settle down to [his] real work," that is, restrict himself to the discipline of his academic appointment as he moved toward tenure.

Another risk of interdisciplinary collaboration is that your results may be unwelcome. One researcher offered another researcher substantial insight that enabled him to create and execute experiments with unanticipated results. The experimenter drafted a paper, but its results did not support the insightful collaborator's theory, so she declined to coauthor. The author published the paper alone, and warned "you can't come [to interdisciplinary collaboration] with an agenda."

Of course, interdisciplinary collaboration offers remarkable rewards as well. An expanded perspective is virtually guaranteed. One respondent noted that "the difference in perspective you gain [by looking at a problem from vantage point of another discipline] is liberating." Another praised interdisciplinary collaboration because "[It] forces me to think in very different ways." One psychologist, working with a team of computer scientists, found that their questions inspired the design of new psychological experiments even as hers inspired new simulation techniques. Substantial results are often the case; in 12 years of work together, a linguist and a computer scientist coauthored more than 30 papers. Another researcher described a tremendous synergy that was "multiplicative, not additive."

With or without publication, there can be powerful intellectual benefits. Most collaborative results are coconstructions; but even better is appropriation, "the incorporation of jointly constructed ideas into one's own thinking"(John-Steiner, 1998). "Unless it changes you [interdisciplinary collaboration] remains just a division of labor ... you need to appropriate some of [the other collaborators'] knowledge ... individuals become more effective in formulating problems in ways that their own disciplines do not." This kind of integration is not necessarily sustained throughout the entire period, but it is a hallmark of the best collaborations.

In our specialized world, many people are doing similar, related work. One respondent noted that "putting their results together is a big win." Interdisciplinary collaboration often provides enrichment and information to other fields. Another researcher enthused that "Great things can happen that would never be possible otherwise." That includes close personal friendships. One scientist said it very well: "[one recent interdisciplinary collaboration has been] very intense intellectually ... the most stimulating thing I've ever done."

ACKNOWLEDGMENTS

My thanks to Vera John-Steiner, who generously shared her enthusiasm and expertise on this topic. I am also indebted to each of the following for their continued interest, enthusiastic support, candid conversations, and thoughtful writing. They are, indeed, superb collaborators.

- David Aha, Navy Center for Applied Research in Artificial Intelligence
- Liz Bates, Department of Psychology and Department of Cognitive Science, University of California at San Diego
- B. Chandrasekaran, Department of Computer and Information Science, The Ohio State University
- Tony Cohn, School of Computer Studies, University of Leeds
- Kenneth D. Forbus, The Institute for the Learning Sciences, Northwestern University
- Jeff Elman, Department of Cognitive Science, University of California at San Diego
- Robert Dufour, Department of Psychology, Swarthmore College
- Jack Gelfand, Department of Psychology, Princeton University
- Dedre Gentner, Department of Psychology, Northwestern University
- Gerd Gigerenzer, Max Planck Institute for Human Development

• Jordan Grafman, Cognitive Neuroscience Section, National Institute of Neurological Disorders and Stroke

• Pat Hayes, Institute for Human and Machine Cognition, University of West Florida

• Jim Hendler, Department of Computer Science, University of Maryland

• Michael Levison, Department of Computer and Information Science, Queens University

• Susan Lederman, Department of Psychology and Department of Computer and Information Science, Queens University

• Dana Nau, Department of Computer Science, University of Rochester

• Mary Jo Ratterman, Department of Psychology, Franklin and Marshall

• Larry Reeker, National Science Foundation

• Paul Thagard, Department of Philosophy, University of Waterloo

• John Tsotsos, Department of Computer Science and School of Medicine, University of Toronto

• Steven Zucker, Department of Computer Science, Yale University

REFERENCES

John-Steiner, V. (1998). *Creativity and collaboration: A sociocultural approach*. Paper presented as the Annual Research Lecture, April 1997, Office of Research Services, University of New Mexico, Albuquerque, NM.

Thagard, P. (1997). Collaborative knowledge. *Noûs, 31,* 242–261.

10

Interdisciplinarity: An Emergent or Engineered Process?

Yvonne Rogers
Mike Scaife
University of Sussex

Antonio Rizzo
University of Siena

Cognitive science should be more than just people from different fields having lunch together to chat about the mind.
—Thagard (1996, p. 7)

There is a widespread view that interdisciplinary research is a good thing. Interdisciplinarity usually means something like the emergence of insight and understanding of a problem domain through the integration or derivation of different concepts, methods, and epistemologies from different disciplines in a novel way. However, it is also widely believed that "true" interdisciplinarity is very difficult to achieve and more often than not remains an elusive goal. In practice, many self-styled interdisciplinary enterprises actually work at the level of being multidisciplinary (or pluridisciplinary): when a group of researchers from different disciplines cooperate by working together on the same problem toward a common goal but continue to do so using theories, tools, and methods from their own discipline and occasionally using the output from each other's work. They remain, however, essentially within the boundaries of their own disciplines both in terms of their working practices and with respect to the outcomes of the work.

In this respect, it is illuminating to look at the case of cognitive science in which interdisciplinarity was explicitly intended to be the defining feature of progress. Despite the aims of its founders (see later) to integrate disciplines in terms of differing levels of description and analyses, cognitive science has been predominantly a multidisciplinary activity with psychologists that continue to work as psychologists, philosophers as philosophers, and artificial intelligence (AI) researchers as AI researchers, each continuing to use their own tools, methods, and terminology. Indeed, at the 1998 Cognitive Science Conference—which had as its central theme interdisciplinarity—there was not one paper presented on it. Instead, the majority of papers apparently reflected research carried out within a single discipline with only a few combining methods from two disciplines, which tended to be biased heavily toward psychology and AI.[1]

Our aim in this chapter is to examine some of the main assumptions that lie behind the ideal of interdisciplinarity and to ask, When is it is really necessary? How might it be realized? What alternatives are there to help cognitive scientists progress research when they find themselves frustrated and thwarted by the limits of their own discipline's conceptual and methodological armory? We pose these questions in the context of first, previous attempts to consider the issue, and second, in terms of the challenge(s) for contemporary cognitive science. Finally, we shall conclude by a brief examination of some of our own work, reflecting on how we dealt with bridging across disciplinary limitations.

---

[1]We also discovered that on the form for reviewers to judge submissions to the conference there was a question that asked which of the contributing disciplines the paper best fitted (e.g., psychology, computer science, philosophy)—with the omission of cognitive science itself from the possible categories.

RECEIVED VIEWS ON THE MULTIDISCIPLINARY–INTERDISCIPLINARY DISTINCTION

The terms *multidisciplinarity* and *interdisciplinarity* are often used interchangeably to refer to researchers from different disciplines or backgrounds coming together to collaborate on a common goal, be it basic or applied research (e.g., Brown, 1990). At a basic level of research, both terms have been used to describe the way a group of academics from different disciplines (e.g., psychologists, anthropologists, economists, and computer scientists) may collaborate to develop a more extensive understanding of a situation or phenomenon, whereas at an applied level, they have been used to describe the process of bringing a team of professionals together from different backgrounds—such as designers, educational technologists, computer scientists, and human factors specialists—to develop a product such as building a software application (e.g., Kim, 1990). Using the two terms interchangeably is not problematic if they are being used simply to refer to some kind of cooperation or collaboration between different people. However, the terms can have quite distinct meanings when used to denote different processes of collaborative activity. For example, bringing together a group of experts from different disciplines or professions to contribute to a single project, which would not be able to be accomplished by any one profession alone, is not necessarily the same process as when a group of researchers from distinct disciplines try to generate novel concepts and integrate different levels of explanation. The former may be considered a form of multidisciplinarity in which each person contributes their expertise to the project; the latter is a form of interdisciplinarity in which novel research questions are addressed (cf. The Royal Society's, 1996, position on the distinction between them). In this sense, the main difference between multidisciplinarity and interdisciplinarity lies in the mechanism of the research process and, relatedly, its outcomes. Interdisciplinary approaches are assumed to derive novel concepts, methods, and theoretical frameworks through the melding of concepts, methods, and theoretical frameworks coming from different disciplines. An example is ecological economics in which scientific aspects of ecological events have been integrated with their social consequences to make objective assessments of ecological aspects. By contrast, multidisciplinary approaches are assumed to evolve new understanding through adapting and modifying existing concepts, methods, and theoretical frameworks within a discipline and occasionally borrowing ideas from others. Here, we would classify the majority of research carried out in cognitive science (see also Schunn, Crowley, & Okada, 1998).

This gloss has the benefits of polarizing the distinction between the two terms and we use it following to consider what is really important when trying to specify good working practices. However, we are well aware that such descriptions as we have offered previously do, of course, beg many questions. For example, the very idea of a "discipline" as something coherent in terms of its methodology and theory and of the assumption of a single agreed internal language is itself suspect. Within psychology, for example, the chasm between the various "branches," such as social and cognitive, makes the assumption of disciplinary inclusiveness an interesting one. We need also to be aware that the practices within a discipline may owe more to internal political agendas than we would like. In short, although we have referred to interdisciplinarity as an ideal, even the glossing of multidisplinarity as being based on positions firmly rooted "within" disciplines is itself an idealization.

WHEN DO YOU NEED INTERDISCIPLINARITY?

Although it is relatively easy to be multidisciplinary because everyone is good at championing their own areas of expertise, it is much harder to achieve interdisciplinarity. There are many epistemological obstacles and cultural differences that prevent cross-fertilization of ideas. These include incommensurability of concepts, different units of analysis, differences in world views, expectations, criteria, and value judgements. For example, a traditional divide is between the social and cognitive sciences: social scientists (e.g., sociologists) do not accept cognitive concepts as having explanatory power, whereas cognitive scientists have ignored the environment and social structures. The main reason behind this obduracy is that each views the other as self-evident and hence simply takes it for granted (Cicourel, 1995). For example, in one of our European project[2] meetings, a heated debate took place between the sociologists and the cognitive psychologists whereby the latter were try-

[2]We were all partners of an European research training network called "Cooperative Technologies for Complex Work Settings" in which cognitive ergonomists, psychologists, linguists, ethnomethodologists, and computer scientists were brought together to develop theoretical and methodological frameworks for analyzing complex work settings for cooperative technologies.

ing to convince the former of the need to understand how cognition works with respect to work practices. One of the sociologists simply retorted, "Why do you need to look inside the brain? It's all out there." There are still other problems, however, for those who are able to see beyond their disciplinary boundaries. Bannon (1992) warned of the danger of attempting to "wed" different conceptual frameworks to develop a new unified one—it inevitably results in the dilution of the contributing specific theoretical frameworks. The added value of such efforts can all too often fail to materialize, and instead a mishmash of ideas, methods, and theory may be the ensuing result.

A critical and ultimately more valuable question might be, therefore, When do you really need interdisciplinarity? In many situations, it is possible to continue progressing research by adopting a multidisciplinary stance. For example, research into the use of language, which has made substantial advances, is an area in which cognitive psychologists, computational linguists, and sociolinguists have come together in the same forum borrowing ideas from each other where they see fit but primarily remaining as separate and distinct approaches. In other situations, however, an impasse may be reached in which researchers find that working by themselves is too limiting, and they are unable to address the problem at hand. Accordingly, they may reach a point in which the constraints of their own discipline prevent them from making any further progress and as a consequence are forced to work at the fringes of their field and in so doing forge new ones (The Royal Society, 1996). It is at these junctures that striving for interdisciplinarity may be most promising. Yet how can we characterize what these junctures are and perhaps recognize the need for a change of tactic from the multidisciplinary approach?

One way forward here is to identify what the impetus has been in cases in which we can see an area that clearly demands input from more than one conventionally defined discipline and in which no one alone has a comprehensive set of theoretical frameworks or methodological tools to deal with it. For convenience, we divide these into cases in which an existing problem has simply seemed too large for a single discipline to cope with by itself and those in which something external to the disciplines has forced itself on their attention. Here, we consider examples of both of these: first, a program to develop a more comprehensive account of cognitive science and second, the evolution of two related applied fields—human–computer interaction (HCI) and computer-supported cooperative work (CSCW).

INTERDISCIPLINARITY AS AN IDEAL

Cognitive science is a classic example of the emergence of a new field that set itself up to be truly interdisciplinary. A main motivation behind its inception was to enable a number of different disciplines to come together to develop a better understanding of how the mind works. An overarching aim was to provide more extensive accounts than was possible from a single discipline, which would comprise interacting levels of description that covered social, behavioral, cognitive, and biological aspects. For example, Norman (1980) argued that the study of cognition needed to be much more far reaching, considering it in terms of interacting aspects of a phenomenon, including social, cultural and internal cognitive factors: "I wish Cognitive Science to be recognized as a complex interaction among different issues of concern, an interaction that will not be properly understood until all the parts are understood, with no part independent of the others, the whole requiring the parts, and the parts the whole" (p. 3).

In the beginning, the disciplines that were brought in to develop the new field of cognitive science were cognitive psychology, AI (computer science), linguistics, philosophy, and neuroscience (Green, 1996; Johnson-Laird, 1988; Thagard, 1996; Von Eckardt, 1993). Since then, a whole host of others have been identified as important contributors including anthropology, sociology, engineering, HCI, and education (Schunn et al., 1998). With so many potential collaborators, the stage seemed set for a range of combined efforts to emerge. Indeed, a number of such collaborations have been reported throughout the potted history of cognitive science. For example, Schunn et al. (1998) cited the early collaborative efforts of Simon, Newell, and Shaw's when building their logic theorist program. It is claimed that their work involved combining ideas from economics, psychology, mathematics, and computer science. Their output was a computer program that arguably had more explanatory power than what would have evolved from within a single discipline. Kosslyn's (1980) work on mental imagery is also viewed as a paradigmatic example of interdisciplinary research—whereby his early empirical research from the 1980s on how mental images work spurred a number of other researchers from different disciplines to extend his research in relation to their own models, perspectives, and empirical findings. For example, Farah (1984) extended Kosslyn's work into the area of imagery deficits in neurologically impaired people, developing a further explanation of the role of imagery in cognition by identifying its physiological localization (Von Eckardt, 1993). More recently, Green et al. (1996) pointed out how neuropsychological research (especially on brain-damaged patients) has provided insight for models of cognitive functioning and in so doing, they claim enabling a better integration of biological and cognitive accounts. Schunn et al. (1998), in a survey of the main publications arising out of cognitive science in the last 20 years, noted how there have been a significant increase in papers citing research carried out in which the methods of computer science and psychology have been combined—the most notable being the presentation of computer simulations of previously published empirical data sets.

Although these examples have been set up as paradigmatic of interdisciplinary research in cognitive science, it is actually quite difficult to determine what form they have taken. In many instances, the main form of collaboration is in terms of borrowing or building on another's findings or ideas to extend or support their own theoretical explanation of some aspect of cognition. Another form is the use of another discipline's methodology, the most common being psychologists using computer programs to simulate some aspect of cognitive behavior (see Schunn et al., 1998). By our criteria, that is really a form of multidisciplinarity. Moreover, it is a far cry from the desideratum proposed by Norman (1980) for interdisciplinary research—that its focus be concerned with developing explanatory accounts that cover complex interactions between different levels of description of the phenomena.

So why has there been so little demonstrable interdisciplinary research? Ten years on from his original thesis of what was needed in cognitive science, Norman (1990) acknowledged that part of the problem for there being no real interdisciplinary progress is that the key issues he thought were important to this kind of research (e.g., learning, memory, thought, language, emotion, consciousness) could actually continue to be studied within a single existing discipline, not requiring any new interdiscipline to frame the research questions and methods needed. Indeed, this appears to have remained predominantly the status quo under the umbrella of cognitive science: Psychologists have continued to study psychology, computer scientists computer science, and linguists linguistics. Part of the problem for this resistance is that one can carry out research within a single discipline with much more certainty. Methodologies have been tried and tested and can thus be relied on for carrying out the research in a rigorous and systematic fashion. Furthermore, there are still many unanswered questions within the separate disciplines to keep researchers busy for a long time. To move beyond the boundaries of one's own discipline into an unknown territory and to try integrating unfamiliar terms and concepts is much more of a high risk enterprise in which the process and outputs are always uncertain. Those who have tried even multidisciplinary collaboration will often report how difficult it is to reach any level of shared understanding between the different parties with regard to the referents and terms each is using. Such frustrations can leave researchers wondering whether the costs involved in such ventures outweigh the benefits of doing so. For example, in their survey of multidisciplinary research in cognitive science, Schunn et al. (1998) found that multidisciplinary collaborations were not rated as being any more successful than monodisciplinary collaborations. One of the biggest complaints was that multicollaborations generated too many different ideas. Similarly, Scaife, Curtis, and Hill (1994) noted that one of the key problems arising from their collaborative research project, with partners from cognitive psychology, design, and computer science, respectively, was the difficulty of communicating and knowing what to do with the different ideas generated between them.

Problems can also arise when groups try to agree as to what constitutes interdisciplinarity. For example, when the European Commission (EC) tried to encourage interdisciplinarity within one of its funded programs, it failed. In their first call for a proposal of a new Training and Mobility Research Program, they established a panel of evaluators coming from different backgrounds with the aim that they should assess proposals in terms of their interdisciplinarity. This was the first time that such a panel had been established to evaluate a research program. However, all the projects submitted to this panel were awarded B grades, which meant that they were rejected. No single project passed the review due to the difference of evaluation in the panel. The following year, the EC eliminated the interdisciplinary panel.

As mentioned previously, we are trying to identify occasions when there has been an effective impetus for interdisciplinary research to break out and make any headway. Cognitive science was "founded" because of the perception that we had reached a point in which no further progress could be made in understanding the mind because of the limitations and restrictions of the parent disciplines. However, pursuing interdisciplinarity for its own sake in the hope that a better understanding would result has proven to be a route that very few researchers have been prepared to take. Instead, most researchers in cognitive science have preferred to take an egocentric stance, remaining within the confines of their own discipline, importing ideas and methods from other disciplines when they see them as being useful for supporting their own research.

If true interdisciplinarity is ever to take off, then what is needed is a paradigm shift whereby a whole set of new issues and research questions are framed that force new ways of conceptualizing and working. Norman (1990), in his critique of cognitive science 10 years on, recognized this dilemma and with hindsight suggested that there are now such a set of interdisciplinary concerns emerging—that cannot be addressed within the confines of a single discipline as could the previous issues he had identified 10 years earlier. These included the need first, to develop an applied cognitive science, and second, a way of overcoming the deficiencies of a disembodied theory of cognition.

The need for an applied cognitive science stems from a recognition that there are "massive gaps in our scientific knowledge ... because there has not been sufficient study of real, naturally occurring behavior" (Norman, 1990, p. 4). We come back to this later, but first we briefly consider the external (to cognitive science itself) push for disciplinary collaboration generated by technological advances in computing and telecommunication technologies. These have provided us with much scope for new forms of collaboration, communication, and computational support including the ability to manipulate and interact with information in a multitude of ways, together with interacting with each other in remote and virtual spaces. In turn, there has been a growing expectation within the system design community that cognitive science should and could have practical application for understanding these developments. Existing tools, theories, and methods from within the contributing disciplines, especially cognitive psychology, however, have proven to be largely disappointing, being inappropriate and largely unusable (e.g., see Barnard, 1991; Rogers, 2004). Here, therefore, was an opportunity for a breakaway group of researchers frustrated by the limits of their existing disciplinary knowledge to come together and create a new field that could evolve new knowledge and methods that could be applied to practical problems.

INTERDISCIPLINARITY: FORMING APPLIED FIELDS

The perceived need for a new form of interdisciplinarity was very much the driving force behind the emergence of two new applied fields—HCI and CSCW in the late 1970s and 1980s, respectively. In both cases, evolving technological advances created new theoretical and applied challenges, which existing tools, methods, and frameworks from the separate disciplines of psychology, computer science, engineering, human factors, and design were considered inadequate to deal with by themselves. In HCI, the goal was originally to bring together psychology and computer science to design more effective human–computer interfaces for single user applications. In CSCW, the goal shifted toward bridging the gap between the social sciences and computer science to develop more usable and useful collaborative computer systems for multiuser settings.

What we can see, however, from these kinds of more applied endeavors that have tried to attain interdisciplinarity is that the process is very much an uphill struggle to break away from a multidisciplinary mindset. The jury is still out as to whether either HCI or CSCW have in fact been able to achieve any significant level of interdisciplinarity. In a critique of the interdisciplinary accomplishments of the two fields, Bannon (1992) argued that although there have been several laudable attempts to develop new frameworks that allow for a family of theories and different concepts to be incorporated (e.g., Kuutti & Bannon, 1993), there has yet to be any convincing research projects reported in which different disciplines have genuinely wedded together and made mappings across concepts that have resulted in the development of a common unified theory. However, rather than see this as a failing of such enterprises, Bannon argued that the goal of true interdisciplinarity in these contexts is fundamentally flawed because the worldviews, backgrounds, research traditions, perspectives, and so forth of each of the contributing disciplines are often so different that they are simply not commensurable with each other. It is a mistake, therefore, on these grounds to assume that it is even possible to develop an interdisciplinary theory per se. Attempts to build such hybrid frameworks are destined to fail. Recent examples such as Mantovani's (1996) model of social context—in which he took a wide range of concepts and research findings from the social and the cognitive sciences, combining top-down with bottom-up approaches for the purpose of analyzing social norms and mental models together—are witness to this. Different terms, ontologies, and methods are mixed together in one big melting pot making it difficult to make any sense of the various strands and levels presented in the framework. The outcome is all too often confusing rather than insightful.

INTERDISCIPLINARITY: CHANGING THE UNIT OF ANALYSIS

One of Norman's (1980) other proposals for progressing cognitive science was to study cognition as it occurs in the real world rather than the traditional model of isolating and controlling it in a laboratory setting. The reason for this is based on a growing acknowledgment that the assumptions behind laboratory-based cognition do not necessarily hold true in the real world: "In the tradition of disembodied intellect, the person simply cogitates. The assumption is that the person starts with full and complete knowledge of the world-state relevant to the issue at hand, selects a course of action, then plans and executes it. I argue that this is neither what people do nor is it possible" (Norman, 1990, p. 6).

Thus, although a psychologist can try to study the behavior of participants in an experimental laboratory—observing them interacting with environments that embody knowledge they can control—they cannot understand the behavior of, say, operators in a control room because they cannot extrapolate from the former setting to the latter. This is because they have no real understanding of the knowledge embodied in the external representations that the operators create and use in their work. The continuous interplay of internal and external representations is completely out of the psychologist's range of investigation unless they begin to study, together with engineers, physicists and others, the way in which artifacts are actually used in the control room work.

Several researchers within cognitive science have taken up the challenge of studying cognition as practiced in different cultural settings, providing alternative explanations that reconceptualize cognition as situated within its cultural, social, and environmental context (e.g., see Greeno's special issue of Cognitive Science, 1993, on situated cognition). Such attempts have tended to adapt and assimilate concepts from other fields to contextualize their existing theories about cognition. As such, the process of evolving a new understanding arises through local adaptation. A more extensive form of adaptation is to seek ways of developing a new understanding by reconceptualizing a domain area using a new unit of analysis. An example of this more global strategy is Hutchins' (1995) distributed cognition approach in which he broadened the mainstream cognitive science unit of analysis—which focuses exclusively on the properties and processes inside the mind of a single person—to one that extends to a family of cognitive systems. As well as continuing to allow for a unit of analysis that accounts for the workings of the individual mind, Hutchins proposed that we really need to begin studying other more extensive kinds of cognitive systems such as an individual interacting with a set of tools and even more comprehensively, sociocultural systems comprising a group of individuals interacting with each other and a set of artifacts over a historical period of time (e.g. the collaborative activity of flying a plane be described in terms of the interactions between the pilots and the air traffic controllers and the pilots and their interactional use of the instruments in the cockpit).

The rationale for developing the distributed cognition approach was largely motivated by a deep dissatisfaction with previous efforts in the disciplines of cognitive science and anthropology to explain cognition and culture. Hutchins (1995) noted how both disciplines have marginalized the role of culture in cognition and, conversely, the role of cognition in culture to such an extent that "history and context and culture will always be seen as add-ons to the system, rather than as integral parts of the cognitive process, because they are by (*their*) definition, outside the boundaries of the cognitive system" (p. 368). To put matters right, Hutchins argued that researchers need to study cognitive systems in "the wild" by carrying out a kind of "cognitive ethnography." This involves analyzing the processes and properties of a particular cognitive system in terms of the propagation of representational state across the different media in that system—which may be inside or outside of the individuals. In other words, Hutchins advocated continuing to use the conceptual currency of classical cognitive science but in a manner that is modified to enable a much more flexible unit of analysis. Hutchins acknowledged, though, that broadening the scope of one's enterprise makes the work of carrying out the research much more difficult and the ensuing outcomes much more uncertain. However, Hutchins argued that one of the benefits of adopting this more precarious stance, is that it can result in a more accurate picture of the functional specification of cognition and culture—something that has been largely missing in existing accounts of cognition and culture. In particular, it is suggested that one outcome of analyzing different kinds of cognitive systems is that it can reveal cognitive properties that cannot be reduced to those of individual persons.

Here is an attempt, then, to obtain new understanding of a phenomenon by reconceptualizing the domain of interest through using a modified unit of analysis. Such an approach is not new in itself. Indeed, a similar reconceptualization of the unit of analysis was proposed by

Vygotsky (1978) in his cultural-historical approach for analyzing psychological processes. Vygotsky argued that to understand cognition, we need to go beyond analyzing the isolated mind in the laboratory to studying our everyday practices in relation to the nature and evolution of cultural artifacts in use. Further, Vygotsky claimed that the more pervasive such artifacts are in our everyday life the more they mediate cognition.

What is significant about these kinds of theoretical developments is that they have made advances in our understanding through extending and adapting concepts rather than through creating novel ones from an interdisciplinary approach. In setting itself more modest objectives, the distributed cognition approach has been able to develop a more integrated view of human cognition that arguably overcomes the shortcomings of previous attempts (i.e., cognitive science and cognitive anthropology) to account for the relation between cognition and culture.

INTERDISCIPLINARITY AS PRACTICED: THE CHALLENGE OF WORKING TOGETHER

Whether one is discussing interdisciplinary or multidisciplinary collaboration, there is a problem that is common to both: the evolution of a common ground that will allow the coordination of concepts and efforts. There is surprisingly little written about this process perhaps because the lessons of such collaboration are usually of the "we could have done better" variety. However where there are accounts they tend to come from applied research in which the situation can arise when difficult problems need to be solved and individual disciplines are not adequately equipped to do so by themselves. Here, multidisciplinarity as basic working practice—in which novel outputs are yielded through different individuals coming together and working as a team—is generally viewed as desirable. To achieve this, however, requires the various individuals becoming more open to new ideas and ways of communicating with each other. It also means learning about and accepting the other discipline's way of working. It is becoming increasingly apparent, however, that to enable this kind of mutual understanding to occur also requires some kind of lingua franca (Green et al., 1996; Rogers, 2004). In particular, what is needed is a way of representing and talking about new concepts that can be readily exchanged between the participating disciplines.

A good example of where researchers from different disciplines can work together and develop a new method is a project carried out by a team of sociologists and software engineers at Lancaster University (Sommerville, Rodden, Sawyer, Bentley, & Twidale, 1993). Sommerville et al. were interested in developing systems for multiple end users, in particular, for the domain of air traffic control. Sommerville et al.'s starting point was to acknowledge that the conventional software engineering approaches to requirements capture and analysis were inappropriate in their current form for use in the design of these kinds of collaborative systems. This was because the software engineering methods—developed originally to support formal structures—were seen as being unable to cope with the dynamic and informal ways of working that groups of people invariably adopt in different work settings. An important step for the group was then to determine how groups actually work together and to then work out a way of using this information to inform the design of new systems. It was here when it was considered that the sociologists could help by providing accounts of group working from their ethnographic studies. However, a dilemma emerged. Members of the two disciplines recognized a language problem: The sociologist's detailed descriptions did not fit in with the structured way of working required to do software engineering. They needed to determine, therefore, how the software engineers could usefully take on board the sociologist's findings and analyses. Their solution was to design a new tool that could enable the sociologists to represent their findings in a more structured form that the software engineers could more easily relate to and use when designing software.

However, there are a number of persistent difficulties preventing multidisciplinary teams from successfully communicating ideas between each other as can be seen in another research study, this time on producing a tool for fashion designers (Scaife et al., 1994). This was a study of collaboration between computer scientists and a field research team with backgrounds in psychology/cognitive science. They encountered a number of problems: how to relate observation of work practices to design decisions, how to manage responsibilities within the project, what priorities to adopt in development, and how to involve users in prototyping. Many difficulties stemmed from the differing backgrounds and concomitant assumptions about good practice and what the correct way of working should be. Goldstein and Alger (1992) made a useful distinction between (a) software development, which involves a model of how the world operates, and (b) software development methodology, which is a model of how people behave in a software project. Just this kind of dual perspective was operating in the project. The field workers were attempting to map their model of the (fashion designer) world into a series of incomplete prototypes. They were not thinking in detail about software issues. The software developers, with their own model of how development should occur, were resistant to this process. Thus, initial debates about the completeness/functionality of the software conflated the issue of development methodology with the issue of what constituted an adequate model of the fashion design process itself. This is important because as Bond (1992) pointed out, one problem of collaboration in joint design is that each party may have a "private justification language" (p. 463). In this project, the field team produced summary models and lists, believing that this was a format that software developers would be happy with. However, there were problems with these summary formulations because they were still very close to the original data: They reflected the designers' terminology and concepts. This is likely to be an issue in any study in which the need to provide abstractions for system design competes with the requirement to phrase models in a way that users can understand and comment on. This is a separate problem from that of understanding terminologies across disciplines.

What the preceding examples demonstrate, therefore, is that even multidisciplinarity in an applied domain can be highly problematic. It suggests that for multidisciplinary teams to have the best chance of succeeding, then both the nature of the problem space has to be clear to all and that all parties concerned are prepared to act on it by being willing to change how they do their research. This may mean taking a radical departure from what is prescribed in their parent discipline, but in breaking away, novel solutions can emerge.

INTERDISCIPLINARITY AS EMERGENT: UNDERSTANDING HOW EXTERNAL REPRESENTATIONS WORK IN RELATION TO HUMAN COGNITION

We have shown something of the problems (and opportunities) that can occur in the process of collaboration and communication in multidisciplinary team work. However, in the cases we've discussed, the aim was to produce a new artifact. What occurs when the goal is the more nebulous one of "promoting understanding" or "evolving new ideas" within a domain? We can examine something of this process by looking at the efforts with a particular research problem: the role of external representations in cognition (e.g., Marti, Rizzo, Rogers, & Scaife, 1997). We consider ourselves as cognitive scientists, and so we briefly present this work through the lens of Norman's (1980, 1990) desiderata for cognitive science that we have quoted previously. Among other things, we will recall, he argued for the necessity of understanding interactions between issues and for a more applied and situated orientation. One question here is whether in so doing we can therefore allow ourselves the label of interdisciplinarists?

The impetus for our research was the lack of any generalizable theories in this domain. By this, we mean explanations that could enlighten us on how people interact with different kinds of external representations—be they diagrams, animations, multimedia, or virtual reality—for a variety of cognitive activities (e.g., learning, problem solving, reasoning). In an extensive review of the literature including cognitive science, education, psychology, instructional science, HCI, and art history, we discovered a fragmented and poorly understood account of how graphical representations work, thereby exposing a number of assumptions and fallacies (see Scaife & Rogers, 1996). The main reasons for this state of affairs seemed to be twofold. First, prior empirical work had a strongly parochial flavor, with investigations largely limited to the efficacy of this or that representation within a particular context such as a pump animation for teaching high school physics. Second, such mainstream cognitive theories as were available focused primarily on internal representations (e.g., Bauer & Johnson-Laird, 1993; Hegarty, 1992), missing out much of the cognitive processing that goes on when interacting with external representations (Rizzo, Marti, Veneziano, & Bagnara, 1999). Consequently, there was little or no integration of levels of description in such process models as were offered. The result of this situation was a chorus of disaffection from the community of designers and educational advisers that there was little of general relevance coming out of the research to date.

This led us to recognize that we needed to (a) explain more adequately the interplay between internal and external representations and (b) use this analysis to better understand, design, and select graphical representations, which are appropriate for the learning environment, problem-solving task, or entertainment activity in question. In particular, we believed that the value of adopting this alternative approach would be to focus our attention more on the cognitive processing involved when interacting with graphical representations, the properties of the internal and external structures, and the cognitive benefits of different graphical representations. What we were not sure of, however, was how best to characterize and operationalize this relation between the internal and external. In essence, we needed a new set of concepts that were generalizable in both a theoretical and applied context.

We were fortunate enough to discover that we were not alone in our endeavor, that others in the field of cognitive science had begun to recognize the need to broaden and situate the base from which to explain cognitive behavior (e.g., see special edition of the journal of *Cognitive Science,* 1993). A few researchers had also specifically been giving external representations a more central functional role in relation to internal cognitive mechanisms (e.g., Cox & Brna, 1994; Kirsch & Maglio, 1994; Zhang & Norman, 1994). Others, too, had begun putting forward alternative concepts like rerepresentation and expressiveness—originating from philosophy and logic—to explain why certain graphical representations were more effective than others (e.g., Stenning & Tobin, 1997). Finally, we were much inspired by Green's (1989, 1990) work on cognitive dimensions in which he has sought to develop a set of high-level concepts that are easy to use by academics and designers alike for evaluating the design and assessment of informational artifacts such as software applications.

On the basis of these ideas and Rogers and Scaife (1997), we identified novel concepts that provide a useful analytic framework from which to explicate aspects of external cognition. These are computational offloading, rerepresentation, graphical constraining, and temporal and spatial constraining. Each of these refer to different aspects of the relation between internal and external representations and which can be operationalized further in terms of empirical predictions and specific design decisions. For example, these include design concepts of visibility, explicitness, and annotatability (see Fig. 10.1 for more details of this operationalization). Our approach, therefore, has been to develop a better understanding of a domain by rejecting "old" computational theories that focus exclusively on the internal mind and constructing an alternative framework that conceptualizes the problem space using a different kind of framework. This, we believe, conforms at least partially to what Norman (1990) would consider a broader approach. It tries to map between levels of description/analysis; has a "purely" cognitive level (computational offloading) and an applied (design) level; and allows the analysis of situated use of external representations, for example, by identifying trade-offs between design decisions for clarity of information presentation.

At the highest conceptual level cognitive interactivity refers to the interaction between internal and external representations when performing cognitive tasks (e.g. learning). At the next level this relationship is characterized in terms of the following dimensions:

- **computational offloading**—the extent to which different external representations reduce the amount of cognitive effort required to solve informationally equivalent problems

- **re-representation**—how different external representations, that have the same abstract structure, make problem-solving easier or more difficult

- **graphical constraining**—this refers to the way graphical elements in a graphical representation are able to constrain the kinds of inferences that can be made about the underlying represented concept

- **temporal and spatial constraining**—the way different representations can make relevant aspects of processes and events more salient when distributed over time and space.

For each of these dimensions we can make certain predictions as to how effectively different representations and their combinations work. These dimensions are then further characterized in terms of design concepts with the purpose of framing questions, issues and trade-offs. Examples include the following:

- **explicitness and visibility**—how to make more salient certain aspects of a display such that they can be perceived and comprehended appropriately

- **cognitive tracing**—what are the best means to allow users to externally manipulate and make marks on different representations

- **ease of production**—how easy it is for the user to create different kinds of external representations, e.g. diagrams and animations

- **combinability and modifiability**—how to enable the system and the users to combine hybrid representations, e.g. enabling animations and commentary to be constructed by the user which could be appended to static representations.

FIG. 10.1. A theoretical framework of cognitive interactivity.
Note. Adapted from Rogers & Scaife (1997).

CONCLUSION

So, are we interdisciplinary? Well we're not really sure how to answer this and whether it is really important to do so. As researchers, we are all currently working within several environments (cognitive psychology, cognitive ergonomics, multimedia), moving back and forth between fields. De facto we are doing interdisciplinary work because we are constantly having to translate between disciplinary-based concepts and, perhaps most important, producing a framework that is explicitly aimed at being intelligible to people from several disciplinary areas. In practice, we believe, it may be of far greater practical significance to recognize the moment for a new "look" (our search for "impetus") than to worry too much about precisely how we engineer it. Much of the writing on taxonomies of types of disciplinary collaboration (multidisciplinary, transdisciplinary, pluridisciplinary, interdisciplinary, etc.) reflects an abstraction that, although useful in an analytic sense, may have little use value in the laboratory. The emergence of radically new forms of knowledge in individuals is a persistent problem for epistemology, but what is generally accepted is that collective cognition can proceed very effectively through the mutual awareness of other points of view. It is in this sense that we see the commonality of forms of disciplinary collaboration.

ACKNOWLEDGMENT

We gratefully acknowledge support from the European Commission Training and Mobility of Researchers programme, project COTCOS (Cooperative Technologies for Complex Work Settings), Number ERBFMRXCT960014.

REFERENCES

Bannon, L. (1992). Interdisciplinarity or interdisciplinary theory in CSCW? *Workshop Proceedings of CSCW '92. Workshop on Interdisciplinary Theory for CSCW Design,* October 31, 1992, Toronto, Canada.

Barnard, P. (1991). Bridging between basic theories and the artifacts of human–computer interaction. In J. Carroll (Ed.), *Designing interaction: psychology at the human–computer interface* (pp. 103–127), New York: Cambridge University Press.

Bauer, M. I., & Johnson-Laird, P. N. (1993). How diagrams can improve reasoning. *Psychological Science, 4,* 372–378.

Bond, A. H. (1992). The cooperation of experts in engineering design. In L. Gasser & M. N. Huhns (Eds.), *Distributed AI, Pitman,* 463–483.

Brown, G. D. A. (1990). Cognitive Science and its relation to psychology. *The Psychologist, 8,* 339–343.

Greeno, J. (Ed.). (1993). Situated action [Special issue]. *Cognitive Science, 17*(1), 1–133.

Cicourel, A. V. (1995). Cognition and cultural belief. In P. Baumgartner & S. Payr (Eds.), *Speaking minds: Interviews with twenty eminent cognitive scientists* (pp. 47–58). Princeton, NJ: Princeton University Press.

Cox, R., & Brna, P. (1994). *Supporting the use of external representations in problem-solving: The need for flexible learning environments* (Research Paper No. 686A). Edinburgh, Scotland: Department of AI, University of Edinburgh.

Farah, M. H. J. (1984). The neurological basis of mental imagery: A componential analysis. *Cognition, 18,* 245–272.

Goldstein, N., & Alger, J. (1992). *Developing object-oriented software for the Macintosh.* Addison-Wesley.

Green, D. W. (1996). *Cognitive science: An introduction.* Oxford, England: Blackwell.

Green, T. R. G. (1989). Cognitive dimensions of notations. In A. Sutcliffe & L. Macaulay (Eds.), *People and computers V* (pp. 443–459). Cambridge, England: Cambridge University Press.

Green, T. R. G. (1990). The cognitive dimension of viscosity: A sticky problem for HCI. In D. Diaper, D. Gilmore, G. Cockton, & B. Shakel (Eds.), *Human–computer interaction—INTERACT '90* (pp. 79–86). North Holland: Amsterdam.

Hegarty, M. (1992). Mental animation: Inferring motion from static displays of mechanical systems. *Journal of Experimental Psychology: Language, Memory and Cognition, 18,* 1084–1102.

Hutchins, E. (1995). *Cognition in the wild.* Cambridge, MA: MIT Press.

Johnson-Laird, P. N. (1988). *The computer and the mind.* Cambridge, MA: Cambridge University Press.

Kim, S. (1990). Interdisciplinary collaboration. In B. Laurel (Ed.), *The art of human computer interface design* (pp. 31–45).

Kirsch, D., & Maglio, P. (1994). On distinguishing epistemic from pragmatic action. *Cognitive Science, 18,* 513–549.

Kosslyn, S. (1980). *Image and mind.* Cambridge, MA: Harvard University Press.

Kuutti. K., & Bannon, L. (1993). Searching for unity among diversity: Exploring the interface concept. In S. Ashlund, K. Mullet, A. Henderson, E. Hollnagel, & T. White (Eds.), *Proceedings of INTERCHI '93* (pp. 263–268). New York: ACM.

Mantovani, G. (1996). Social context in HCI: A new framework for mental models, cooperation and communication. *Cognitive Science, 20,* 237–269.

Marti, P., Rizzo, A., Rogers, Y., & Scaife, M. (1997). *External representations* (Technical Document (WP1.2) for project Cooperative Technologies for Complex Work Settings). Retrieved from www.sv.cict.fc/cotcos/pjs/cotcos.htm

Norman, D. (1980). Twelve issues for cognitive science. *Cognitive Science, 4,* 1–32.

Norman, D. (1990). *Four (more) issues for cognitive science* (Cognitive Science Tech. Rpt. No. 9001). Department of Cognitive Science, University of California, San Diego.

Rizzo, A., Marti, P., Veneziano, V., & Bagnara, S. (1999). Engaging with organizational memories. In J. Bliss, P. Light, & R. Saljo (Eds.), *Learning sites: Social and technological contexts for learning* (pp. 110–119). Oxford, England: Elsevier Science.

Rogers, Y. (2004). Beyond the cognitivist crisis: What is the value of recent theoretical developments in HCI for system design? New Theoretical Approaches for Human Computer Interaction. *Annual Review of Information, Science and Technology, 38,* 87–143.

Rogers, Y., & Scaife, M. (1997). How can interactive multimedia facilitate learning? In J. Lee (Ed.), *Intelligence and multimodality in multimedia interfaces: Research and applications.* Menlo Park, CA: AAAI Press.

Scaife, M., Curtis, E., & Hill, C. (1994). Interdisciplinary collaboration: A case study of software development for fashion designers. *Interacting With Computers, 6,* 395–410.

Scaife, M., & Rogers, Y. (1996). External cognition: How do graphical representations work? *International Journal of Human-Computer Studies, 45,* 185–213.

Schunn, C. D., Crowley, K., & Okada, T. (1998). The growth of multidisciplinarity in the cognitive science society. *Cognitive Science, 22,* 107–130.

Stenning, K., & Tobin, R. (1997). Assigning information to modalities: Comparing graphical treatments of the syllogism. In E. Ejerhed & S. Lindstrom (Eds.), *Logic, action and cognition.* Dordrecht, The Netherlands: Kluwer.

Sommerville, I., Rodden, T., Sawyer, P., Bentley, R., & Twidale, M. (1993). Integrating ethnography into the requirements engineering process. In A. Finklestein & S. Fickas (Eds.), *IEEE Symposium on Requirements Engineering* (pp. 165–173). New York: ACM.

Thagard, P. (1996). *Mind.* Cambridge, MA: MIT Press.

The Royal Society. (1996). Interdisciplinarity—Transport and the environment. Retrieved December 12, 1996, from http://www.royalsoc.ac.uk/st_pol08.htm

Von Eckardt, B. (1993). *What is cognitive science?* Cambridge, MA: MIT Press.

Vygotsky, L. S. (1978). *Mind in society: The development of higher psychological processes.* Cambridge, MA: Harvard University Press.

Zhang, J., & Norman, D. A. (1994). Representations in distributed cognitive tasks. *Cognitive Science, 18,* 87–122.

11

Cognitive Science: Interdisciplinarity Now and Then

Christian D. Schunn
University of Pittsburgh

Kevin Crowley
University of Pittsburgh

Takeshi Okada
Nagoya University

Cognitive science is a relatively young field. Its beginnings are typically dated in the 1950s, and it did not become a large-scale activity (with a society, meetings, and degree programs) until the late 1970s or early 1980s. As in most young fields, one of the salient features of cognitive science is its interdisciplinarity. All definitions of cognitive science make important mention of this feature, although they may vary slightly in which disciplines they include as contributing to cognitive science (e.g., Collins, 1977; Hardcastle, 1996; Simon, 1982; Simon & Kaplan, 1989; Von Eckardt, 2001).

Why did researchers from several different disciplines come together to form cognitive science? More generally, why do researchers from different disciplines ever come together to form new disciplines? The institutional and disciplinary barriers to interdisciplinary work can be quite formidable. Our goal in this chapter is to examine this general question using the case of cognitive science.

To examine this question, we use a historical analysis methodology. In particular, we analyze the state of interdisciplinarity in cognitive science both historically and recently. Two questions will be at the forefront of our investigation. First, which disciplines have taken part in cognitive science? Second, what factors account for the relative presence or absence of each discipline in cognitive science? It is our belief that understanding how and why members of existing disciplines take part in the formation of another will provide an important insight into the process by which new disciplines are formed.

In studying the field of cognitive science, we use a case study approach. In particular, we focus on the Cognitive Science Society, an influential force in the field. The Society has two main social worlds: its production world, the journal *Cognitive Science*, and its communal world, the *Annual Meeting of the Cognitive Science Society*. At the end of the chapter, we briefly consider data from sources outside of the Society.

This chapter extends similar data analyses found in Schunn, Crowley, and Okada (1998). The 1998 paper should be consulted for more detailed analyses of the functionality of collaborations. On the topic of the interdisciplinarity of cognitive science, in this chapter, we go beyond the earlier paper both in the kinds of analyses conducted and the scope of the data sets analyzed.

THE JOURNAL *COGNITIVE SCIENCE*

Background

Cognitive Science is a high quality journal that has citation impact levels among the upper tiers of social science journals. The journal was first published in 1977. At that time, its subtitle was *a Multidisciplinary Journal of Artificial Intelligence, Psychology, and Language*. Thus, we see early evidence of both interdisciplinarity and focus on a particular set of disciplines.

In 1980, it was given to the Cognitive Science Society to become the Society's official journal. Since then, all members receive a subscription as part of their membership dues. In 1984, *Cognitive Science* merged with the journal *Cognition and Brain Theory*. For several years thereafter, the journal's subtitle was *Incorporating Cognition and Brain Theory*. *Cognition and Brain Theory* had a stronger focus on neuroscience and philosophy (Ringle & Arbib, 1984; Waltz, 1985). Therefore, we see an apparent growth of *Cognitive Science* to include these other disciplines as well.

In 1988, reflecting this growth in disciplines, the journal subtitle became *A Multidisciplinary Journal Incorporating Artificial Intelligence, Linguistics, Neuroscience, Philosophy, Psychology*. Of particular note, the order of the disciplines is now alphabetical, reflecting a desire for equal contributions from each of the disciplines. In 1997, the journal subtitle added anthropology and education to the subtitle list of disciplines. The alphabetic ordering was maintained.

The journal has typically been published quarterly, with three to five articles per issue; in 2001, it moved to bimonthly. The small number of articles in each issue allows for longer articles. These longer articles facilitate presentation of interdisciplinary work because multiple methodologies often require extra space to be adequately described, especially to members of the other disciplines.

But How Interdisciplinary Is It Really?

The descriptions of the journal suggest that the journal should be quite interdisciplinary. However, official descriptions and actual content are not always the same. With the goal of examining how interdisciplinary the journal actually is, we examine three aspects of the journal. First, we examine the participation of each discipline with an analysis of departmental affiliations of article authors. Second, we examine the reliance on past work from each discipline with an analysis of references found in articles. Third, we examine the use of methodologies taken from each discipline with an analysis of the primary methods used in each article. To examine the evolution of these measures throughout the history of the journal, we compared articles published in the first 5 years of the journal (1977–1981), articles published in the mid 1980s (1984–1988), articles published in the early 1990s (1991–1995), articles published in the late 1990s (1996–1997), and articles published in early 2000s (2002). Interrater agreement for all coding exceeded 90%.

Department Affiliations

One measure of disciplinary membership is departmental affiliation. To examine disciplinary participation in the journal, we coded the departmental affiliations of the first authors of all articles excluding editorials, commentaries, and special issues. Only first authors are included because *Cognitive Science* formatting lists the department affiliations of only the first author. We code all such articles in each of the time periods.

Figure 11.1 presents the percent of affiliations in each of the categories in each of the time periods. From this figure, it is clear that the journal has always been dominated heavily by psychologists and computer scientists. At its inception, the journal had a near majority of computer scientists (41%), whereas this has shifted to a majority of psychologists (64%) more recently. Of particular note is the minimal participation of philosophers, linguists, and neuroscientists (means each less than 5%). Turning to the latest additions to the journal (education and anthropology): There have always been some authors from Schools of Education publishing in the journal, although the levels have never been very high (< 7%). There has also been no increase since the change in the subtitle. If the affiliations are accurate, no author from anthropology department has yet been published in *Cognitive Science*.

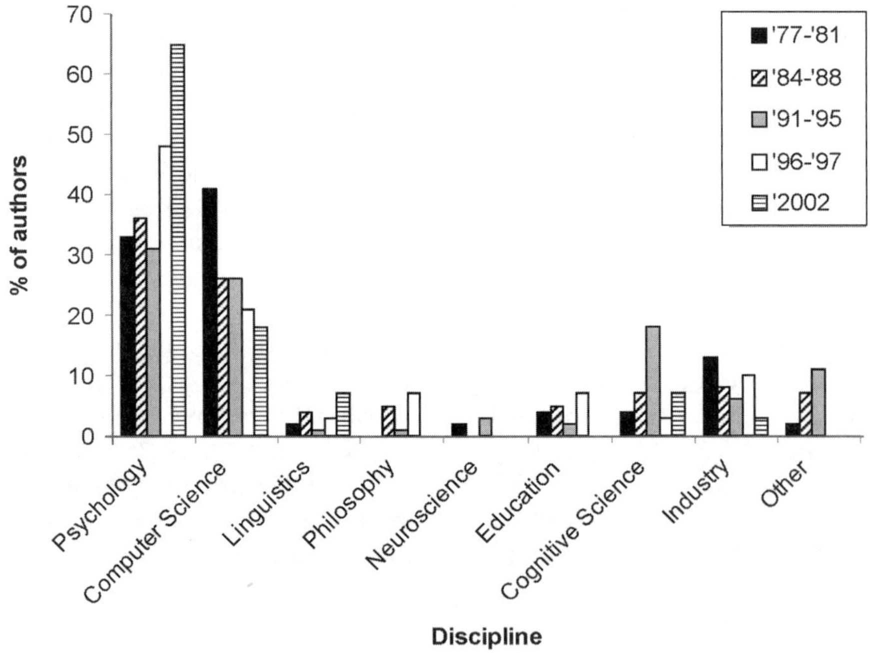

FIG. 11.1. Percentage of first-author affiliations for each discipline in the journal *Cognitive Science*.

It is likely that the drop in number of computer scientists is due to the formation of the American Association for Artificial Intelligence in the early 1980s. It is unclear whether the recent drop in number of researchers with cognitive science affiliations is a real effect or simply noise due to the smaller number of articles included in the most recent time period.

It is worth noting that first-author affiliations can be misleading, especially in the case of interdisciplinary work. Computer scientists or anthropologists can hold positions in psychology departments, for example. Moreover, even when training degrees match departmental affiliations, some of the people could hold undergraduate or masters in other fields. Finally, because the journal only listed first-author departmental affiliation, it is possible that the other disciplines participated more heavily in lower status author positions. For all of these reasons, it is important to look beyond just listed affiliations.

Citation of Disciplinary Work

Another measure of interdisciplinary activity is the extent to which new work builds on previous work in other disciplines. One measure of reliance on previous work is the frequency of citations. We see citations rates as approximate measures of a discipline's impact on cognitive science articles. Clearly, it is possible for a single seminal article to have a greater intellectual impact on a research project than, for example, any 10 other citations from the reference list combined. However, in the absence of any better way to assess the relative impact of previous research, the overall proportion of references to each discipline serves as a functional estimate of how influential various disciplines have been in aggregate on the work published in *Cognitive Science*.

For each article, we coded every reference for the discipline it represented: psychology, artificial intelligence, linguistics, neuroscience, philosophy, education, cognitive science, or other. As the reference coding involved significant labor, we coded references from 3 years within each of the three large time periods (i.e., 9 years total). These were selected by taking the first, middle, and last year within each 5-year age range, producing 5,349 coded references.

The disciplines of journal articles were coded using the classifications in Ulrich's international periodical directory (1994). The disciplines of conferences were coded according to the name of the conference. The disciplines of technical reports were coded according to the department publishing the report. It proved much more difficult to find external validation for coding books and book chapters. They were coded according to a best guess of their content. When a book or book chapter was too ambiguous to code, coders were instructed to exclude that item. This occurred for 39% of book references. The three most common reasons that references were not coded were non-English titles, very short book titles, and missing departmental affiliations from references to dissertations and technical reports. Although the dropout rate for books and book chapters is quite high, it is important to note that when the data were analyzed excluding all books and book chapters, the overall pattern of results did not change. Moreover, careful investigation of a subset of these ambiguous references suggested that there were not large differences in the proportion of references to each discipline between the nonambiguous and ambiguous references.

Figure 11.2 presents the mean percentage of citations to each discipline within each of the time periods. The overall pattern is similar to what was found with analyses of affiliations. First, just as with the affiliation data, there is a dominance of psychology and computer science (means of 37% and 25%, respectively), beginning with more computer science and ending with more psychology. However, there is one important difference from the affiliation analyses: Linguistics was cited more frequently (mean of 12%) than linguists took part in the journal as first authors (mean of 2%). Thus, it appears that linguistics is considered relevant to cognitive science by other cognitive scientists. We return to possible causes of this asymmetry in participation and apparent relevance later. It appears, however, that there is no asymmetry for philosophy, neuroscience, anthropology, and education: Not only do researchers from those disciplines not publish in *Cognitive Science*, it appears that research from those disciplines is not typically cited either (means all below 3%).

Methodologies

A third measure of interdisciplinarity of the journal involves the methodologies of the work reported in the journal articles. Because the previous analyses revealed that the journal is dominated by psychology and computer science related work, one could reasonably focus on the primary methodologies from those two disciplines: empirical studies of behavior and computer simulation. We originally coded each article

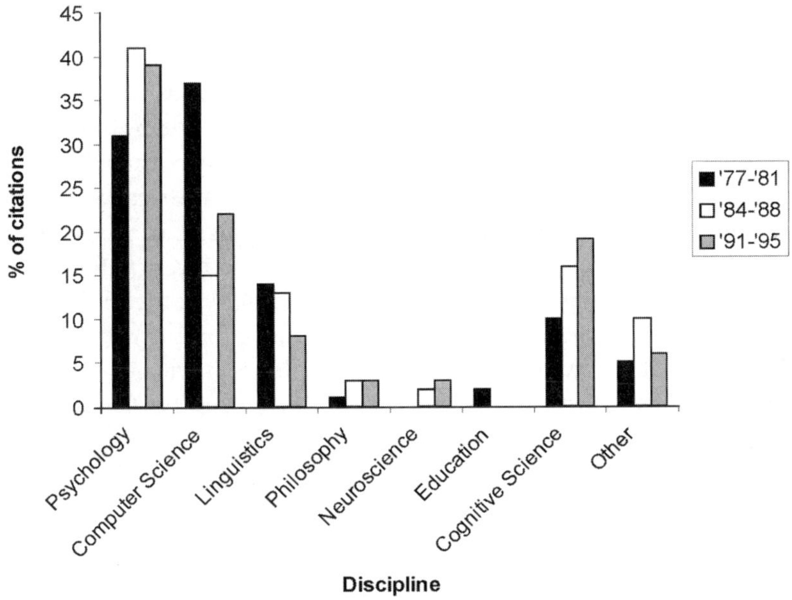

FIG. 11.2. Percentage of citations to work in each discipline within *Cognitive Science* journal articles.

from the 1977 to 1981, 1984 to 1988, and 1991 to 1995 periods into one of five categories: simulation only, empirical study only, simulation + empirical study, simulation of data, or neither simulation or empirical study. The simulation of data category was used for articles presenting simulations of empirical data sets that had been presented elsewhere.

These analyses found that just under one third of the articles published in *Cognitive Science* used empirical studies, and one fourth used simulations only. Thus, at least half of the work in the journal reflects within-discipline activity, and those disciplines are computer science and psychology. However, one sixth of the articles involved either empirical + simulation studies or simulations of data, suggesting that there is some integration of psychological and computer science activities within cognitive science.

The remaining one third of the articles presented neither simulations nor empirical studies. What kind of research did these articles represent? Is it possible that here a strong philosophical or linguistic presence could be seen? To investigate this issue, we coded all articles in the 1996 and 1997 issues of the journal *Cognitive Science* using finer coding categories. The categories and the percentage of articles coded into each category were psychological empirical, 34%; running simulations, 34%; linguistic analyses, 0%; philosophical argumentation, 3%; neurological data involving brain damaged patients or brain imaging data, 0%; new theory/frameworks, 7%; combinations, 17%; and other, 3%. All but one of the combinations involved psychological empirical and simulations—the remaining combination involved linguistic analysis and running simulation. Analyses of methodologies in 2002 articles found a similar trend but a larger proportion of psychological empirical. Thus, it appears that psychological and computational methodologies remain the dominant forces within the journal—philosophical argumentation, linguistic analyses, and neurological data can only be found in very small quantities.

Summary of Findings

From the analyses of departmental affiliations, discipline citations, and research methodologies, a consistent picture of the journal *Cognitive Science* has emerged: It has been dominated by psychology and computer science. The other disciplines thought to belong to cognitive science—linguistics, philosophy, and neuroscience—appear to be unequal partners in the journal. Interestingly, cognitive science as a discipline of its own with its own communal and production worlds is becoming increasingly more common in the journal. We now turn to examining the communal world associated with the Cognitive Science Society: its annual meeting.

ANNUAL MEETING OF THE COGNITIVE SCIENCE SOCIETY

Background

The annual meeting was first convened in 1979. In 1983, it became peer reviewed. The reviewing selection criteria have become stricter each year since then. In the past several years, approximately 40% of submitted papers are accepted. Over the years, the conference has also grown steadily in size until it stabilized in the early 1990s at approximately 140 papers and posters and 500 attendees. In the middle 1990s, a category of member abstracts was added in which any member of the society could have a poster and a one page abstract in the proceedings without having to go through the review process. Although the analyses presented following do not include data from member abstract posters, the results do not change noticeably when they are included.

Motivation for Examining the Conference

There are three reasons that one might expect to see a different picture of cognitive science when examining the annual meeting. First, the journal *Cognitive Science* is highly selective. Yet, interdisciplinary work tends to be more novel and speculative in nature. Therefore, it is possible that much interdisciplinary work has occurred in cognitive science settings but has not appeared within the more conservative journal setting. If one were to examine a somewhat less selective setting, such as a conference, one might find a higher proportion of interdisciplinary work.

Second, it is possible that the discipline of psychology places a heavier emphasis on journal publications than do some of the other disciplines. For example, whereas conference publications might be considered of little value on a psychologist's vita, conference publications can be viewed quite favorably for computer scientists. Thus, one may find less of a dominance of psychology in a conference setting.

Third, in contrast to the journal articles, the affiliations of all authors are presented for the conference papers. It is possible that members from disciplines other than computer science and psychology are taking part in cognitive science activities but only in conjunction with psychologists and computer scientists. Thus, focusing on the first authors in the journal analyses may have partially masked the presence of the other disciplines.

Issues Examined

The first goal of examining the annual conference is to see whether the journal results replicate. As we just argued, there are several reasons why one might expect to find different levels of interdisciplinarity in the annual conference. Toward this goal, we again present analyses of departmental affiliations of paper authors, although this time also including more than just the first authors. However, we do not rely exclusively on departmental affiliations as a measure of disciplinary participation. We also include results from questionnaires studies in which the authors where asked about their training backgrounds. This allows us to test whether departmental affiliation measures used in the journal analyses were good measures of disciplinary participation.

The second goal of examining the annual conference is to investigate the collaborations found at the conference. In an interdisciplinary field, not only is it desirable for researchers from many different disciplines to participate, but it is also desirable for researchers from different disciplines to collaborate on research projects. The self-reported backgrounds of article authors are used to test for the presence or absence of such interdisciplinary collaborations.

Affiliation Analyses

Figure 11.3 presents the percentage of departmental affiliations to each of the disciplines for first authors. As we see, the relative frequencies of each discipline at the conference are similar to those found in the journal. If anything, philosophy, linguistics, and neuroscience play an even smaller role in the conference (means each 2% or below).

In 1994 and 1995, a short questionnaire was sent to the authors of all multiauthored papers at the annual meeting (see Schunn et al., 1998 and Schunn, Okada, & Crowley, 1995 for the details of these questionnaires). The questionnaires included questions about each of the au-

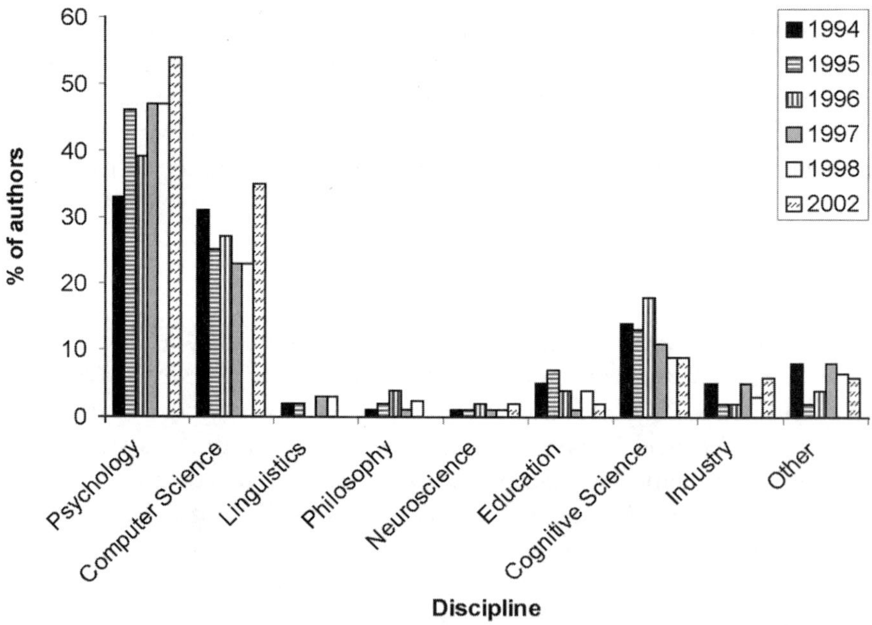

FIG. 11.3. Percentage of first-author affiliations for each discipline at the cognitive science conference.

thors' training backgrounds. Responses were obtained for over 60% of the papers, providing a good opportunity to examine whether the affiliation measure was a good surrogate measure for training background. Figure 11.4 presents the percentage of authors in each discipline according to the department affiliations and training backgrounds. As can been seen, there was a relatively similar distribution of listed affiliations and training backgrounds. The primary difference was that there were almost no authors who listed cognitive science as their training background. It is interesting to note that psychology and computer science were listed as training backgrounds more often than as affiliations, suggesting that psychology and computer science trained researchers are working in other settings (such as industry, education, and, most notably, cognitive science departments).

Another question one might ask is whether examining only first-author affiliations presents an accurate picture of participation. It is possible that other disciplines participate together with psychologists and computer scientists but only as second, third, or fourth authors on the papers. To examine this issue, we coded the affiliations of all authors for

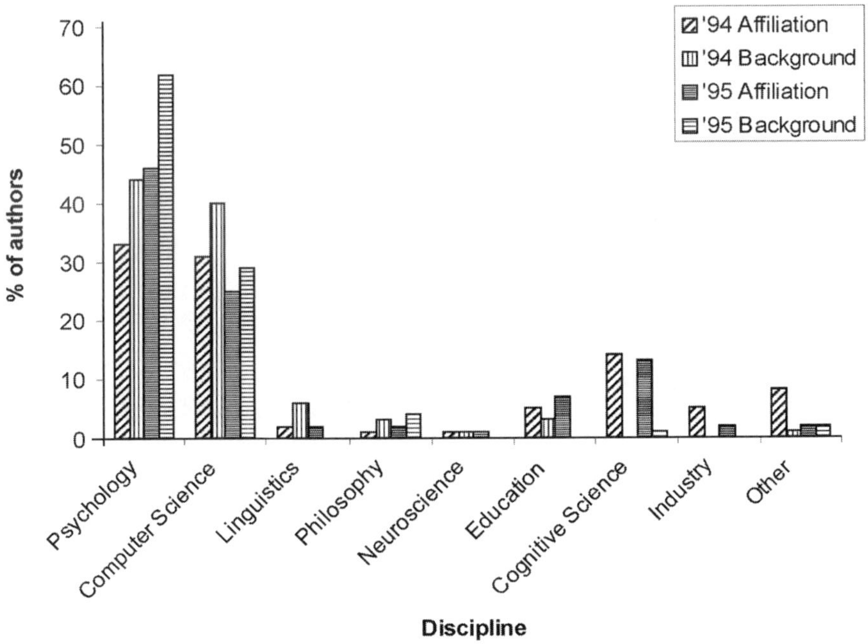

FIG. 11.4. Percentage of first authors at the annual conference with affiliations or training backgrounds in each of the disciplines.

the 1994, 1995, and 1996 annual conferences. Whereas the majority of the papers had more than one author, a minority had three authors, and very few had four or more authors. Therefore, only the first through fourth authors were included in the analyses, and the results were collapsed across the 3 years. Figure 11.5 presents the percentage of authors in each discipline as a function of authorship order. The percentages are remarkably stable across the different authorship positions, with the possible exception that the proportion of neuroscientists is slightly higher in the third and fourth positions. The slight rise in proportion of neuroscientists is not remarkable: Neuroscience projects often require more researchers (and hence more authors) than do projects in psychology, computer science, philosophy, linguistics, and education. It may also be that as brain research has advanced, there has been increasing interest on the part of funding agencies, researchers, and the public at large in discovering the implications of this work for education and other applied fields, encouraging cross-disciplinary collaboration with neuroscientists (e.g., see Bruer, chap. 8, this volume).

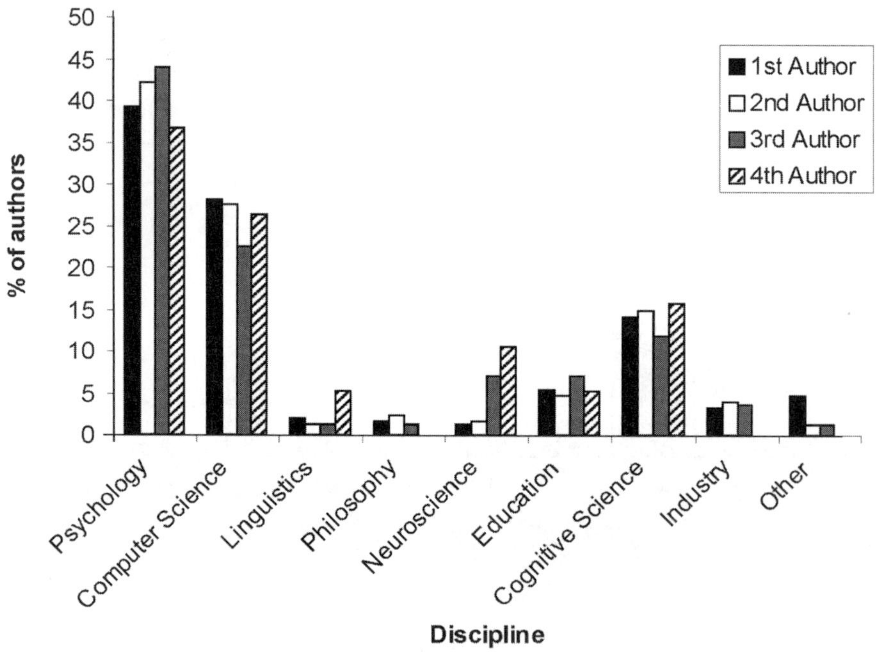

FIG. 11.5. Percentage of authors at the 1994, 1995, and 1996 annual conferences with affiliations in each of the disciplines as a function of authorship order.

Comparing listed affiliations for all authors directly against self-rated training backgrounds (excluding the ambiguous affiliations of industry laboratories, government laboratories, and cognitive science), we found 80% agreement. In other words, rated training backgrounds usually matched the affiliations that were listed. The authors in cognitive science departments were for the most part trained as psychologists (68%) and computer scientists (21%), with a few trained as linguists (5%). Thus, we see a dominance of psychology and computer science even within the new cognitive science institutes and departments.

The 20% mismatches in affiliation present an interesting piece of data. They could be viewed as mistakes in the authors' ratings. More likely, however, is that they are cases of interdisciplinary hires. How are these hires distributed? The two most common categories are computer scientists in psychology departments (31%) and psychologists in computer science departments (23%). Then there are a few philosophers in computer science departments (10%). The remaining cases consist of computer scientists and psychologists in schools of medicine or schools of education (5% frequency of each case). Thus, we see primarily cross-fertilization of psychologists and computer science with some additional spreading of psychologists and computer scientists to other disciplines.

In sum, there is a very consistent picture of disciplinary participation at the annual conference. As with the journal, the conference is dominated by psychology and computer science, with little participation of the philosophers, linguists, educators, and neuroscientists. These results are true across the years and across authorship order. They also do not depend on whether self-reported background or listed departmental affiliations are used.

Proportion of Interdisciplinary Collaborations

The second goal of examining the annual conference was to investigate the collaborations found at the conference. Most of the papers were multiauthored. How many of these multiauthored papers involved collaborations of researchers from different disciplines? To test for the presence or absence of such interdisciplinary collaborations, we used the 1994 and 1995 questionnaire data.

As we described earlier, these questionnaires were sent out to the authors of all multiauthored papers at these two conferences. The authors were asked about the primary training background of each of the authors. Using these responses, each paper was classified into intradisciplinary (all from the same background) or interdisciplinary (at least one author from a different background). The authors were also asked about the relative status of each author (e.g., faculty, postdoctoral associate, graduate student, undergraduate student, etc.). Responses were used to classify each paper into a peer collaboration (authors of same status) or apprenticeship collaboration (authors of different status). Of interest was whether interdisciplinary collaborations happened primarily in peer or apprenticeship collaborations.

At the 1994 conference, 47% of multiauthored papers were interdisciplinary. At the 1995 conference, 57% of multiauthored papers were interdisciplinary. Thus, although the conference is dominated by psychology and computer science, there is a high proportion of interdisciplinary collaboration. This level of interdisciplinary collaboration proved to be equally prevalent for peer and apprenticeship collaborations. At the 1994 conference, 50% of peer collaborations were interdisciplinary and 38% of apprenticeship collaborations were interdisciplinary. At the 1995 conference, the corresponding numbers were 50% and 59%, respectively. Thus, interdisciplinary collaborations neither seem to require that all participants have equal status, nor do they seem to require an apprenticeship relationship. Okada, Crowley, Schunn, and Miwa (1996) found similarly high levels of interdisciplinary collaborations within the Japanese Cognitive Science Society meetings (41% of peer collaborations and 54% of apprenticeship collaborations). This suggests that interdisciplinary collaboration is a central part of cognitive science more generally rather than just the activities of the Cognitive Science Society.

Did all disciplines participate in these interdisciplinary collaborations? To answer this question, we analyzed how often each discipline was paired with each other discipline at the 1994 and 1995 conferences. For simplicity, only the first two authors were considered. Almost all of the interdisciplinary collaborations involved either cognitive psychologists or computer scientists. Table 11.1 presents the frequency of each discipline combination (the first three columns) as well as the frequency of intradisciplinary collaborations for comparison (the rightmost column). The order of authorship is not represented. Interestingly, all of the disciplines except for cognitive psychology had at least as many or more interdisciplinary than intradisciplinary collaborations.

Many readers of our work have found our reported proportion of interdisciplinary collaborations surprisingly high. The barriers to interdisciplinary collaboration are formidable (e.g., see Epstein, chap. 9, and Klein,

TABLE 11.1

The Frequency of Interdisciplinary and Intradisciplinary Collaborations Among the Disciplines (Between First and Second Authors) for the 1994 and 1995 Annual Conferences

	Interdisciplinary Collaborations			*Intradisciplinary Collaborations*
Discipline	*Cognitive Psychology*	*Comp Science*	*Other*	
1994				
Cognitive psychology	—	13	8	24
Computer science	13	—	7	14
Educational psychology	3	0	0	1
Philosophy	1	2	0	0
Linguistics	2	3	0	2
Other	2	2	0	1
1995				
Cognitive psychology	—	12	6	20
Computer science	12	—	7	8
Developmental psychology	4	3	0	0
Philosophy	0	2	1	0
Linguistics	1	1	0	0
Other	1	1	2	1

chap. 2, this volume). What forces would lead so many researchers to collaborate with researchers from other disciplines? One hypothesis is that interdisciplinary collaborations result in more successful work. To investigate this issue, the participants of the 1995 conference were asked how successful they thought the work reported in their paper was. Interdisciplinary collaborations were not rated more successful than intradisciplinary collaborations (58% vs. 56% rating their projects very successful, respectively). As a converging piece of evidence, interdisciplinary collaborations were no more likely to be a paper (vs. a poster), which is determined by reviewer ratings, than were intradisciplinary collaborations (65% vs. 67%, respectively). Finally, as a follow-up 3 years later, we sent e-mail to the authors of the 1995 conference papers asking them whether the work reported at that conference had been published elsewhere in the form of a book chapter or journal article. It is possible that the authors and reviewers were not able to immediately evaluate the successfulness of interdisciplinary work, which may require a longer time to come together. We received responses from over 65% of our original set of papers, a surprisingly high rate given that this was an e-mail survey and that many of the original authors had changed institutions. However, once again, there was no difference in the successfulness of interdisciplinary versus intradisciplinary collaborations (61% vs. 63% reporting subsequent publications elsewhere, respectively).

What about the details of the collaborations? Participants of the 1995 conference were asked to estimate how often communication had occurred within the collaboration, the means of communication (face-to-face meetings, e-mail, etc.), the mesh or clash of the collaborators' background knowledge and intellectual styles, and the benefits and frustrations of the collaboration. Differences between interdisciplinary and intradisciplinary collaborations were observed on several dimensions. First, participants were more likely to say that their coauthors frequently came up with alternative hypotheses in an interdisciplinary collaboration situation (45% vs. 28%). Second, the interdisciplinary collaborators were more likely to say they had a different research style than their coauthors (44% vs. 28%). Third, interdisciplinary collaborators were more likely to say that they had an equal status relationship (86% vs. 65%). This result held even when the relationship was inherently one of unequal status (100% vs. 88% reporting equal status for peer collaborations, and 82% vs. 58% for apprenticeship collaborations). Finally, three benefits were rated as occurring more often in the interdisciplinary collaborations than in intradisciplinary collaborations: having different ideas (75% vs. 43%), having a stimulating relationship (81% vs. 43%), and challenging each other's ideas (56% vs. 39%).

Was there evidence that having different backgrounds contributed to the structure of the collaboration? For the 1994 conference, we asked the first authors to report which author played a primary role in each phase of the research: selecting the research question, designing the study or simulation, providing the resources for the study or simulation, running the studies or simulations, and writing the paper. Comparing interdisciplinary and intradisciplinary collaborations, we found no differences in the number of roles to which the first author contributed (mean number of 3.9 vs. 4.0, respectively). However, the second author of interdisciplinary collaborations contributed to more roles than did the second author of intradisciplinary collaborations (means of 2.8 vs. 1.8, respectively). This pattern held for both peer and apprenticeship collaborations.

We conducted a similar analysis for the 1995 conference. For that conference, we had asked all authors (rather than just first authors), and we had asked each author only about themselves (rather than to report on the contributions of the other authors). Moreover, rather than ask who contributed to which roles, we asked the authors to rate the percentage of the work that they did for each of the roles and overall. Figure 11.6 presents the mean ratings for overall percentage of work. Once again, we see no differences for the first author and greater work levels for the second and third authors for interdisciplinary collaborations relative to intradisciplinary collaborations. Turning to the different roles, this pattern of differences held primarily for the roles of designing the study/simulation and analyzing the data. There were no differences between intradisciplinary and interdisciplinary collaborations for coming up with the research questions and writing the paper.

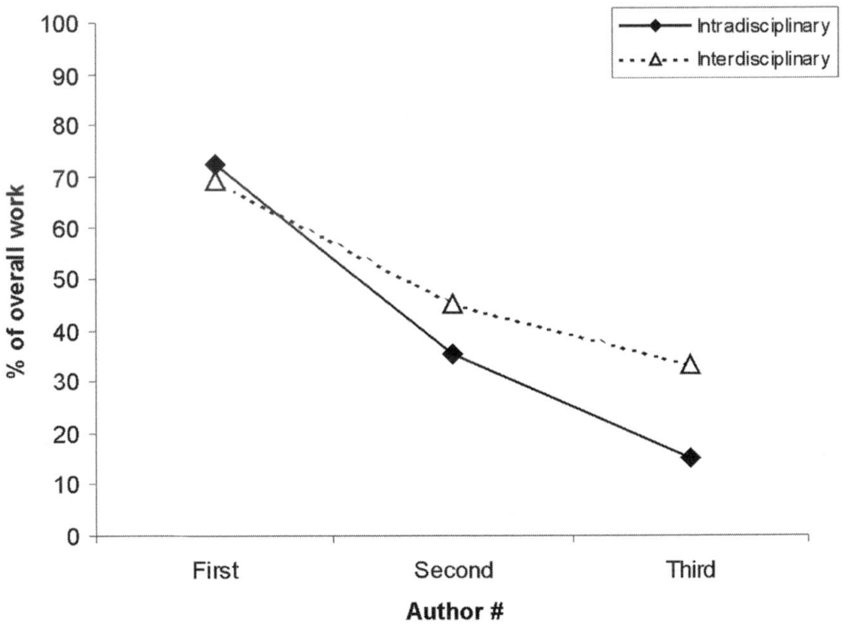

FIG. 11.6. Mean self-rated percentage of the overall work to which each author contributed as a function of interdisciplinary and intradisciplinary collaboration for the 1995 conference.

Summary of Findings

The state of interdisciplinarity in the cognitive science society can be viewed as the proverbial glass: half empty or half full. As the glass half full, we saw a high proportion of interdisciplinary collaborations. This very high rate of interdisciplinary collaborations seemed related to better divisions of labor and other details of the structure of the collaboration rather than differences in the successfulness of interdisciplinary collaborations. Another positive feature of interdisciplinarity in the Cognitive Science Society was a historical trend for an increasing frequency of individuals trained in cognitive science per se and using multiple methodologies. As the glass half empty, we saw a domination by psychology and computer science as well as the presence of pure psychology (e.g., psychologists working together presenting only data from psychology experiments) and pure computer science. These features have been true of the Cognitive Science Society from its inception. In sum, cognitive science can be seen as interdisciplinary now and then rather than completely intradisciplinary or completely interdisciplinary.

BEYOND THE COGNITIVE SCIENCE SOCIETY

Perhaps this domination by psychology and computer science is just a problem in the Cognitive Science Society outlets, not a factor in cognitive science more generally. To explore this possibility, we examined two other journals that are commonly considered prominent outlets of cognitive science research: *Cognition* and *Behavioral and Brain Science (BBS)*. *Cognition* is a bimonthly journal with approximately two to six articles per issue. It has been published since 1972. Its subtitle is *International Journal of Cognitive Science*. *BBS* is a quarterly journal that typically publishes two to four large articles in each issue. Each article presents in detail a particular author's thesis, typically a controversial one. The article usually reviews the authors' previous research rather than presenting new evidence. The article is then followed by 10 to 50 commentaries by other researchers followed by a response by the author to each of the comments. *BBS* was first published in 1978 and describes itself as a journal for research in psychology, neuroscience, behavioral biology, and cognitive science. Thus, it actually has a scope larger than that of cognitive science.

We coded the departmental affiliations of article first authors and the methodologies presented in each article in 1996 and 1997. Figure 11.7

presents the percentage of authors with each discipline affiliation for these two journals as well as *Cognitive Science* for comparison. As can be readily seen, *Cognition* is even less well balanced than *Cognitive Science*. It is almost completely dominated by psychologists (69%). The levels of neuroscience are slightly higher than those found in *Cognitive Science*, but the levels are still well below 10%. Most important, there is no computer science to be found in *Cognition* during the years examined. Examinations of *Cognition* articles from the early 2000s suggests this trend of psychology domination continues.

BBS presents yet another picture of cognitive science. Although psychologists are the largest plurality, they are not a majority in this journal (37%). Moreover, neuroscience plays a large role in this journal (26%). However, there continues to be little presence of linguistics (5%), philosophy (11%), and education (0%). Moreover, very few computer scientists take part in *BBS* (5%).

Turning to the methodologies used in each of these journals, Fig. 11.8 presents the percentage of articles using each of the methodologies for *Cognitive Science* and *Cognition*. The same categories as discussed earlier were used. *BBS* is not included in this analysis because

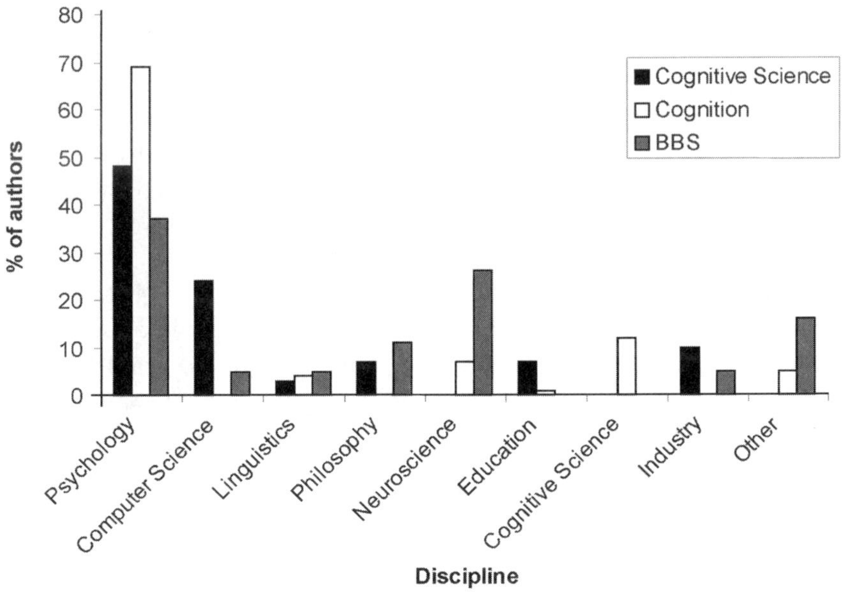

FIG. 11.7. Percentage of first-author affiliations for each discipline in the journals *Cognitive Science, Cognition,* and *Behavioral and Brain Sciences* for 1996 and 1997.

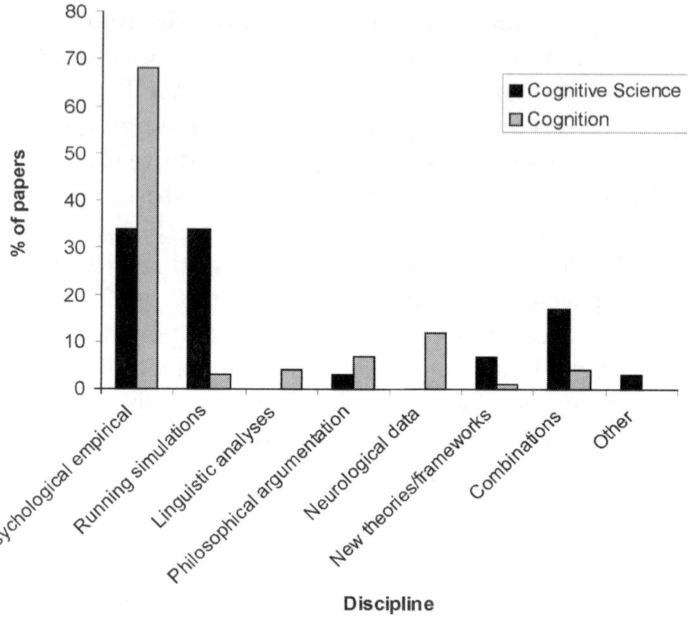

FIG. 11.8. Percentage of articles using each of the disciplinary methodologies for the journals *Cognitive Science* and *Cognition* in 1996 and 1997.

articles do not typically present new research. From this figure, it is clear that *Cognition* is heavily dominated by psychological empirical studies. There is, however, slightly more linguistic analysis and philosophical argumentation in *Cognition*, although still in very small quantities. Other than psychological empirical studies, it is only the neurological data methodology that exceeds 10% in *Cognition*.

In sum, although *Cognition* and *BBS* may present slightly different perspectives of cognitive science, they still do not have an equal balance among the disciplines. The question remains: Why do linguistics, philosophy, neuroscience, and education not play larger roles in cognitive science?

WHERE DID LINGUISTICS, PHILOSOPHY, AND NEUROSCIENCE GO?

Why have linguists, philosophers, and neuroscientists not taken a greater role in cognitive science? The answer to this question is likely to provide insights into the creation and evolution of a new discipline.

Here, we consider several different possible factors in turn, beginning with the simplest factors. It is likely that no one factor is responsible for influencing so many disciplines, and the important factors are likely to vary by discipline.

Problems at the Top or at the Bottom

One simple explanation for the disciplinary distributions within cognitive science is that it is the result of explicit or implicit editorial practices in the journal reviewing and conference reviewing and organization. However, many factors argue against this hypothesis. First, the explicit editorial policies of the journals and the annual conference clearly invite submissions from the other constituent disciplines of cognitive science. Second, in discussions of this issue with many of the past conference organizers, editors, and society presidents, the majority of them stated that they made several efforts over the years to include the other disciplines. Christian Schunn, author of this chapter, was chair of the annual cognitive science conference in 2002 and can report first hand that psychology dominated the conference despite many attempts to include other discipline participation and use discipline-neutral reviewing criteria. Third, there is concrete evidence of these efforts. For example, there were several special invited issues highlighting activities from other disciplines in the journals *Cognitive Science* and *Cognition*. At several of the annual conferences, there were special invited symposia highlighting activities from other disciplines. Thus, it is unlikely that the absence of linguistics, philosophy, and neuroscience is the result of explicit editorial and organizational biases (see Greeno et al., 1998).

Size of the Disciplines

Another very simple explanation of the disciplinary distributions within cognitive science is the relative size of each discipline. In other words, psychology and computer science may simply have more active researchers than philosophy, linguistics, and neuroscience. In the case of neuroscience, this explanation is simply humorous. Neuroscience is itself a combination of several disciplines (e.g., psychiatry, neuroanatomy, neurophysiology, neuropsychology, neurochemistry, etc.). Its main conference, the Society for Neuroscience, attracts about 40,000 attendees—more than the largest psychology and computer science conferences put together.

The case of philosophy and linguists may not be as implausible, as there are fewer research philosophers and linguists than research psychologists. However, available overall productivity data do not support this hypothesis for these disciplines either. Simple online searches were conducted using the WorldCat, Journal, Proceedings, and Conference databases available in FirstSearch of work produced from 1990 to 1995. These searches revealed that there were approximately as many psychology conference papers as philosophy papers, and linguistics produced over 25% more conference papers than did psychology. Moreover, although linguistics produced fewer journal articles and books than psychology, philosophy produced as many. Thus, although differential size of the disciplines may contribute, the productivity data suggests it is not likely to be the source of differential participation in cognitive science.

Recognition of Cognitive Science by Mainstream Elements

We found that linguistics has historically been the third most cited of the constituent disciplines behind psychology and computer science. This suggests that linguistics is considered relevant to cognitive science by cognitive scientists. Yet, for both the journal and the conference analyses, participation by linguists has never exceeded 4% of all articles or conference papers. Why do the linguists not participate?

One possible explanation is that linguistics does not consider cognitive science to be relevant. To investigate this explanation, we conducted several citation analyses using the Social Science Citation Index. First, we counted the number of citations to the journal *Cognitive Science* in the journal *Linguistic Inquiry* for all main articles (ignoring news items, remarks, and reply articles) for the years 1980, 1987, and 1994. For comparison, we also conducted the same analysis for the journal *Psychological Review*. Both *Linguistic Inquiry* and *Psychological Review* are the top mainstream journals within each of those two disciplines as measured by citation index impact factor. As can be seen in Fig. 11.9, *Cognitive Science* articles were regularly cited in *Psychological Review* but were never cited in *Linguistic Inquiry*. Thus, mainstream linguistics articles appear not to cite *Cognitive Science* articles.

As a second citation analysis, we examined how often *Cognitive Science* articles by linguists are cited by anyone in contrast to *Cognitive Science* articles by psychologists and computer scientists. The citation analysis was done for papers from 1980, 1986, and 1991, years se-

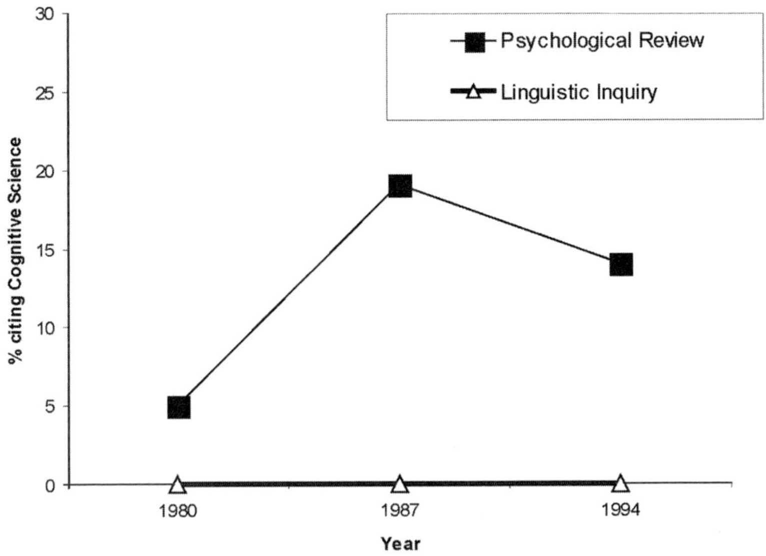

FIG. 11.9. Percentage of articles in *Psychological Review* and *Linguistic Inquiry* citing work in the journal *Cognitive Science*.

lected randomly from the three time periods used for the other journal analyses. Citations were counted 2, 4, and 8 years after publication. Figure 11.10 presents the mean number of citations collapsed across publication year. Psychology articles (N = 14) received the largest number of citations (M = 3.4 per year per article). Computer science articles (N = 10) were next most cited (M = 2.1 per year per article). Linguistic articles, what few there were (N = 2), were the least cited (M = 1.3 per year per article). Perhaps it is because mainstream linguists do not read *Cognitive Science* articles that the citation rates are so low for linguistics papers when they do appear in *Cognitive Science*. Whatever the cause, there are clear external reinforcers for linguists not to publish in *Cognitive Science*.

It is difficult to do corresponding analyses for philosophy, both because philosophers publish even less frequently in *Cognitive Science* and because philosophy does not have a central, high-impact journal. However, the existence of the vibrant Society for Philosophy and Psychology suggests that many psychologists consider philosophy to be relevant to psychology and many philosophers consider psychology to be relevant to philosophy.

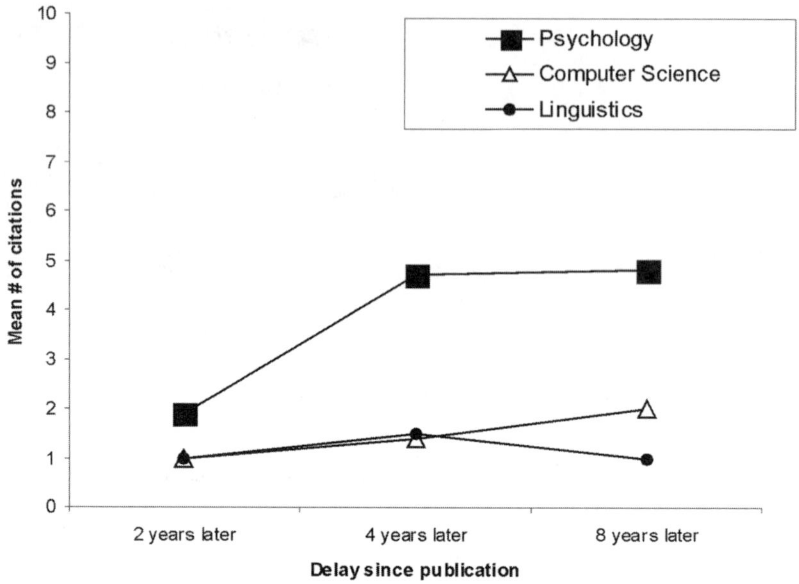

FIG. 11.10. Mean number of citations to *Cognitive Science* articles published by psychologists, computer scientists, and linguists 2, 4, and 8 years after publication.

Where Is the Money?

It is likely that availability of funding has an impact on where people receive training, to which departments they chose to belong, what kinds of research is conducted, and who can afford to attend conferences. Various public and private funding sources have provided large quantities of money to cognitive science research over the years (e.g., National Science Foundation, Defense Advanced Research Projects Agency, Office of Naval Research, Air Force Office of Scientific Research, James S. McDonnell Foundation, A. W. Mellon Foundation, and Spencer Foundation). Many of these funding initiatives have had an applied bent, particularly toward education and training. It may be that this applied bent is part of the reason that researchers from different disciplines were brought together—applied problems tend not to reside nicely within only one discipline and usually require contributions from multiple disciplines. However, it is beyond the scope of this chapter to speculate on how funding patterns may or may not have excluded certain disciplines. Moreover, the relation between funding and research is typically bidirectional—funding for particular approaches spurs productivity in the area, and at the same time, new productive approaches to an area tend to draw more funding (see Bruer, chap. 8, this volume for examples of how a grant agency contributed to the growth of interdisciplinary work).

Competition for Time and Money

In addition to limitations imposed by funding trends, there are resource allocation issues providing a general pressure against the creation and acceptance of new interdisciplinary journals and conferences. Academics typically have limited resources of both time and money when it comes to attending conferences and subscribing to journals and societies. As long as there are sufficiently relevant and interesting conferences, journals, and societies within a discipline for a researcher, they may be reluctant to devote extra resources to an interdisciplinary conference, journal, or society. For example, participation of computer science in cognitive science decreased in the early 1980s when the American Association for Artificial Intelligence formed, creating a new journal and annual conference.

The funding issue is not likely to play a large role in the case of neuroscience. In this case, there are many other conferences (Neuroscience, Cognitive Neuroscience, Computational Neuroscience, and others) and journals (*Journal of Neuroscience*, *Journal of Cognitive Neuroscience*, *Cognitive Neuropsychology*, *Neurocomputing*, and others) that well fit the interests of researchers interested in the intersection of psychology and neuroscience or computer science and neuroscience. The journal *Cognition* competes with *Cognitive Science* for the attention of philosophers and linguists interested in cognitive science. The annual meeting of the Society for Philosophy and Psychology competes for the attention of philosophers interested in cognitive science.

The funding resource issue may be more critical in the case of disciplines that are less well funded. For any particular researcher, this issue has both direct effects (Can I afford to attend this conference?) and indirect effects (Will my colleagues in my discipline notice my work in this interdisciplinary setting?). However, future research is required to assess whether this factor actually played a role in the lack of participation of certain disciplines in the Cognitive Science Society, particularly because the Cognitive Science Society journal and conference are not all that expensive.

Micro Cognitive Sciences

There is also no particular reason why the many constituent disciplines and possible pairwise combinations among constituent disciplines of cognitive science should cohere at the level of one big cognitive science social world. There are journals that correspond to most if not all of the possible pairwise combinations of psychology, computer science, linguistics, philosophy, anthropology, education, and neuroscience. Thus, there may be many micro cognitive sciences that reflect different pieces of the overall cognitive science. For example, research on education-related cognitive science is often published in *Cognition & Instruction* or the *Journal of the Learning Sciences*. Linguistics-related cognitive science is often published in *Cognition* or *Computational Linguistics*. Neuroscience-related cognitive science is often published in the *Journal of Cognitive Neuroscience*. Because the combination of psychology and computer science has "owned" the Cognitive Science Society outlets from the beginning, it may be for this reason that it is the only micro cognitive science that tends to take part in that social world. Even within the combination of psychology and computer science, there are other cognitive science social worlds. For example, researchers interested in human–computer interaction have their own large conference and journals. Thus, it may be misleading or naïve to think of cognitive science as a unitary discipline or to expect that it should be one.

Epistemological and Methodological Stances

The preceding section does not address the issue of why different pieces of cognitive science might not cohere together. One factor that is likely to play a very large role is differences in epistemological and methodological stances. If individuals cannot agree on what counts as an interesting question, what counts as acceptable data, and/or what counts as acceptable theory, they are unlikely to participate in the same social worlds. In our questionnaire survey of authors of cognitive science conference papers, we asked about the frustrations of collaboration. The only frustration that was listed more often with interdisciplinary collaborations than intradisciplinary collaborations was having ideas that were too different. Okada et al. (1995) interviewed 21 cognitive scientists about factors that are important in their collaborations. When one of the researchers was asked how he evaluates whether a person would make a good collaborator, he responded, "(The important thing is) whether or not a person shines his eyes at the right times when we are talking about interesting ideas. As a joke, we call it the 'shining eye test.' ... It indicates whether or not we can share interesting problems." Another researcher was more blunt: "If we don't share interests, I cannot work with him."

With respect to the Cognitive Science Society, it tends to emphasize empirical and/or applied work that advances theory from within computer science, psychology, education, and linguistics. That is, the models should actually run rather than exist only on paper, the psychological and educational studies should involve rigorous experimental methodology with careful data analyses, and the theories should address issues of performance and learning rather than just competence. Linguistics, philosophy, education, and anthropology research tend to clash with the Cognitive Science Society approach on several of these dimensions. There are also epistemological clashes. In the great debates of situation cognition versus symbolic processing (e.g., Anderson, Reder, & Simon, 1996, 1997; Greeno, 1997; Vera & Simon, 1993), for example, the social worlds of the Cognitive Science Society have tended to favor the symbolic processing side.

Although the organizers of the Society have often tried to be open to alternative approaches as evidenced by journal editorial statements (e.g., Greeno et al., 1998) and calls for conference participation, it may be difficult to overcome this fundamental clash. Reviewers will continue to apply the standards from their own epistemological/methodological approach, and researchers will not participate when they find the work being presented to be opaque and/or uninteresting from their own perspective.

CONCLUSION

In this chapter, we have examined the historical and recent state of interdisciplinarity in cognitive science. We have found it to be a case of interdisciplinarity now and then: One often finds researchers from different disciplines working together using methodologies from multiple disciplines, but the field continues to be dominated by two constituent disciplines, psychology and computer science. Although much of this chapter has focused on the glass half empty, the positive findings regarding interdisciplinarity in cognitive science should not be forgotten. We found significant evidence for many researchers overcoming large barriers to actually work together with researchers from other disciplines or at least read literature and learning methodologies from other disciplines. Moreover, we described some of the benefits of these interdisciplinary collaborations.

Turning back to the glass half empty, many questions remain to be answered: Is this domination hindering progress in cognitive science, or is it a necessary state of affairs? Would deep interdisciplinarity lead to more innovative work, or will the best science emerge from limited interdisciplinary exchanges around shared epistemological stances? How should the next generation of cognitive scientists be trained: deeply in a single discipline or broadly across several? What kinds of infrastructure innovations will lead to true interdisciplinary work and what kinds will simply reinforce existing disciplinary boundaries? In an academic climate in which junior faculty are often advised that tenure comes from publishing conservative research in traditional discipline-specific journals, are we minimizing the contributions of exactly the scientists with the best chance of forging true links between disciplines? These are empirical questions, but they have rarely been the subject of empirical research. We hope our work will provide data that may help guide the continued emergence of the field of cognitive science.

ACKNOWLEDGMENTS

Christian Schunn and Kevin Crowley are in the Learning Research and Development Center, University of Pittsburgh; Takeshi Okada is in the School of Education, Nagoya University. This work was supported by a grant from the Mitsubishi Bank Foundation to all three authors. Portions of the data have been presented in partial form in Schunn, Okada, and Crowley (1995), and Schunn, Crowley, and Okada (1998).

REFERENCES

Anderson, J. R., Reder, L. M., & Simon, H. A. (1996). Situated learning and education. *Educational Researcher, 25*(4), 5–11.

Anderson, J. R., Reder, L. M., & Simon, H. A. (1997). Situated versus cognitive perspectives: Form versus substance. *Education Researcher, 26*(1), 18–21

Collins, A. (1977). Why cognitive science. *Cognitive Science, 1,* 1–2.

Greeno, J. G. (1997). On claims that answer the wrong questions. *Educational Researcher, 26*(1), 5–17.

Greeno, J. G., Clancey, W. J., Lewis, C., Seidenberg, M., Derry, S., Gernsbacher, M. A., Langley, P., Shafto, M., Gentner, D., Lesgold, A., & Seifert, C. M. (1998). Commentary: Efforts to encourage multidisciplinarity in the cognitive science society. *Cognitive Science, 22,* 131–132.

12

Being Interdisciplinary: Trading Zones in Cognitive Science

Paul Thagard
University of Waterloo

By the early part of the 20th century, academia in the English-speaking world had stabilized (or ossified!) into a set of scientific and humanistic disciplines that still survived at the century's end. The natural sciences have such disciplines as physics, chemistry, and biology, and the social sciences include economics, psychology, and sociology. These disciplines provide a convenient organizing principle for university departments and professional organizations, but they often bear little relation to cutting-edge research, which can concern topics that cut across or occur at the boundaries of two or more of the established disciplines. When this happens, productive research and teaching must be interdisciplinary.

Cognitive science is the interdisciplinary study of mind, embracing psychology, artificial intelligence, philosophy, neuroscience, linguistics, and anthropology. It is undoubtedly one of the major interdisciplinary successes of the 20th century with its own society, journal, and textbooks and with more than 60 cognitive science programs established at universities in North America and Europe. This chapter is an attempt to answer the question, What are the factors contributing to the success of the interdisciplinary field of cognitive science?

My discussion is organized around the metaphor of the trading zone, a novel and fertile analogy that Gallison (1997) developed for his rich and detailed discussion of the practices of 20th-century physics. To understand the diverse groups of experimenters and theoreticians, Galison (1997) presented their interactions in terms of the trading zones described by anthropologists:

> Subcultures trade. Anthropologists have extensively studied how different groups, with radically different ways of dividing up the world and symbolically organizing its parts, can not only exchange goods but also depend essentially on those trades. Within a certain cultural arena—what I call in chapter 9 the "trading zone"—two dissimilar groups can find common ground. They can exchange fish for baskets, enforcing subtle equations of correspondence between quantity, quality, and type, and yet utterly disagree on the broader (global) significance of the items exchanged. Similarly, between the scientific subcultures of theory and experiment, or even between different traditions of instrument making or different subcultures of theorizing, there can be exchanges (coordinations), worked out in exquisite local detail, without global agreement. (p. 46)

Gallison uses this analogy to depict the interactions of theory and experiment in a way that appreciates the importance of both to the development of physics. Klein (chap. 2, this volume) also compares developing an interdisciplinary perspective to entering another culture.

What are the trading zones in cognitive science? Inevitably, there are difficulties of communication and cooperation faced by researchers from the particular fields of cognitive science as they attempt to work with people from other fields. However, just as traders from different cultural groups have often succeeded in overcoming their differences, so cognitive scientists have frequently surmounted disciplinary barriers. In this chapter, I describe how successful interdisciplinary work in cognitive science has been possible because of important people, places, organizations, ideas, and methods. I begin with a description of some of the key people in the early days of cognitive science in the 1950s and show how the fact that each of them had strong interdisciplinary interests was important for getting the field underway. I then describe how a number of universities in the 1960s and 1970s provided fertile places where cognitive science work could develop and recount how the Cognitive Science Society and the journal *Cognitive Science* began to contribute to interdisciplinary work. However, the point of this paper is not merely sociological, for I also describe some of the ideas and methods of cognitive science that make the field hold together as more than just a bunch of people getting together to chat about the mind. As a more specific example of interdisciplinary research in cognitive science, I describe how understanding of analogical thinking has improved dramatically as the result of people, places, organizations, ideas, and methods. Finally, I conclude with a summary of what the discussion of trading zones in cognitive science contributes to understanding of the past successes and future prospects of cognitive science.

Interestingly, an anthropological metaphor has already been used by cognitive science educators quite independently of Galison's (1997) account of trading zones in physics. Janet Kolodner (1994) discussed pidgin and creole languages, which emerge when cultures trade, in her report on a workshop held to promote cognitive science education. Kolodner (1994) summarized some remarks by Paul Smolensky:

> Paul's underlying concern was how do we produce the next generation of cognitive scientists—the ones who will take cognitive science its next step forward into a unique and identifiable interdisciplinary endeavor? Based on an earlier comment by Angel Cabrera, a graduate student in the audience, he made the analogy to naturalistic language evolution where speakers from a variety of different language backgrounds, when living together in the same community, seem to develop an impoverished language, called a pidgen [sic], that allows them to communicate with each other. The new generation born to the community picks up the pidgen and develops a new language from it, called a creole. Creoles are real languages, as structured and expressive as other languages. One can, however, see the roots, in creoles, of the languages they derived from.
>
> Now for the analogy. We are currently a group of researchers from a variety of different disciplines trying to communicate with each other. We have developed pidgens to allow us to communicate and collaborate. Few of us, however, are native speakers of Cognitive Science. Most of us come first from an associated discipline. The argument based on language theory goes like this: If Cognitive Science is to become an autonomous discipline, with its own language and methods, then we will need to have offspring who are born into Cognitive Science, offspring for whom Cognitive Science is their first language, for whom the natural environment is one with members from the variety of disciplinary communities. The new generation will evolve the many pidgen dialects into a creole, a distinct discipline, with its own methods and issues.

> Those of us who have been in cognitive science for some time had hoped that our interdisciplinary collaborations and the pidgen dialects we developed to communicate across the disciplines would evolve into a creole—a distinctive, real, hybrid discipline—but it hasn't happened yet. Why not, and how can we aim towards a creole.

Even if cognitive science has not developed such an integrated language, it has had considerable success in tying together disparate disciplines. I now look at some of the trading zones that have fostered the development of cognitive science.

PEOPLE

There is no canonical list of the "founders" of cognitive science, but such a list could not omit the following figures who were active in the mid-1950s eruption of ideas that provided the intellectual origins of the field: Noam Chomsky, George Miller, Marvin Minsky, Allan Newell, and Herbert Simon. My aim is not to retell the history of cognitive science (Gardner, 1985; Thagard, 1992, chap. 9) but to highlight the origins of the field in the intense interdisciplinary interests of some its founders.

Chomsky's theories of grammar revolutionized linguistics in the 1950s and 1960s and contributed mightily to the downfall of behaviorist theories of language use. Chomsky's linguistic theories diverged radically from those of his teacher, Zellig Harris, and displayed the influence of diverse intellectual sources including the logicians and philosophers whom Chomsky read avidly from an early age (Barsky, 1997). Chomsky's early work was inspired in part by such philosophers as Bertrand Russell and Nelson Goodman. Chomsky (1953) published his first article, "Systems of Syntactic Analysis," in the *Journal of Symbolic Logic*. Before receiving his PhD in linguistics from the University of Pennsylvania, Chomsky spent several years in Harvard's interdisciplinary Society of Fellows. Although Chomsky's ideas were subsequently to have a great impact on cognitive psychology and computer science, he does not seem to have been directly influenced by these fields. Nevertheless, his early combination of linguistic and philosophical ideas shows that his research was interdisciplinary from the start.

Miller's 1956 article, "The Magical Number Seven, Plus or Minus Two," is generally considered to be one of the seminal works in cognitive psychology. Miller's introduction of Shannon's information theory into psychology was only one of several interdisciplinary innovations that he produced (Hirst, 1988; Shannon & Weaver, 1949). Miller,

Galanter, and Pribram (1960) published what is probably the first book in modern cognitive science, ***Plans and the Structure of Behavior***. The Miller et al. (1960) book replaced behaviorist notions of reflexes and associative links with the concept of a plan, a "hierarchical process in the organism that can control the order in which a sequence of operations to be performed, ... essentially the same as a program for a computer" (p. 16). Influenced in part by the work of Newell and Simon, Miller et al. described the psychological advantages of computational ideas and computational simulations. In the 1960s, Miller collaborated with Chomsky to bring ideas about transformational grammar to the attention of psychologists, and in the 1970s, Miller coined the term "cognitive neuroscience" to describe the emerging relevance of brain research to cognitive psychology. Miller's own history exhibits the fertility of combining psychological, mathematical, computational, linguistic, and neurological interests.

Minsky was a participant in the 1956 conference at Dartmouth that inaugurated artificial intelligence (AI) and his contributions to that field and cognitive psychology have been legion. As an undergraduate at Harvard, he had three laboratories of his own in biology, physics, and psychology where he worked with Miller (Bernstein, 1981; McCorduck, 1979). Minsky's early interests ranged from mathematics to electronics to psychology, and he did his PhD at Princeton on the mathematics of neural networks. Minsky's (1975) AI paper on frames influenced and was influenced by psychological work on schemas, and Minsky's (1986) later Society of Minds theory shows some Freudian influences. It is clear that Minsky would not have been drawn to artificial intelligence if he had not had from the beginning a strong multidisciplinary interest in the nature of mind.

Newell and Simon were also at the 1956 Dartmouth AI conference, and their interests were more avowedly psychological than Minsky's. Simon's PhD was in political science, but he had strong early interests in mathematics and psychology. As a consultant at the RAND corporation, he met a young mathematician, Newell, who was interested in adding intelligence to the primitive computers of the day. With Shaw, Newell and Simon produced the first AI program, which was also intended to be a model of human thinking. From the General Problem Solver through influential later projects, Newell and Simon (e.g., Newell, 1990; Newell & Simon, 1972; Simon, 1991) combined computational and psychological research. Newell also made important contributions to computer hardware and the field of human–computer interaction, and Simon's accomplishments include a Nobel prize in economics and valuable philosophical work on causality. As with Chomsky, Miller, and Minsky, these two founders of cognitive science were thoroughly interdisciplinary in themselves.

Just as cultural trading zones require people who learn enough of another culture and language to be able to initiate trade with strangers, interdisciplinary fields require individuals who can get them going by each working in more than one field. I do not know of any cognitive scientist who can claim to have worked in all six of the constituent disciplines of cognitive science, but the five seminal figures I have discussed each operated in two, three, or four of them. The obvious lesson for interdisciplinary work is if you want to start an interdisciplinary field, start with people themselves whose interests and abilities are already interdisciplinary. Table 12.1 summarizes the interests of Chomsky, Miller, Minsky, Newell, and Simon, none of whom has had much to do with anthropology.

PLACES

The development of an interdisciplinary field requires more than a few brilliant individuals who generate ideas at the intersection of established disciplines. It also requires institutions that provide opportunities for interdisciplinary contacts and collaborations. In its early days, before the term *cognitive science* was coined in the 1970s, cognitive science benefited from several places in which interdisciplinary work flourished. In this section, I describe the impact of two important institutions, the Graduate School of Industrial Administration at the Carnegie Institute of Technology and the Center for Cognitive Studies at Harvard University.

TABLE 12.1
Interdisciplinary Interests of Some of the Founders of Cognitive Science

Founder	*Artificial Intelligence*	*Linguistics*	*Neuroscience*	*Philosophy*	*Psychology*
Chomsky		v		v	v
Miller	v	v	v		v
Minsky	v		v		v
Newell	v				v
Simon	v			v	v

In 1955, Newell went to Pittsburgh to work and do a PhD with his collaborator Simon. Simon was a professor in the Graduate School of Industrial Administration at what was then Carnegie Tech and is now Carnegie Mellon University. This school was sufficiently flexible that Simon's students, who later included Edward Feigenbaum and many other important early contributors to AI, could receive PhD degrees for computational models of human thinking. Simon was instrumental in the reconstitution in the early 1960s of Carnegie's psychology department as a major concentrator on cognition and in the creation in 1965 of its computer science department, which to this days retains ties with psychology through joint appointments such as John R. Anderson. Simon's efforts in an unlikely location—a business school in a technical university—produced four decades of influential work in cognitive science.

In 1960, Miller and Jerome Bruner founded the Center for Cognitive Studies at Harvard University with support from the Carnegie Corporation and Harvard University. According to Bruner (1988, p. 91):

> There was undoubtedly a suspicion abroad that the old disciplinary boundaries, though they had once been useful in shaping the division of scholarly labors, were no longer the natural joints of the enterprise. In circles where this general view prevailed, psychology was believed to be too narrowly focused on a few traditional problems to deal interestingly with the nature and uses of the human mind, a view shared by many inside psychology, who felt that the old behavior was a hopelessly wrong epistemological base from which to view the higher functions of the mind.

Fellows and visitors at the center included an amazing group of established and beginning scholars from linguistics (e.g., Roman Jakobson, Chomsky), philosophy (e.g., Nelson Goodman), and psychologists (e.g., Donald Norman, Peter Wason) as well as other fields. Miller and Chomsky collaborated on developing a formal theory of grammar, and at the center, Chomsky (1965) completed his influential book, *Aspects of a Theory of Syntax*. Weekly colloquia brought in a broad and distinguished series of speakers from many disciplines, although there does not seem to have been any direct connection with the AI group that Minsky and John McCarthy started at Massachusetts Institute of Technology (MIT) in 1957.

According to Allan Collins (personal communication, June 5, 1998), the term *cognitive science* was created by Daniel Bobrow for their interdisciplinary book, *Representation and Understanding: Studies in Cognitive Science* (Bobrow & Collins, 1975). Explicit cognitive science programs came into being in the late 1970s when the Sloan Foundation poured millions of dollars into new ventures at such institutions as Yale, MIT, the University of Pennsylvania, the University of California at San Diego, and the University of Michigan. Another important source of funds was the Systems Development Foundation, which established the interdisciplinary Center for the Study of Language and Information at Stanford University and supported research at other universities. Today, although there are still very few actual departments of cognitive science in universities, there are numerous cognitive science programs in the United States, England, Germany, Canada, and other countries.

My own intellectual trajectory was dramatically affected by cognitive science programs that I participated in at the University of Michigan in the early 1980s and at Princeton University later in that decade. Each institution provided an exciting interdisciplinary intellectual environment along with computational and other resources. Like Harvard in the 1960s and Carnegie Mellon from the late 1950s until today, the cognitive science programs at the University of Michigan and Princeton brought together people from several disciplines both inside and outside the host institution. Cynics had remarked of the influx of Sloan Foundation money in the late 1970s that some sciences are theory driven and others are data driven, but cognitive science is money driven. However, skeptical predictions that cognitive science programs were transitory results of financial incentives have been refuted by the large number of thriving programs at the end of the millennium.

Thus, an interdisciplinary field needs not only brilliant people to get it going but also places where they can work together and where interdisciplinary research is fostered and encouraged. Through its relatively short history, cognitive science has seen some programs flower then fold (e.g., Harvard, Yale), whereas a few programs have developed into full-scale departments (e.g., University of California at San Diego, Johns Hopkins University). Other programs have shifted their emphases as faculty come and leave. However, there can be little doubt that places such as Carnegie Tech, Harvard's Center for Cognitive Studies, and the major centers that arose in the late 1970s contributed greatly to the development of cognitive science as an interdisciplinary field.

At a more local level, interdisciplinary work can take place in particular research groups independent of the umbrella of a general cognitive science program. Hall, Stevens, and Torralba (chap. 5, this volume) describe some of the social and cognitive processes involved in interdisciplinary groups.

OTHER ORGANIZATIONS

Universities, with their departments, centers, and programs, are not the only trading zones that produce interdisciplinary contacts. As part of the mid-1970s jump of interest in what was by then called cognitive science, the journal *Cognitive Science* began publishing interdisciplinary work in 1977. The three original joint editors were Roger Schank (AI), Collins (psychology), and Eugene Charniak (AI). The initial editorial board had 29 members, more than half of them from AI; the rest were psychologists, except for a couple of linguists. The 1998 editors included three from psychology and one from AI; and the editorial board has now shifted so that 15 out of 32 members are psychologists, with 8 from AI, 4 from linguistics, 3 from philosophy, and 1 each from anthropology and neuroscience. This classification is somewhat misleading, however, because many of the current members of the editorial board do research that crosses over into other disciplines, and several have appointments in departments of cognitive science. In the early years, as today, the journal and the proceedings of the annual conference consisted predominantly of articles that are psychological and computational, although papers oriented more toward linguistics, neuroscience, and philosophy occasionally appear (for an insightful analysis, see Schunn, Crowley, & Okada, 1998).

The Cognitive Science Society actually followed the journal, originating in 1979, although the journal was later given to the Society by its publisher, Ablex. The organization began with a meeting at the Dallas airport initiated by Collins, Norman (who did not want to travel to the East coast), and Schank (who did not want to travel to the West coast). The attendees at this meeting were the following:

- Daniel Bobrow, AI, Xerox Palo Alto Research Center.
- Eugene Charniak, AI, Brown University.
- Allan Collins, Psychology, Bolt Beranek and Newman.
- Edward Feigenbaum, AI, Stanford.
- Charles Fillmore, linguistics, University of California, Berkeley.
- Jerry Fodor, philosophy and psychology, MIT.
- Walter Kintsch, psychology, University of Colorado.
- Donald Norman, psychology, University of California, San Diego.
- Zenon Pylyshyn, psychology, University of Western Ontario.
- Raj Reddy, AI, Carnegie Mellon University.
- Eleanor Rosch, psychology, University of California, Berkeley.
- Roger Schank, AI, Yale.

It is interesting that the 12 founding members of the executive committee included 5 AI researchers, 5 psychologists, a philosopher, and a linguist. All but 4 of them were on the original editorial board of the journal *Cognitive Science*. Since then, the society executive committee (now called the governing board) has tilted more toward psychology and away from AI, reflecting the evolution of the society. The 13 1998 members of the governing board included 8 psychologists, 3 AI researchers, a philosopher, and a linguist. It is notable, however, that the philosopher (Thagard), the linguist, and one of the psychologists each works with computational models. The 2004 governing board is similar.

According to the minutes of the meeting recorded by Charniak and Norman, the two main issues discussed were the nature of the membership of the organization and the role of AI in it. It was decided that there should be two membership categories, "Fellow" and "Member," with fellows being carefully selected on the basis of significant contributions to cognitive science beyond the PhD dissertation. The main reason for making this distinction seems to have been to eliminate the need for refereeing papers at the projected annual conference, following a model used by the Psychonomic Society. Later, the categories were changed to "Member" and "Associate Member," and eventually the distinction was dropped altogether, and refereeing of conference papers began. Some members of the first executive committee thought that the Cognitive Science Society should be an artificial intelligence society and should try to host an annual AI conference. However, such close identification was resisted by other members of the committee, and in 1980, the American Association for Artificial Intelligence was formed and began its own annual meeting. The Cognitive Science Society Executive Committee met again on August 12, 1979, just before the first conference of the society at the University of California at San Diego. Present were Bobrow, Collins, Norman, Pylyshyn, Reddy, Rosch, and Schank, who agreed to organize the 1980 conference at Yale.

Over the past 20 years, annual meetings of the Cognitive Science Society have provided the primary site where researchers can gather to present research of interdisciplinary interest and gain some idea of what is happening in other fields. Typically 400 to 500 people attend out of the approximately 1,000 people who belong to the Society, and the conference proceedings include hundreds of papers and abstracts. A standard feature of the conference is a set of symposia that have speakers from more than one discipline. The content of the conference can vary greatly from year to year, reflecting the different interests of the organizers who are largely drawn from the host institution. It would be very difficult for any one conference to cover the multitude of topics of interest to the highly diverse membership of the Cognitive Science Society, but substantial diversity is assured over the course of successive meetings. One problem for the society is that involvement by AI researchers has dropped off somewhat over the past two decades, reflecting the trend in AI toward engineering rather than cognitive modeling approaches. On the other hand, involvement by philosophers seems to be increasing, but linguists and neuroscientists attend their own disciplinary meetings.

The journal *Cognitive Science* has far fewer participants than the conference because only about 25 articles appear in it annually. It is, however, not the only interdisciplinary journal in cognitive science, as the following partial list demonstrates: *Behavioral and Brain Sciences*, *Cognition*, *Computational Linguistics*, and *Mind and Language*. Moreover, in addition to the annual meetings of the Cognitive Science Society, there are other conferences where researchers can pursue questions at the intersection of such fields as linguistics and computation, philosophy and psychology, cognition and neuroscience, and so on. The Society for Philosophy and Psychology and Cognitive Neuroscience Society are two of the organizations that serve to forge links at a more local level than the entire field of cognitive science. In addition, every year there are special-topic conferences on particular aspects of the mind that are geared toward interdisciplinary participation, on topics such as text processing, computer–human interaction, and AI and education.

For the past two decades, then, cognitive science has benefited from having extrauniversity organizations that foster its development, particularly the Cognitive Science Society with its annual conference and journal. Conferences are probably the closest analog to intercultural trading zones, as people from various disciplines and countries gather to exchange ideas.

One would get, however, a feeble anthropological understanding of trading zones if one concentrated only on the people and places where they meet. Just as the point of economic trading zones is the exchange of goods, so the point of intellectual trading zones is the exchange of ideas, and I have said little so far about the ideas and methods that make interdisciplinary work in cognitive science possible and desirable. Understanding the interdisciplinary character of cognitive science requires much more than biography and sociology, so I turn to a discussion of the intellectual content of cognitive science.

IDEAS

For an interdisciplinary field to have an intellectual purpose, it must involve ideas that cut across disciplinary boundaries. For cognitive science, the most important ideas have been mental representation, computational procedures, and the brain as a representational-computational engine. My aim here is to describe how each of these has helped to make possible trading zones in cognitive zones; fuller accounts of the history and content of these ideas can be found in other sources such as Johnson-Laird (1988), Churchland and Sejnowski (1992), and Thagard (1996).

The concept of mental representation is ancient, evident in the writings of philosophers such as Plato, Locke, and Kant. However, in the early 1950s, especially in American psychological circles, the concept of mind had become suspect, a metaphysical construction incompatible with the positivist and behaviorist prescriptions of the time. Chomsky's work in linguistics and Miller's work in psychology was revolutionary in that they allowed and required the discussion of mental representations such as rules, plans, and schemas. From its beginnings, AI was representational, writing programs using computer structures assumed to be analogous to ones that underlie human thought. Cognitive theorizing has postulated various kinds of mental representation to explain intelligent behavior including sentences expressed in logical formalism, rules, concepts, analogs, visual images, and distributed representations in artificial neural networks (see Thagard, 1996, for a survey). Discussion of these representations has been at the center of interdisciplinary debates involving psychologists, AI researchers, philosophers, linguists, neuroscientists, and anthropologists. Although there is by no means general agreement on which kinds of representation are most important for explaining mental capacities, it is striking that the discussion of representation is at the core of interdisciplinary discourse. Heideggerians and social constructivists who completely reject the concept of mental representation operate only at the fringes of cognitive science. Trading zones do not require complete agreement or a universal vocabulary, but they do require an overlapping conceptual core among the cultures or disciplines that participate in them. For cognitive science, the idea of mental representation is a crucial part of that core.

Although cognitive science merely revived and enriched the idea of mental representations, it also had from the start a core idea that was much more original. To explain intelligent functioning, it is necessary to postulate not only mental representations but procedures that operate on them to produce performance. Before computational ideas came along in the 1940s, philosophers and psychologists were limited in the kinds of processes they could discuss, for example, association of ideas and logical inference. Moreover, it was not at all evident how such processes could be understand mechanistically or how the brain could carry them out.

By the early 1950s, however, the first computers were in use, and computation was becoming understood both theoretically and practically. The pioneers of AI quickly saw the potential for understanding thinking as a kind of computation, and by 1956, , Shaw, and Simon had produced the first computational model of human problem solving, the Logic Theorist, which performed logical proofs. Although Chomsky has never embraced the computational view of mind because he contends that linguistics need only explain competence and can ignore performance, the view of thinking as analogous to or even as a kind of computation has united many other linguists, most psychologists, some philosophers, and even cognitive neuroscientists who understand the brain as a computational device. It is not an exaggeration to see cognitive science as a spin-off from a technological development—the invention of digital computers in the 1940s. In particular, the rapid growth of cognitive psychology in the 1960s and 1970s employed a view of thinking as information processing that heavily employed computational ideas and metaphors.

The major development in cognitive science in the 1980s was the growth of connectionist models using artificial neural networks, and the most striking expansion of the 1990s has been in work on cognitive neuroscience using brain scanning methods I discuss in the next section. Through this work, the computational approach to thinking has been enriched by thinking of the brain as a representational-computational machine and using what is known about the brain to enhance ideas about representation and computation. The result has been a new set of ideas that cross disciplinary boundaries including distributed representations and parallel processes. Increasingly, the brain and what is rapidly becoming known about it are furnishing topics for interdisciplinary discourse.

Although concepts involving representation, computation, and the brain are at the center of the cognitive science trading zone, there are other more local concepts that provide intersections for particular pairs of disciplines. For example, psychology, philosophy, and AI share a concern with inference, although philosophy and AI are often concerned more with normative issues of how people and machines should infer than with descriptive psychological issues about how people actually do make inferences. Concepts of culture, long a staple of anthropological investigation, are starting to make inroads into social and cognitive psychology. It would be interesting to compile a complete list of ideas at the intersection of two or more of the six disciplines that constitute cognitive science.

There is no reason to suppose that an interdisciplinary field such as cognitive science should be limited to a fixed set of contributing disciplines. Just as new cultures can arrive to contribute to an anthropological trading zone, so new disciplines can emerge as relevant to an interdisciplinary field. At its inception in the 1950s, cognitive science was mostly a mixture of psychology, AI, and linguistics, and only later was the strong relevance of neuroscience, philosophy, and anthropology recognized. The early emphasis on mental representation led to neglect of matters that have received more attention in recent cognitive science such as the role of the human body in cognition and the importance of the physical and social environments in which cognition takes place. However, the embodiment and situatedness of cognition do not provide reasons for abandoning the representational-computational theory of mind, only for expanding and supplementing it (Thagard, 1996). My guess is that the next major addition to the interdisciplinary mix of cognitive science will be molecular biology as knowledge increases dramatically about the genetic and chemical basis of neurological processes. Bruer (chap. 8, this volume) discusses some of the potential interconnections between genetic studies and cognitive science. The ebb and flow of contributions of different disciplines to an interdisciplinary field can not be managed by any central body such as the Cognitive Science Society but depends on the unpredictable course of theoretical and experimental developments.

The journal *Cognitive Science* currently lists education as one of the areas of cognitive science in addition to the six disciplines that I have been discussing. Education is an extremely important area of application of cognitive science but is not a contributing discipline in itself. Like other applied areas such as computer–human interaction and expert system development, education has provided challenging problems for cognitive scientists to work on from an interdisciplinary perspective (chap. 8, this volume). Yet education is primarily a borrower of ideas and methods rather than a disciplinary contributor to understanding of how the mind works.

METHODS

A discipline is constituted not only by its ideas but by its methods. Typically, for example, psychologists run experiments, AI researchers write computer programs, linguists analyze languages, and neuroscientists record brain operations. An interdisciplinary field requires methods that cross disciplinary boundaries, and there are two such methods that have had the greatest impact on work in cognitive science: computer simulation and brain scanning. I briefly describe the nature of these two methods to show how the cognitive science trading zone involves not only ideas but also activities of an interdisciplinary nature.

When computers began to become available in the 1940s, scientists quickly realized their potential for investigating physical processes. Even when a physical system has a mathematical description, it is often not possible to work out its behavior in any detail because the equations that describe it may have no tractable solution. However, if programmable equations can be written that approximate its behavior, then running a computer program can provide predictions about behaviors too complex to be worked out by direct mathematical methods. Galison (1997) described how computer simulations became a standard part of the practice of physics in the 1950s, and today, computer simulations are widely used in disciplines as diverse as economics and evolutionary biology.

I have already described how cognitive science pioneers such Newell, Simon, Miller, and Minsky recognized in the 1950s the potential for computational simulation of human thought, and such simulations have been at the core of theoretical developments in cognitive science ever since. In fact, computer simulations are even more central to cognitive science than to other disciplines by virtue of the theoretical identification of thinking as a kind of computation. Computer simulation not only offers cognitive science the benefit of complex calculation found in computer modeling in such disciplines as physics, economics, and biology; it also provides a major theoretical impetus. The structures and procedures in the computer model of mind are hypothesized to be analogous to the mental representations and procedures that underlie human thinking.

As in other disciplines in which computer models are useful, one of the merits of computational models of cognition is that they serve to draw out the unforeseen empirical consequences of cognitive theories and display their limitations. The assessment of cognitive models should address questions such as the following (Thagard, 1998, p. 57):

1. Genuineness. Is the model a genuine instantiation of the theoretical ideas about the structure and growth of knowledge, and is the program a genuine implementation of the model?
2. Breadth of application. Does the model apply to lots of different examples, not just a few that have been cooked up to make the program work?
3. Scaling. Does the model scale up to examples that are considerably larger and more complex than the ones to which it has been applied?
4. Qualitative fit. Does the computational model perform the same kinds of tasks that people do in approximately the same way?
5. Quantitative fit. Can the computational model simulate quantitative aspects of psychological experiments, e.g. ease of recall and mapping in analogy problems?
6. Compatibility. Does the computational model simulate representations and processes that are compatible with those found in theoretical accounts and computational models of other kinds of cognition?

When such questions are addressed, computational models of human cognition can provide valuable insights into the nature of the mind and potential applications to areas such as education.

Computer simulation is an interdisciplinary method for two reasons. First, computational modeling is not normally part of the training of psychologists, philosophers, neuroscientists, linguists, or anthropologists, and second, it usually draws on ideas about structures and algorithms that are part of the branch of computer science called AI. Yet computer simulation is obviously not just part of computer science and AI because knowledge of psychology, philosophy, language, or neuroscience is crucial for determining what to simulate. The method of computer simulation requires either (a) interdisciplinary collaboration between computer scientists and members of other interdisciplinary fields or (b) the acquisition by individuals from a particular discipline of ideas and skills from the other. A great deal of cognitive modeling has been accomplished by psychologists who have stepped outside the typically empirical orientation of their discipline to acquire computational skills to perform computational simulations. More rare are AI researchers who have acquired sufficient knowledge of psychology or linguistics to produce computational models in these areas, and rarer still are philosophers who have adopted computational modeling as a methodology.

Another interdisciplinary method has become important to cognitive science in recent decades. Brain imaging began in the early 1970s when X-ray computed tomography was developed (Posner & Raichle, 1994). Soon, developments such as positron emission tomography (PET) and magnetic resonance imaging (MRI) made it possible to image the brain's changing blood flow during sensory stimulation and cognitive operations. These instruments depended on many technological advances, including the availability of computers to collect data and produce interpretations of brain activity. In the 1980s, cognitive psychologists such as Michael Posner began to use PET and MRI to determine the operations that the brain performs when people are given experimental tasks that had been used in experiments over the preceding three decades. Edward Smith (1997, p. 72), another cognitive psychologist turned neural imager, reported that cognitive psychologists are turning to neuroscience for several reasons. First, neuroscientists have found out a great deal about the neural bases of memory and are now able to use PET and MRI to observe brain changes while an organism is engaged in various tasks. These results place constraints on cognitive theories. Second, neuroimaging techniques may eventually provide more directly interpretable information than that obtained in strictly cognitive experiments. Third, cognitive neuroscience can also suggest new ways of dividing cognition into meaningful areas of study. In recent years, many leading cognitive psychologists have shifted their research in neuroscientific directions.

Thus, brain scanning is a new method that ties together cognitive psychology and neuroscience and that is beginning to yield results of interest to linguists and philosophers as well. Like computer simulation, it is an inherently interdisciplinary method because it requires the knowledge and skills of both experimental psychologists and neuroscientists. This new intersection has spawned new journals such as *Cognitive Neuroscience* and a new organization, the Cognitive Neuroscience Society, with its own annual meeting. Neural imaging is potentially a complementary method to computer simulation because the information it provides can contribute ideas and constraints on computational models of how the brain processes information. For example, Kosslyn (1994) used imaging studies and computational models in a complementary fashion to defend strong theoretical claims about visual imagery. We can expect these two interdisciplinary methods to continue to work together as cognitive science continues. I now look at a more specific case of interdisciplinary research.

CASE STUDY: ANALOGY RESEARCH

Over the past two decades, research on analogical thinking has been one of the most successful areas in cognitive science, and it well illustrates the benefits of interdisciplinary collaboration. I do not try to survey that research (see Holyoak & Thagard, 1995) and certainly want to avoid any kind of partisan defense of my own views over those of other analogy researchers. Rather, I want to describe how interdisciplinary research on analogy has benefited from trading zones comprised of people, places, organizations, ideas, and methods.

Before 1980, analogy was primarily a topic discussed by philosophers such as Hesse (1966), but cognitive research has since flourished. Ignoring many independent researchers, we can divide the most active participants into four main camps:

1. Structure mapping theory comprising Dedre Gentner, Ken Forbus, and numerous collaborators.
2. Multiconstraint theory comprising Keith Holyoak, Paul Thagard, and numerous collaborators.
3. Case-based reasoning comprising Kolodner, Chris Hammond, Colleen Seifert, and numerous other researchers inspired by Schank.
4. Fluid analogies research group comprising Douglas Hofstadter, Melanie Mitchell, and other collaborators.

Notably, the first three groups involved a mixture of psychologists whose research consists primarily of experiments (Gentner, Holyoak, Seifert) and AI researchers who produce computer programs (Forbus, Thagard, Kolodner, Hammond). All three of these projects have involved interconnected work on both psychological experimentation and computer modeling. Hofstadter's group has not been so explicitly psychological but has been interdisciplinary in its own way involving people with backgrounds in computer science, philosophy, and physics. None of these analogy researchers produces both computational models and psychological experiments alone, but all have willingly expanded beyond their initial training disciplines.

Places were crucial in the initial constitution of all of these groups. Holyoak and I got together at the University of Michigan in the early 1980s. Gentner and Forbus began collaborating at the University of Illinois in the 1980s and have continued together at Northwestern University in the 1990s. The case-based reasoning group were mostly graduate students at Yale in the 1980s when an active cognitive science program formed by Schank and Robert Abelson brought together students from both psychology and computer science. Hofstadter's group began at the University of Indiana, moved to Michigan, then back to Indiana. Cognitive science programs have been active at all four institutions crucial to the rise of analogy research in the 1980s—Michigan, Illinois, Yale, and Indiana.

Other organizations also helped move research along. Annual meetings of the Cognitive Science Society provided occasions for debate and exchange of information. For example, a symposium in 1993 on cognitive models of problem solving included presentations by Gentner, Forbus, Holyoak, Thagard, Seifert, and Hammond. Gentner, Holyoak, Seifert, Thagard, and Forbus have all been elected to the governing board of the society. Funding organizations have been crucial for fostering interdisciplinary research. From 1986 to 1992, Holyoak and I were funded by the Basic Research Office of the U.S. Army Research Institute, and the Office of Naval Research has provided funding for collaborative projects by Gentner and Forbus and by Hammond and Seifert.

All four analogy research projects described previously have worked within the fundamental hypothesis of cognitive science that thinking consists of computational processes operating on mental representations. Although there has been much dispute concerning the particular nature of the processes and representations used in analogical thinking, the different approaches all share fundamental ideas about the nature of mind operations. Similarly, analogy researchers all take for granted the value of combining multiple methods involving both psychological experimentation and computational modeling.

Application of multiple methods requires broader collaboration as evident in the case of the multiconstraint theory: Fig. 12.1 displays analogy collaborations of Holyoak and Thagard up to 1995. Most of Holyoak's collaborators were involved in psychological experiments, although Hummel and Melz in particular contributed computational models. In my research, Cohen, Gochfeld, Hardy, Nelson, and Fleck all worked on computational modeling, Buchanan and Joordens developed psychological experiments, and Barnes and Shelley helped with philosophical analyses. Similar diagrams could be produced for the other analogy research groups.

What have the numerous groups of people, places, organizations, ideas, and methods contributed to the understanding of analogical thinking? In contrast to the situation before 1980, there is now a wealth

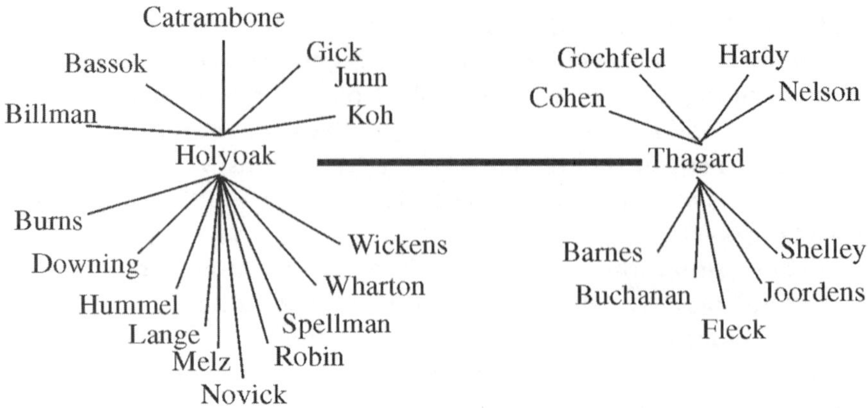

FIG. 12.1. Collaborators of Holyoak and Thagard between 1980 and 1995. *Note.* From Thagard, P. (1997). Collaborative knowledge. *Noûs, 31,* 242–261. Copyright 1997 by Blackwell Publishers. Reprinted with permission.

of experimental data on how people use analogies and rich theoretical explanations of how minds think analogically. Theoretical advances have involved intense interaction between psychological experiments and computational models. For example, after Holyoak (1995) and his co-workers performed experiments on analogical problem solving, he and I set out to produce a computational model of such thinking. Our first attempt, the PI (process of induction) model of analogy, failed to convince even us, and we were impelled to produce the multiconstraint models of analogical mapping and retrieval, which in turn led to further psychological experiments (see Holyoak & Thagard, 1995, for the whole story). Similarly, Gentner, Forbus, and their collaborators have benefited from alternation and interpenetration of psychological experimentation and computational modeling. The neurological study of analogical thinking using brain scanning is just beginning.

CONCLUSION

Cognitive science has been successful as an interdisciplinary field because of the establishment of fertile trading zones at the intersections of its six constituent disciplines. I have described how these trading zones have been constituted by people, places, organizations, ideas, and methods. Cognitive science has flourished due to the presence of the following:

- People, both at the inception and in the maturity of the field, who are willing and eager to cross disciplinary boundaries.
- Places where interdisciplinary communication and communication is encouraged.
- Organizations such as societies and journals that foster interdisciplinary communication.
- Ideas that provide bridges between disciplines and show that problems cross disciplines.
- Methods that require participation of people trained in more than one discipline.

These factors have enabled cognitive science to have the kind of overlaps between disciplines recommended by Campbell (chap. 1, this volume) with his fish-scale model of knowledge.

I suspect that the same factors have been crucial to the success of other interdisciplinary fields. For example, history and philosophy of science emerged in the late 1950s and early 1960s and has benefited from:

- Pioneers such as N. R. Hanson, Thomas Kuhn, and Stephen Toulmin whose work was both philosophical and historical.
- Places such as Princeton University, Cambridge University, Indiana University, and the University of Pittsburgh that encouraged interdisciplinary work.
- Journals such as *Studies in History and Philosophy of Science* that invite work from more than one discipline.
- Methods combining historical interpretation and philosophical analysis.

It would be interesting to attempt to apply my fivefold analysis of interdisciplinary trading zones to other fields. It would also be interesting to discuss the question of how undergraduate and graduate education can foster future work in cognitive science. One consequence of the preceding analysis is that training in cognitive science should involve not only acquisition of the representational-computational ideas that connect disciplines but also training in the methods that operate at the core of interdisciplinary research. Such training should enable future students to thrive in the 21st-century trading zones of cognitive science.

ACKNOWLEDGMENTS

I am grateful to Allan Collins, Alan Lesgold, Donald Norman, and Andrew Ortony for historical information; to Colleen Seifert for providing access to the minutes of the early meetings of the Cognitive Science Society; and to Sharon Derry for very helpful comments on a previous draft.

REFERENCES

Barsky, R. F. (1997). *Noam Chomsky: A life of dissent*. Cambridge, MA: MIT Press.
Bernstein, J. (1981, December 14). Profile of Marvin Minsky. *New Yorker, 57,* 48–126.
Bobrow, D. G., & Collins, A. (Eds.). (1975). *Representation and understanding: Studies in cognitive science*. New York: Academic.
Bruner, J. (1988). Founding the center for cognitive studies. In W. Hirst (Ed.), *The making of cognitive science: Essays in honor of George A. Miller* (pp. 90–99). Cambridge, England: Cambridge University Press.
Chomsky, N. (1953). Systems of syntactic analysis. *Journal of Symbolic Logic, 18,* 242–256.
Chomsky, N. (1965). *Aspects of a theory of syntax*. Cambridge, MA: MIT Press.
Churchland, P. S., & Sejnowski, T. (1992). *The computational brain*. Cambridge, MA: MIT Press.
Galison, P. (1997). *Image & logic: A material culture of microphysics*. Chicago: University of Chicago Press.
Gardner, H. (1985). *The mind's new science*. New York: Basic Books.
Hesse, M. (1966). *Models and analogies in science*. Notre Dame, IN: Notre Dame University Press.
Hirst, W. (Ed.). (1988). *The making of cognitive science: Essays in honor of George A. Miller*. Cambridge, England: Cambridge University Press.
Holyoak, K. J., & Thagard, P. (1995). *Mental leaps: Analogy in creative thought*. Cambridge, MA: MIT Press/Bradford Books.
Johnson-Laird, P. N. (1988). *The computer and the mind*. Cambridge, MA: Harvard University Press.
Kolodner, J. (1994). Workshop on Cognitive Science Education: An Idiosyncratic View. Retrieved July 14, 2004, from http://www.cc.gatech.edu/aimosaic/cognitive-science-conference-1994/education-workshop-review.html
Kosslyn, S. M. (1994). *Image and brain: The resolution of the imagery debate*. Cambridge, MA: MIT Press.
McCorduck, P. (1979). *Machines who think*. San Francisco: Freeman.
Miller, G. (1956). The magical number seven, plus or minus two: Some limits on our capacity for processing information. *Psychological Review, 63,* 81–97.
Miller, G. A., Galanter, E., & Pribram, K. (1960). *Plans and the structure of behavior*. New York: Holt, Rinehart & Winston.
Minsky, M. (1975). A framework for representing knowledge. In P. H. Winston (Ed.), *The psychology of computer vision* (pp. 211–277). New York: McGraw-Hill.
Minsky, M. (1986). *The society of mind*. New York: Simon & Schuster.

Newell, A. (1990). *Unified theories of cognition*. Cambridge, MA: Harvard University Press.

Newell, A., & Simon, H. A. (1972). *Human problem solving*. Englewood Cliffs, NJ: Prentice Hall.

Posner, M. I., & Raichle, M. E. (1994). *Images of mind*. New York: Freeman.

Schunn, C., Crowley, K., & Okada, T. (1998). The growth of multidisciplinarity in the Cognitive Science Society. *Cognitive Science, 22,* 107–130.

Shannon, C., & Weaver, W. (1949). *The mathematical theory of communication*. Urbana: University of Illinois Press.

Simon, H. A. (1991). *Models of my life*. New York: Basic Books.

Smith, E. E. (1997). Infusing cognitive neuroscience into cognitive psychology. In R. L. Solso (Ed.), *Mind and brain sciences in the 21st century* (pp. 71–89). Cambridge, MA: MIT Press.

Thagard, P. (1992). *Conceptual revolutions*. Princeton, NJ: Princeton University Press.

Thagard, P. (1996). *Mind: Introduction to cognitive science*. Cambridge, MA: MIT Press.

Thagard, P. (1998). Computation and the philosophy of science. In W. Bynum & J. C. Moor (Eds.), *The digital phoenix: How computers are changing philosophy* (pp. 48–61). Oxford, England: Blackwell.

Thagard, P. (1999). *How scientists explain disease*. Princeton, NJ: Princeton University Press.

Author Index

A

B

C

G

H

I, J

Subject Index

Note: Page number followed by *n* indicates note.

D

M

Q

R

S

U

V

W

X

Y